Orchid Cultivation and Management — Recent Advances

NIPA® GENX ELECTRONIC RESOURCES & SOLUTIONS P. LTD.
New Delhi-110 034

www.nipaers.com
Browse, Search, Read & Buy...
Print Books
eChapters
eBooks
Publishers
Forthcoming
Subject Catalogues
New Titles
Online Resources on:
✓ Current Affairs
✓ Reasoning, Logic & Aptitude
✓ Competitive Examinations
✓ English Language Lab
✓ Writing & Pronunciation Tools
✓ Personality Development
✓ Online Programmes for Professionals
✓ Interview Preparation
Pay Using
PhonePe
Google Pay
paytm
PayPal
UPI
CC Avenue

Orchid Cultivation and Management — Recent Advances

Authors

Ajai Kumar Tiwari

and

Ramesh Kumar

Directorate of Floricultural Research
Indian Agricultural Research Institute Campus
New Delhi-110012

NIPA® GENX ELECTRONIC RESOURCES & SOLUTIONS P. LTD.
New Delhi-110 034

NIPA® GENX ELECTRONIC
RESOURCES & SOLUTIONS P. LTD.

101,103, Vikas Surya Plaza, CU Block
L.S.C. Market, Pitam Pura, New Delhi-110 034
Ph : +91 11 27341616, 27341717, 27341718
E-mail: newindiapublishingagency@gmail.com
web: www.nipabooks.com

For customer assistance, please contact
Phone: + 91-11-27 34 17 17
Fax: + 91-11- 27 34 16 16
E-Mail: feedbacks@nipabooks.com

ISBN: 978-81-19235-93-3

Composed and Designed by NIPA®.

Preface

Ornamental Horticulture assumes great importance in improving total environment and checking visual pollution by way of beautification. The flowers have been eulogised by poets and artists for generations not merely for their ornamental value but for their versatile curative properties. The flowers have a unique impact on the human mind, the colour adds beauty, fragrance keeps the mind and body afresh but the harmony lies in the art of flowers arrangement. Human association with flowering plants is prehistoric as they have always been attracted by beautiful and fragrant flowers.

Orchids are widely considered to be the most highly evolved of all flowering plants. They are amongst the most beautiful flowers of the entire plant kingdom, combining exotic looks with a diverse set of characteristics. Orchid plants are capable of growing indoors and outdoors, and are no doubt unique difficult to grow successfully. Among flower crops, orchids play a key role in Indian floral economy as they share sizeable export. The vast area and the varied agro-climatic conditions of our country, ranging from tropical to temperate region make possible to grow all kinds of orchids.

In this context, a need of simple book on recent advances in cultivation and management of orchid is highly demanded by students of Horticulture & Forestry and florists of India. This book is thus brought about mainly to fulfil the above requirement and is also the outcome of our teaching, research and field experiences of several years. We describe the morphology, propagation methods, cultivation, breeding, post harvest management, diseases and insect pest management of orchids in details in this book. Effort has been made to present the subject matter in a simple and lucid style and highlighting recent available information. The illustrations have been drawn from own specimens with photographs.

A number of text books, periodicals, bulletins etc. have been consulted during the preparation of this manuscript. We therefore; express our sincere gratitude to authors, scientists and publishers who have generated the aforesaid scientific literatures.

Since, this is the first edition of book and which has been published within a very short time and hence, there may be possibilities of several errors. We would therefore, feel pleasure if critical comments by the readers are provided with regard to same and other limitations in the present edition.

The keen interest and the sincere efforts of the publisher of this book "New India Publishing Agency, New Delhi", needs a special appreciation for publication within a short span of time. We also express our sincere thanks to our colleagues and friends who have contributed and encourage us to bring out this book.

Authors

Contents

Chapter - 1

Introduction to Orchid Flowers

Theophrastus gave the name "ORCHIDS" to the group of bizarre plants on the basis of the resemblance of paired underground tubers of these plants to masculine anatomy. This resemblance was also responsible for the mistaken belief that the orchids posses aphrodisiac properties and eating of underground tubers might "provoke venus" and they may be get male children. The appreciation of orchid beauty has a very long history in both the Western and Eastern cultures. Much of this is attributed to the diverse form and structure of orchids and the large number of species in the orchid family. Orchidaceae is amongst the most diverse family of the flowering plants consisting of 35,000 species under 880 genera (Kumar and Sheela, 2007, Fay and Chase, 2009). Orchids belong to the highest level of organization and are distinguished by the complicated biology of reproduction with long developmental cycle (Vakhrameeva *et al.,* 1991, Sherif *et al.,* 2012). They are beautiful and exotic perennial herbs grown in both north and south hemispheres. They exhibit an incredible range of diversity in size, shape and colour of flowers. Its flowers have lustrous, translucent petals rich in variegated colors and fragrances. Generally Orchids can grow at different elevations (Kumar and Sheela, 2007).

The cultivated orchids are native of tropical countries and occur in their greatest diversity in humid tropical forest of South and Central America, Mexico, India, Ceylon, Burma, South China, Thailand, Malaysia, Philippines, New Guinea and Australia. Brazilian *Cattleyas,* Mexican *Laelias* and Indian *Dendrobiums, Cymbidiums* and *Vandas* have played a major role in the development of modern orchid industry in the world. Each type of orchid has its own peculiarity. *Dendrobiums* are often used as cut flowers in floral arrangements, while *Cymbidium* was thought as the Mothers of orchids. Because of their simplicity and elegance, *Phalaenopsis* (Butterfly) orchids are the first choice for modern flowers arrangements, however, during 40s and 50s, *Cattleya* orchids ranks the first as far as the choice of brides are concerned. Some orchids are edible and used as garnish in culinary presentations. *Vanilla* is one of the examples of orchids being used for food.

Orchids in India

Majority of the cultivated orchids are native of tropical climates and are found in abundance in India in the states of Assam, Meghalaya, West Bengal, Karnataka, Sikkim and Kerala. Kalimpong, Thiruvananthapuram, Yercaud and almost all the coastal areas of India are the places most suitable for the cultivation of orchids (Kumar and Sheela, 2007). They have one of the richest orchid floras with over 1100 species and 163 genera occurring in all the major phytogeographical regions especially in peninsular India, north eastern region and the Himalayas. Out of the 1100 species, about 314 are endemics, 215 endangered and 14 nearly extinct (Shukla *et al.,1998*).

The Himalaya, with vast expanse and varied climatic regimes, is home to nearly 880 orchid species in 150 genera. Such a rich diversity is attributed to the prevalence of congenial conditions for large scale mutations (frequent seismic activity), wide matings (proximity to several immigrant taxa from adjoining areas of Bhutan, Burma, China, Nepal, Tibet, *etc.*), and better growth and development (high humidity, thick vegetation). The eastern Himalaya support more of epiphytic orchids whereas the western Himalaya are dominated by the ground orchids. The North West Himalaya (28 degrees 45'-36 degrees N latitude and 71 degrees 31'-80 degrees 40'E longitude), extends from Kumaon (India) to Chitral (Pakistan) but is phytogeographically indistinct from the other sections of Himalaya. Nearly 60% of the 300 orchid species estimated to grow, in the Indian part of the range, are ground growing in habit. They represent all the major tribes, are essentially an extension of the East Himalayan elements, and are primarily distributed in sub-tropical to temperate climates (1200-2700 m). The ground orchids dwell in a variety of ecological niches like open grassy and/or gravely slopes; semi-shaded to shaded, humus-rich, and moist forest floors; and disturbed situations along forest borders and road embankments where they often grow in close association with ferns, mosses, liverworts, and lichens. The Orchideae (72 species) is represented the best whereas Epidendreae

(57 species) and Neottieae (54 species) find almost equal representation. The number of species in each tribe and their frequency of occurrence diminishes westwards with the decline in rainfall. *Habenaria* with 17 species is the largest genus and 18 genera are represented by a single species each. *Epipactishelleborine, Goodyera repens, Herminium lanceum, Malaxis acuminata, Satyrium nepalense*, and *Spiranthes sinensis* have vast ecological amplitude, whereas *Brachycorythis obcordata, Cephalanthera ensifolia, Cypripedium cordigerum, Goodyera biflora, Liparis paradoxa, Neottia listeroides, OreOrchis indica, etc.* are more or less localized in distribution. The ground orchids are cytologically more dynamic than their epiphytic counterparts and diversification due to polyploidy occurs at higher elevations. Several therapeutically important species like *Dactylorhiza hatagirea, Epipactishelleborine, Eulophia dabia, Habenaria edgeworthii, H. intermedia, Malaxis acuminata, and M. muscifera* are faced with unregulated commercial collection pressures, while others are progressively losing their delicately balanced habitats. Loss of pollinators resulting in poor seed set, alongwith the natural calamities like heavy landslides and forest fires has also added to the woes of these habitat specific plants. As a result, many species *(AphyllOrchis gollanii, Cypripedium elegans, C. himalaicum etc.*) are no more available in their original habitats, while the size and frequency of populations, in others, has been detrimentally affected (Pathak *et al.*, 2005).

North-eastern India owing to its peculiar gradient and varied climatic conditions contains largest group of temperate and sub-tropical orchids. In about 132 wild genera nearly 287 species distributed over 71 genera are endemic (Rao, 1998). However, Lama *et al.*, 2012 have been recorded a total of 311 Orchids species under 85 genera from the Darjeeling Himalaya of India. Of them, 17 species with 6 genera are semi saprophytic, 77 species with 29 genera are terrestrial and the rest 217 species with 52 genera are epiphytic. 42 species found in the regions are medicinally important. Out of 184 genera and about 1200 species reported from India, 115 genera and about 496 species are represented in Sikkim (Shukla *et al.*, 1998). Roychowdhury (2004) reported that the orchids are prolific in North-East India particularly Darjeeling hills due to existing conducive climatological and phytogeographical conditions. There were 322 orchid species belonging to different genera reported from Darjeeling hills. By an evaluation programme on 21 *Dendrobium* spp. at Orchid Research Station, Kalimpong he revealed that *D. moschatum* followed by *D. fimbriatum and D. nobile var. alba* had the maximum plant height. Shoot production was more in *D. jenkinsii* followed by *D. gibsoni* and *D. ochreatum*. The number of leaves per plant were highest in *D. chrysotoxum* followed by *D. farmeri* and *D. nobilescens*. The largest flower in size was produced by *D. jamesonium* followed by *D. nobile*. The number of flowers per spike were highest in *D. densiflorum* followed by *D. fimbriatum* and *D. chrysanthum*. The highest no. of spikes per plant was obtained in *D. jenkinsii* followed by *D. fimbriatum, D. ochreatum* and *D. capillipes*. The longest spikes were produced by *D. nobile* var. *alba* followed by *D. moschatum, D. chrysanthum* and *D. nobile*. The flower spikes remained fresher in *D. nobile* followed by *D. fimbriatum* and *D. williamsoni*.

A taxonomic survey of the plants of Orchidaceae was made in the Dibrugarh District of Assam, India, during 2006-2010. Total 111 species under 50 genera are reported in this communication (Borah *et al.*, 2012). Whereas, Borah *et al.*, 2010 made a taxonomic survey of the plants of Orchidaceae in the Jokai reserve forest of Dibrugarh District of Assam, during 2006-2008 and reported a total 31 species under 22 genera in their communication. A survey was also conducted during 1999-2002 on the epiphytic orchids of Chirang Reserve Forest under Haltugaon Forest Division of Kokrajhar district of Assam, India, and 42 species within 22 genera were recorded (Basumatary and Sarma, 2004). A new orchid species from Assam, *Ornithochilus cacharensis barbhuiya*, is described by Barbhuiya *et al.*, 2012. It differs from the other species in the small genus *Ornithochilus* (Lindl.) Benth in floral morphology and in the uniformly red-purple flowers.

A report on the threatened orchids of Manipur was published by Ramakantha,1995 and mentioned that Only 3 of the 34 orchid species from north-east India listed as threatened have previously been reported as occurring in Manipur, namely *Dendrobium bensoniae, Renanthera imschootiana* and *Vanda coerulea*. However, 3 more threatened species have been observed, *Paphiopedilum spicErianum, Pleione hookEriana* and *Galeola falconeri. P. spicErianum* and *P. hookEriana* are being successfully grown in the Manipur state orchid gardens as a part of an ex situ conservation programme. The orchid genus *Kalimpongia* and 3 further species, *Kalimpongia narajitii, ScheonOrchis manipurensis* and *Ascocentrum ampullaceum* var. *aurenticum* are endemic to Manipur and should be listed as threatened. Singh *et al.*, 1990 published reports and the results of surveys in the state of Mizoram, formerly Lushai Hills, record 246 orchid taxa in 74 genera. Of these, 67 taxa (including saprophytes) are terrestrial and 179 are epiphytic. *Bulbophyllum parryae, Eria lacei* and *Sterogyne lushaiensis* are endemic. *Dendrobium* with 41 species and one variety, is the largest genus. Forty genera, 3 being monotypic, are represented by a single species. About 55% of the total taxa appear to be endangered.

A total of 123 orchid taxa (including 4 new combinations) in 42 genera, collected in Orissa during the past 2 decades, are listed together with details of their habit, distribution, and flowering period (Misra, 1989). The genus *Habenaria* Wild is represented in Maharashtra with about 21 species. Among them, *Habenaria stenopetela* Lindl. is seems to be a rather very rare orchid of Maharashtra State. It was reported from Mumbai, Pune and Thane district of Maharashtra by the earlier workers. There is no recent authentic report on its occurrence from the State (Bachulkar *et al.*, 2011).

Andaman and Nicobar Islands have a vast variety of orchids. There are about 138 orchids distributed in different Islands. Many of the species found here are rare and threatened. Four genera *viz., Grossourdya, Macropodanthus, Malleola* and *Plocobglottis* from these islands have so far not been reported from any other parts of India (Singh and Medhi, 2006). However, eighteen orchid species belonging to 12 genera (*Agrostophyllum, Bulbophyllum, Coelogyne, Cymbidium, Dendrobium, Eria, Eulophia, Luisia, Oberonia, Papilionanthe, Rhynchostylis* and *Vanilla*) were recorded by Singh *et al.*, 2002 in Andaman Islands.

Most of the *Paphiopedilum* (lady's slipper) species are restricted to N.E Himalayas except for *P. druryi* which has been found in Kerala. The principal genera involved in orchid export include *Pleione, Dendrobium, Cypripedium, Paphiopedilum, Coelogyne, Cymbidium, Phalaenopsis* , *Vanda, Calanthe* and *Aerides*, and the principal importing countries include Japan, USA, UK, Netherlands and Germany. While the genus *Pleione* is predominantly exported to Japan, the genus *Dendrobium* is exported to the USA (Chakrabarti, 1995)

All wild growing orchids are banned or restricted for export from India. The Convention of International Trade in Endangered Species of Fauna and Flora (CITES), ratified by India, places all species of Orchidaceae under Appendix II, meaning thereby that their trade will be only through export permits. Orchid sanctuaries are established in Sikkim at Singtom and Deorali and in Arunachal Pradesh at Tapi. National Orchidarium and Experimental Garden, Yercaud houses a total of 158 orchid taxa, being propagated and maintained as germplasm. Out of these, 42 taxa under 28 genera are rare and 39 taxa under 15 genera are endemic to Indian region (Satyanarayana, 2010). The genus *Taeniophyllum* is represented by 8 species in India. Of these, *T. crepidiforme* (King & Pantl.) King & Pantl. appears to be extremely rare and needs immediate measures for conservation (Kumar and Lal, 1995).

History and Importance

Arditti (1992) had given an excellent historical account of orchids in Asia, Africa, Europe, New Guinea and Australia. It was thought that orchids were ancient flower species, possibly in existence at the time of the dinosaurs in the Late Cretaceous period. It began to appear in Chinese botanical manuscripts by the third century B.C. and in the West, orchid flowers were recorded by Theophrastus in manuscripts, in about 300 B.C. By the mid 1700 many orchid species were brought to Europe from more exotic shores by worldwide plant hunters. The establishment of collected orchid species began at the Royal Botanical Gardens in Kew, London, England; over the years orchid cultivation and orchid hybridization have developed, resulting in many more variations of the orchid species. National Orchid Garden at Singapore holds one of the best collections of cultivated orchid species in the world.

During the Victorian era, orchid symbolism shifted to luxury, and today this sense of magnificence and artful splendour continues with orchids representing rare and delicate beauty. When Confucius came to know the Asian *Cymbidium* he likened it to the virtues of education and he stated that a solitary Orchid stands adorning the side of a mountain and perfumes the air even in the absence of appreciation. He termed the Asian Orchid as the "king of fragrance". Orchids suggested virility to the ancient Greeks. *Cattelya* orchid represents mature charm and pink orchids convey pure affection at wedding anniversary. The *Cymbidium hookErianum* was considered a delicacy in Bhutan where it was traditionally cooked in a spicy curry or stew and called "olatshe" or "olachoto".

The only food obtained from the orchid family is a flavouring substance known as *Vanilla* (Jianping and Webster, 2011,Naidu *et al.*, 2012), which is extracted from the ripe, seed-bearing fruits (beans) of the *Vanilla* orchid (*Vanilla fragrans*), a climbing genera native to the West Indies and Mexico (Nakazawa, 1987, Brochado *et al.*, 2010). The discovery of *Vanilla* is not easy to explain, because the fully ripened , uncurved capsules are devoid of the active principle, vanilline. Vanilline (Frenkel *et al.*, 2006, Lavine *et al.*, 2012) is formed during sweating of the beans which brings through enzyme action. During sweating an aromatic white crystalline derivation of an aldehyde, vanilline (C_8 H_8 O_8), is formed due to changes from glycoside (*Pardio et al., 2009,* Naidu *et al.*, 2012). Natural vanillin is enriched in deuterium and 13°C as compared to synthetic vanillin. This enrichment probably results from differences in the synthetic pathways of this compound. In particular, carbon isotope fractionation is associated with photosynthesis; this fractionation lowers the 13C:12C ratio by about 20%, relative to atmos. CO_2 for the Calvin cycle in photosynthetic CO_2 fixation, which occurs in trees. The fractionation factor is even lower, 13%, for the orchid *Vanilla* planifolia Andrews in which the Crassulacean acid metabolic pathway probably operates (Frenkel *et al.*, 2004). *Vanilla* still remains one of the most important and widely used flavours in the food industry and is also extensively employed by fragrance and pharmaceutical manufacturing companies. Natural *Vanilla* flavour, extracted from the pods of the tropic orchid *Vanilla,* is considerably more expensive than synthetic vanillin (Greule *et al.,* 2010). Production of phenolic flavour compounds, including vanillin, in cell cultures of *Vanilla* spp. is reviewed by (Doernenburg and Knorr, 1996). A comparative study was made by Rey *et al.*, 1980 of 3 widely used *Vanilla* spp., viz. *Vanilla planifolia, V. tahitensis* and *V. pompona*. The content of vanillin is higher in *V. planifolia* than in the other 2 spp. detailed analysis of *V. pompona* revealed the presence of anisic aldehyde, not previously recorded. Presence of piperonal, could not be confirmed in any of the 3 spp. or for any methods of extraction and analysis. Long before the arrival of Europeans in Mexico, the Aztecs or their predecessors had discovered *Vanilla* (Freitag, 2006). This widely used spice and flavouring agent probably served the Mexican natives first as a medicine of, but by the time of the Conquest it was most extensively employed as a flavouring agent for chocolate (Freitag., 2006). The Aztecs were said to drink a mixture of the *Vanilla* orchid and chocolate to give them power and strength, and the Chinese believe that it can help to cure lung illnesses and coughs. Natural *Vanilla* is still prized over the chemically pure vanilline which is employed where low cost rather than flavor is the governing factor. An early report of *Vanilla* in a European herbal was that found in Pomet's "Histoire Generale des Drogues ," published in 1694 Pomet said that the oil and volatile salt of the *Vanilla* pod was mixed in chocolate had a properties of a cordial, cephalick, stomachic, carminative, apertive and it alternates vicious humours, provokes urine and women's courses. *Vanilla* still remains one of the most important and widely used flavours in the food industry and is also extensively employed by fragrance and pharmaceutical manufacturing companies. Natural *Vanilla* flavour, extracted from the pods of the tropic orchid *Vanilla,* is considerably more expensive than synthetic vanillin. The disparity of prices between natural vanillin and that derived from other sources has given rise to many cases of fraudulent adulteration, and for more than 30 years, strenuous efforts have been made to authenticate sources of vanillin (Greule *et al.,* 2010).

The first English herbal, published in 1568 by Turner stated of orchids as Dog Stones. He writes of herb that if the greater root be eaten of men it makes the men children and if the root be eaten of women it make the women children. The claims for therapeutic properties of numerous orchids became so extensive that attempt was made to rationalize and explain the reasons for the supposed physiological actions of the plants. This was true especially for the claims of the influence of orchids on sexual phenomena in man. In 1633, English herbalist 'Gerarde' published an orchid classification according to "stones", correlating with this crude classification on morphological character the supposed virtues associated with the plants. Gerarde's classification makes extremely interesting reading to the modern naturalist. According to Gerarde's classification one group of orchid, for example were called Goat Stones because they had a "stinking and loathesome smell "like that associated with goat. Seraphias Stones bore this name because "sundry of them bring forth flowers resembling flies and such like fruitfull and lascivious insects, as taking their name from Serapias, the god of the citizens of Alexandria in Egypt, who had a most famous Temple at Canopus, where he was worshipped with all kinds of lascivious wantonness, songs and dances. Tuber-bearing orchids often had vernacular names which called attention to their testiculate underground parts- Dog Stones, Fox Stones *etc.*

Reliance upon orchids for certain therapeutic uses did not die out completely after the middle ages in Europe. An American herbal, Good's "The Family Flora" published in 1845, reported for *Cypripedium pubescence* (Lady Slipper Orchid) that it is one of the most valuable vegetable medicines. He also stated that its operation upon the system appears to be in harmony with the laws of animal life, giving tone to the nervous system; and it is therefore useful in all cases of nervous irritability, hysterical affections, spasms, fits and all derangements of the functions of the brain, such as madness, delirium *etc.*, and in all cases of inability to sleep, particularly in fevers consumption, *etc.* Lady Slipper is still found for sale in herb shops in Europe and the United States even though there is no scientific basis upon which its medicinal use may be defended. Country folk in some parts of the United States still rely on *Aerthusa bulbos* to relieve toothache and *Corallorrhiza odontorrhiza* as a febrifuge. The Indians of North America continue to use orchids medicinally, such as *Epipactisgiganteum* for insanity. The native people of the West Indies drink the liquid in which pseudobulbs of *Bletia purpurea* have been boiled to relieve symptoms of ptomaine poisoning from fish.

Old World tropics has made of orchids as therapeutic agents. In Amboina, Indonasia, a pomate made of the pseudobulbs of *Grammatophyllum scriptum* was believed to cure sores and the leaves of *HetaEria obliqua* were used for a similar purpose. Several species of *Dendrobium* find place in Malaya to treat skin eruption and infections, while in tropical Asia generally this genus was valued as cure for dropsy and headache. The Chinese consider *Dendrobium nobile* something of a cure-all. Species of *Nervilia, Vanda* and *Gastrodea* had extensive used as medicines in the Asiatic tropics. A number of orchids were valued as febrifuges in treating malaria; *CorymbOrchis longiflora, Tropidia curculigoides,* and *Acriopsis javanica.*

In Africa it was found that Zulus an infusion of *HabenEria foliosa* and *Ansellia humilis* as an emetic. The Swagi had used *Lissochilus Krebsii* to treat a disease of children. Species of *Eulophia* were thought to prevent miscarriage, cure barrenness and to be valuable for other strange troubles. Medicinal use of orchids amongst the Indian populations was apparently not common in South America. *Epidendrum bifidum* had been employed to expel tapeworm and intestinal parasites, and the Chileans hold that *Spiranthes diuertica* was efficacious as a diuretic.

Hossain (2012) reviews the traditional therapeutic uses of orchids with its recent advances in pharmacological investigations that would be a useful reference for plant drug researches, especially in orchids. According to him Orchids have been used as a source of medicine for millennia to treat different diseases and ailments including tuberculosis, paralysis, stomach disorders, chest pain, arthritis, syphilis, jaundice, cholera, acidity, eczema, tumour, piles, boils, inflammations, menstrual disorder, spermatorrhea, leucoderma, diahorrhea, muscular pain, blood dysentery, hepatitis, dyspepsia, bone fractures, rheumatism, asthma, malaria, earache, sexually transmitted diseases, wounds and sores. Besides, many orchidaceous preparations are used as emetic, purgative, aphrodisiac, vermifuge, bronchodilator, sex stimulator, contraceptive, cooling agent and remedies in scorpion sting and snake bite. Some of the preparations are supposed to have miraculous curative properties but rare scientific demonstration available which is a primary requirement for clinical implementations.

Since the days of Dioscorides, salep had considered as an aphrodisiac. Salep was prepared by tuberous underground parts of any the species of *Orchis*. The roots were collected and scalded in boiling water, and skin was removed. They are then dried to a horny consistency and powdered for use. It had been of great importance in the past as medicine, but its use has now fallen into disrepute except in the country parts of Asia Minor and the adjacent Mediterranean areas. Salep was probably somewhat useful as a food, and as a food that is still employed. Its composition runs about 48 per cent mucilage, 3 per cent starch, 5 per cent nitrogenous material, 2 per cent mineral ashes. In India, the tubers of *Eulophia, Orchis* and *Satyrium* may all be used as salep, it was used like sago and narrow root and was prepared from a large variety of species. Starin (2012) describes the medicinal properties (with emphasis on the aphrodisiac effect) of salepi, a hot, aromatic, sweet, thick drink traditionally made from the powder of dried and pulverized orchid (*Orchis* sp.) tubers, sugar, milk and cinnamon. The dwindling popularity of this beverage in Athens, Greece, and the substitution of orchid tuber powder with other thickeners as a response to the decline in orchid tuber availability. Lawler and Rao (2002) review medicinal uses of orchids in some Asian countries in his review covers information on the medicinal value of 137 species of orchids including the scientific name, plant parts with medicinal value and methods of use to cure different ailments. A brief note on Salep, an aphrodisiac tonic, is presented. Georgiadis *et al.*, 2012 extracted Polysaccharide from ground salep (dried orchid root). These were studied in terms of their composition and their capacity to affect the rheology and stability of emulsions. Successive extractions yielded three separate extracts, namely HBSS (hot buffer soluble solids, extracted at 70°C, (pH = 5.2), CHSS (chelating agent soluble solids, extracted at 70°C , pH = 5.2), and DASS (diluted alkali soluble solids, extracted at 0°C). HBSS comprised of negatively charged macromolecular populations of approx. 600-700 kDa (d approx. 500 nm), one at 50 kDa (d approx. 70 nm), and a third one at 12 kDa (d approx. 20 nm). Upon addition into Tween 20-stabilized emulsions, HBSS caused flocculation and enhanced creaming at low concentrations (0.1%), while at higher concentrations (>=0.7%) it drastically reduced creaming due to the increase of the aqueous phase viscosity. CHSS extracts contained a negatively charged polysaccharide population of approx. 50 kDa (d approx. 140 nm). DASS extracts contained only small molecules. Addition of CHSS or DASS extracts into Tween 20-stabilized emulsions did not affect the viscosity or the creaming stability of the emulsions. Flocculation was induced at 2% CHSS.

Though the Orchidaceae has given man few foods, there are some species which might be worth mentioning as nutrients of local use. The fragrant *Angraecum fragrans* forms the basis of the "faham tea" from Isle de Reunion, of the coast of Africa. It was once rather well known in France. The tea was delicious, but it did not become very widely accepted and had for practical purposes been completely forgotten. The fragrant leaves of *Dendrobium salaccense* were used as a condiment for rice in Malaya, where several species of *Habenaria* were eaten as greens. In the tropics of Asia, roasted tubers of *Gastrodia* are eaten like potatoes. The Brazilians employ the seed capsules of *Leptotes bicolor* to flavour ice cream, and other species of orchids are valued locally throughout the tropics as condiments. Pant and Rinchen (2012) describe the medicinal importance of *Dactylorhiza hatagirea* and mentioned that it is widely used to cure various diseases like dysentery, diarrhoea, chronic fever, cough, stomachache, wounds, cuts, burns, fractures, general weakness, *etc.*, and widely used in modern medicine. Therefore, an attempt has been made to assess the medicinal potential of the species both in traditional as well as in modern medicine system. These properties are supposed to be cured by different active compounds present in the plant.

Bowls for smoking pipes were carved by natives in the West Indies from the hollow pseudo-bulbs of *Laelia Thompsoniana,* and in some parts of tropical Latin America trumpets to call childern were fabricated from the bulbs of *L. tibicinis* -the specific epithet coming from the Latin word for "trumpeter". In South America, glue for the cobbler's art was extracted from pseudobulbs of *Cyrtopodium,* whereas Middle American natives apply the dried gum from these pseudo-bulbs to violin strings. In ancient Greece, orchids were associated with virility. In fact, Greek women believed that if the father of their unborn child ate large, new orchid tubers, the baby would be a boy. If the mother ate small orchid tubers, she would give birth to a girl.

The evolution of orchid culture from hobbyist to commercial grower was very slow as most of the orchids collected from different regions did not adapt to local conditions and died. The methods of propagation were not known and most of the orchids which flowered under utmost care were found to be self sterile. Further, it was difficult to germinate the seeds, which lacks any functional endosperm. The seeds sown in the nursery beds required long period of germination and any disturbance to the soil or physical environment destroyed the whole population.

Conrad Loddiges and sons (1821) started growing orchid plants commercially at their nursery in Hackney, followed by John Daminy of M/s Veitch & Sons who produced the first orchid hybrid using two *Calanthe* species in 1852, which flowered in 1856. This was followed by a large number of manmade hybrids produced by hobbyists and small growers. In 1913, Sun Kee Nursery, Singapore, started the first cut-flower production of *Arachnis* type hybrids.

The major developments in cultivation of orchids in the world have been due to modern scientific technology which has been suitably used in case of orchid seed germination and meristem culture. Today orchids are grown on assembly-line method in extensive glasshouses with controlled environment and the sale of orchid flowers worth in millions of dollars. The modern methods of propagation have bought orchid cultivation *at par* with other commercial crops. Large-scale cultivation of orchid cut-flowers and potted orchids is now the trend. In the past, orchid growers and hobbyists relied solely on the collection of orchid species from the wild because the technique of breeding and selection is not available. Mass cultivation becomes possible with the breakthrough in orchid seed germination. Discovery and development of asymbiotic method of germination of orchid seeds in 1921 by Lewis Knudson revolutionize the orchid industry. The availability of asymbiotic germination and tissue culture has made large-scale orchid cultivation economically feasible. Today, orchids such as *Cymbidium, Dendrobium, Phalaenopsis* and *Oncidium* are marketed globally and the orchid industry has contributed substantially to the economy of many ASEAN (Association of the South East Asian Nations) countries. The market potential for both orchid cut-flowers and potted orchids is very favourable. This is evident from the world market demand of planting materials for orchids grown for cut-flowers and potted plants.

In the Indian Vedic scriptures there is a mention of the plants under the name *"VANDA"*, which has been adapted as a generic name in one of the most beautiful group of orchids. The commercial cultivation orchids in India is not organised and is still in the hands of hobbyist and few dealers who mainly depends on wild collections from forest to meet a large part of their foreign and local demands, due to which some of the orchids-growing areas are now without any orchid and very rare species are now facing the danger of depletion.

Ornamental Indian Species

Some of the Indian orchid species which are of high ornamental value are : *Aerides crispum, A. fieldingii, A. multiflorum, A. odoratum, Anaectochilus roxburghii, Arachnis clarkei, Arundina graminifolio, Bulbophyllum leopardinum, Calanthe masuca, Coelogyne elatn, C. devonianum, Cymbidium pendulum, C. longifolium, C. munronianum, Dendrobium aggregatum, D. aphyllum, D. fimbriatum, D. jenkinsii, D. moschatum, D. nobile, Paphiopedilum faireanum, P. venstum, P. hirsutissium, P. insigne, Phaius wallichii, Pleione praecox, Rhynchostylis retusa, Thunia alba, Vanda cristata, V. coerulea* and *V. coerulescens.*

Orchids with Medicinal Value

In India, orchids are employed for a variety of therapeutic use in different systems of traditional medicines like Ayurveda, Siddha and Unani. *Acampe praemorsa*, an epiphytic orchid is used in rheumatism. Tubers of many *Habenaria's* are being used to treat for unconsciousness, vermicide and as blood purifier. Locals in the form of salep are using *Cymbidium aloifolium* as emetic and purgative. *Dendrobium ovatum* for stomache ache *Eulophia nuda* for tumors and bronchitis, *FlickingEria nodosa* (Purusharatna) is astringent, aphrodisiac, expectorant *etc.* *Luisia zeylanica* is being used as emollient for boils, abscess and burns. *Habenaria acuminata, Nervilia aragoana, Satyrium nepalense, Vanda testacea, Zeuxine strateumatica* are some of the other orchids, which are also medicinally important. A botanical exploration in Palani and Sirumalai hills in Tamil Nadu, India, conducted in 1997-98 by Karuppusamy and Kumuthakalavalli (2002) and identified 4 medicinally important orchids: *Cymbidium aloifolium, Eulophia nuda, Luisia birchia* and *Vanda tesellata.* Aravindhan *et al.,* (2011) highlights the medico-potentiality of 11-orchidaceous species viz., *Acampe praemorsa* (Roxb.) Blatt. & Mc. Cann., *Bulbophyllum acutiflorum* A. Rich., *Bulbophyllum neilgherrense* Wight., *Cymbidium aloifolium* (L.) Sw., *Dendrobium herbaceum* Lindl., *Geodorum densiflorum* (Lam.) Schltr., *Habenaria plataginea* Lindl., *Habenaria roxburghii* Nicolson, *Malaxis acuminata* D. Don, *Oberonia wightiana* Lindl. and *Vanda tessellata* (Roxb.) Hook. f. ex D. Don, collected from the Velliangiri hills, Southern Western Ghats of Tamil Nadu, India.

Orchids represent the largest and most important group of flowering plants in Meghalaya. Besides ornamental importance, some of them are used as medicinal plants. Many of these orchid plants are used to cure a variety of diseases and disorders like stomachache, diarrhoea *etc.* Local people use these medicinal orchids as ayurvedic and Unani treatment of many diseases (Roy *et al.,* 2007). The traditional use, geographical distribution, chemical constituents (alkaloids, flavonoids, anthocyanins, and carbohydrates and glycosides), and conservation status (extinct or endangered) of medicinal orchids in India are described by Miniraj (2010). Twelve species are listed, with vernacular names, distribution and medicinal uses (based on the literature of Ayurvedic and other systems of medicine) by Chauhan (1990). Pathak *et al.,* (2010) describe that Orchids have been used in the local system of medicine, in different parts of the world since time immemorial. Based on the observations, they mentioned an account of 25 species of medicinally important orchids from the Himachal Pradesh along with brief notes on their habit, distribution, and flowering Period and medicinal uses. Dobriyal (2002) described the rare and endangered medicinal orchids of Asthavarga group in Uttaranchal - *Habenaria intermedia* (Virdhii) and *Habenaria edgeworthii* (Ridhii). Ancient Ayurveda of India indicates that Asthavarga is important ingredient of various classical Ayurvedic formulations like Chavyanprasa. Out of eight constituents of Ashtavarga, four have been reported to be orchids (Table 1.1).

Table 1.1 : Orchid plants used in Ashtavarga.

S.No.	Ayurvedic name	Botanical name	Family	Part used
1.	Jivaka	*Malaxis muscifrea*	Orchidaceae	Bulb
2.	Rishbhaka	*Malaxis acuminata*	Orchidaceae	Pseudo-bulb
3	Riddhi	*Habenaria intermedia*	Orchidaceae	Root
4	Vriddhi	*Habenaria edgeworthii*	Orchidaceae	Root

In India work has been carried out on chemical analysis of some medicinally useful orchids such as *Eulophia campestris, Orchis latifolia* and *Vanda roxburgii. Dendrobium macraei* is another important orchid from Ayurvedic point of view as it is reported to be source of Jivanti. *Cypripedium parviflora* is widely used as aphrodisiac and nervine tonic in Western Herbalism. Many medicinal orchids are reported to contain alkaloids. Some of the phytochemicals isolated from orchids along with biological source have been describe in table1.2.

Table 1.2 : Some phytochemicals isolated from orchid

S.No	Name of phytochemical	Phytochemical class	Source	Part used and action
1.	Aeridin	Phenanthropyran	*Aerides crispum* *Aerides affine*	Tubers, AntiBacterial
2.	Agrostophyllinol	Triterpenoid	*Agrostophyllum brevipes*	Tubers.
3.	Agrostophyllinone	Triterpenoid	*Agrostophyllum brevipes* and *Agrostophyllum callosum*	Tubers.
4.	Isoagrostophyllol	Triterpenoid	*Agrostophyllum callosum*	Tubers.
5.	Orchinol, 6-methoxycoelonin, imbricatin,	Stilbenoids	*Agrostophyllum callosum*	Tubers.

Contd.

	flaccidin, oxoflaccidin, isooxoflaccidin, flaccidinin, agrostophyllin, callosin, callosinin, callosumin, callosuminin and callosumidin			
6.	Arundinan	Stilbenoid	*Arundina graminifolia* **Common name:** Bamboo Orchid Rhizome. AntiBacterial	*(Cymbidium bambusifolium* Roxb.)
7.	Cypripedin	1-4 phenanthrene quinone	*Cypripedium calceolus pubescens* **Common name:** Lady's Slipper orchid.	Roots. Antispasmodic, diaphoretic, hypnotic, nervine, sedative, tonic. The plant is used as substitute for *ValEriana officinalis* L. although it is inferior .
8.	Loroglossin	Glucoside	*Orchis latifolia* ***Syn. Dactylorhiza hatagirea*** **Common name:** Salampanja, Marsh *Orchis*, salep orchid.	Roots. Aphrodisiac, expectorant and nervine tonic
9.	Jebantine	Alkaloid	*Dendrobium macraei* **Syn:** *Ephemerantha macraei* (Lindl.) Hunt et Sunmeh, *FlickingEria nodosa* (Dalz.) Seiden f.	Tubers. **Actions**: Tonic. **Therapeutics**: General debility.

Contd.

10.	Gigantol	Bibenzyl	*Dendrobium nobile* **Syn:** *Dendrobium* lindleyanum, *Dendrobium* coerulescens	Whole Plants **Actions**: Antiphlogistic, pectoral, sialogogue, stomachic and tonic . Anti-mutagenic
11.	Moscatilin	Bibenzyl	*Dendrobium nobile*	
12.	Dendrobine	Alkaloid	*Dendrobium nobile*	
13.	Nudol	Phenanthrene	*Eulophia nuda* **Syn:** *Eulophia dabia* (D.Don) Hochr **Common name:** Whitton root, Salep.	Tubers **Actions:** Demulcent and anthemnintic. **Therapeutics:** Worm infestation and scrofula.
14.	Melianin	Glycoside	*Vanda roxburghiiSyn.* ***Vanda tessellata*** **(Roxb.) Common name:** *Vanda.*	**Parts used:** Whole plant. **Actions:** Aphrodisiac, analgesic and nervine tonic **Pre-clinical studies:** Aphrodisiac , anti-inflammatory, anti-arthritic , antimicrobial and wound-healing
15.	Kinsenoside	Glycoside	*Anoectochilus formosanus*	Tubers. Anticancer

Medicinal Orchids of Uttarakhand (Dobriyal, 2002)

1. *Aerides multiflorum* Roxb. Epiphytic on miscellaneous forest of Siwalik found in Maana ranges and sub mountain Himalaya. Its leaf paste applied as poultice on cuts and wounds.
2. *Calanthe tricarinata* Lindley. It is found in syoru damp and shady places in mountain forests as terrestrial plant. Leaf paste applied on sores and eczema. Leaves and pseudo bulbs believed as an aphrodisiac by locals.

3. *Coelogyne cristata* Lindley. This genera is epiphytic on rocks or epiphyte on plants in the area of Gondya, Harjojan mountain forests. Infusion of pseudobulbs given in constipation as also as an aphrodisiac.

4. *Epipactishelleborine* (L.) Crantz. Found as terrestrial plants in the mountain forests under damp shady (Trindrya) places. Infusion of leaves given in intermittent fever. Rhizome regarded as an aphrodisiac.

5. *Eulophia dabia* (D.Don) Hoch. Terrestrial, rare, associated with dense forest (Misri, Salip, floor, Submontane to mountain) Himalaya, Salam mishri, Munjatak. Tubers are given to infants in cough and cold.

6. *Goodyera repens* (L.) R.Br. Terrestrial, in shaded oak forest floor, in Vern: Girwara montane and alpine Himalaya. Plant paste externally applied in syphilis, extract is taken as a blood purifier.

7. *Habenaria intermedia* D. Don. In moist shady places of mountain forest, Vern: RidhilViriddhi (Sans) generally in open grass land. The one of the ingredient of Astavarga of Ayurveda, used as tonic.

8. *Habenaria marginata* Colebr. Terrestrial on grassy slopes, forest Vern- Haldya-Jari cover submontane to montane Himalaya. Thoroughly boiled plant extract taken in flatulence.

9. *Herminium lanceum.* Terrestrial, common in shaded and wet Forest, rock shelters in mountain and alpine Himalaya.Extract of plant given in suppressed urination.

10. *Malaxis acuiminata* D. Don. Terestrial, damp and shady forest Sanskrit name-Jeevak, submontane to alpine Himalaya Vern.- Lasuni. One of the ingredient of "Astavarga" medicine of Ayurveda. Locally the bulbs are used in bronchitis as well as given as a tonic.

11. *Habenaria intermedia* D.Don. In moist shady places of montane forest, Vern: RidhilViriddhi (Sans) generally in open grass land. The one of the ingredient of Astavarga of Ayurveda, used as tonic.

12. *Malaxis muscifera* (Lindley) Kuntze. Terrestrial in moist shady forest grassy Sansk- Rishbhak / jeevak slopes of montane zone. One of the ingredient of "Astavarga" medicine. The underground parts used as tonic.

13. *Pholidota articulata* Lindley. Epiphytic on rock and trees in submontane Sanskrit - Jeevanti to mountain Himalaya. Whole plant used as tonic.
14. *Satyrium nepalense* D. Don. Terrestrial, in open grassy slopes in mountain to Vem-Saleep, salam mishri alpine slopes. Root used as tonic and also used in diarrhoea, tubers edible.
15. *Spiranthes sinensis* (Persoon) Ames. Terrestrial in open grassy slopes Vem-Phirtya submontane to mountain Himalaya. Decoction of plant given in intermittent fever, tubers used as tonic.
16. *Vanda cristata* Linn. Epiphytic in mountain zone. Stem and leaf extract used as a tonic.
17. *Dactylorhiza hatagirea* (D.Don) Terrestrial, Tubers palmate in subalpine Vem-Hathajari, Salampunja and alpine Himalaya in open moist localities. Used as general tonic, as aphrodiasic and in bronchitis.

Orchids Exploitation

About 10 % orchid genera are believed to be endangered in their native habitats. There are two major causes; the primary threat is from the physical destruction of habitat. An analysis of 746 Indian orchids showed that those with large showy flowers were the most vulnerable to commercial exploitation and the most likely to be endangered species (Lokesha and Vasudeva 1994). Habitat destruction not only destroys the places for the plants to live, but also causes loss of the orchid's pollinators, other plants, and fungi that they depend on. The other threat is also man-caused such as over-collection. Despite bans in many countries on the collection and export of native plants, in many cases the damage to the population is already beyond repair and there is still a trade from unscrupulous orchid growers. A large number of orchid species, which were present in plenty in Indian forests, are now at the verge of extinction and some of them have become so rare that a large number of botanical teams were unable to trace them. To cite an example *Paphiopedilum druryi*, which was once found in plenty in Agastaya Hills in south India, is now difficult to locate. Some orchids are endemic to India are so ornamental and in demand that their natural populations have been over exploited. Some species in the genera like *Arundina* , *Cymbidium, Coelogyne, Dendrobium, Paphiopedilum, Renanthera,* and *Vanda* are almost extinct. The provisional list of 150 endangered plants of India includes many orchids like *Acanthephippium sylhetense, Anoectochilus sikkimensis, AphyllOrchis montana, Arachnanthe clarkei, Arundina graminifolio, Cymbidium macrorhizon, Dendrobium densiforum, Didiciea cunninghamii, Eria crassicaulis, Galeola lindleyana, Gastrodia Exilis, Paphiopedilum fairanum, P. druryi, Pleione humilis, Renanthera imschootiana, Vanda coerulea, V. pumila and V.roxburgh.*

Rare Endangered Orchid of Uttarakhand

Rare endangered Orchid of Uttarakhand recorded in Red Data book of Indian Plants (Nair and Shastry, 1987, 1988, 1990):
TAXA RDB Status

1. *AphyllOrchis gallani*
2. *Archinottia microglottis*
3. *Cyperipidium elegans*
4. *Cyperipidium himalaicum*
5. *Diplomeris hirsuta*
6. *Eria occidental*
7. *Eulophia mackinnonii*
8. *FlickingEria hesperis*
9. *AphyllOrchis parviflora*
10. *Calanthe alpina*
11. *Calanthe pachystalix*
12. *Cypripedium cordigerum*

Chapter - 2

Orchid Improvement

Orchid Morphology

Most of the orchids are perennial herbs with simple leaves. Although the specialized flower structure confirms to a standard plan, the vegetative parts are shown great variation, a large number of them being epiphytes, or terrestrial and a few saprophytes are leafless in nature. Orchids can be divided into two groups - monopodial or sympodial depending upon their habit of growth. Monopodial orchids such as *Vanda* have a main stem which continues to grow year after. Sympodial orchids such as *Cymbidium* have a main stem, which terminates growth at the end of each season. A new shoot then grows from the base forming its own bulbous stem called pseudo-bulb, which eventually flowers. The pseudo-bulb or thickened stem is a very useful device for the storage of food and water and function like bulbs. In addition to the epiphytic orchids, there are also grounds orchids or terrestrial orchids, which grow like ordinary plants with their roots in soil. Most of the temperate zone orchids are terrestrial and tropical orchids are epiphytes.

Taxonomy

A number of taxonomists have the opinion that the orchid family is a part of the lily family. The correlation of structural features of a tulip flower to those of an orchid can made it more clear. In a tulip flower and in orchid flowers, there are six floral segments – three outer sepals and three inner petals. In the orchids, one petal is shaped differently from the others to guide the pollinators to the 'right place'. As with tulips, the centre of the flower holds a stem – the continuation of the flower stem/ovary, which in orchids is called the "column". At the end of this stem, in tulips, there is the sticky stigma. In orchids, this stigma is not at the top but slightly beneath the top and it is a little cavity. Surrounding the stigma in tulips, there are six pollen-bearing anthers. In orchids, these anthers are part of the column. The pollen-grains are wrapped in little sacks that sit at the end of the column under the anther cap and are loosely attached to the column by a thin thread called viscidium, which detaches rather easily. Orchid pollen are not dusty but waxy, to ensure safe transfer to the next flower. The lady slippers are built on the same 3+3 system, like tulips with major changes. Main concern in these structural differences is the fact that the stigma is at the back of the column, behind the front-facing staminode.

The most recent classification of the Orchids which widely accepted among botanists is that of Robert Dressler (1981 and 1986). He recognised six subfamilies in the Orchidaceae. However, during recent years, comprehensive molecular phylogenies of this family have supported the division of the Orchidaceae into five monophyletic subfamilies that consecutively have sister relationships to each other. The subfamily Epidendroideae comprises 80 per cent of all orchid species; most of them are epiphytic, although the four most basal subfamilies are terrestrial. *Hypoxis villosa* is supposed as a representative of the closest relatives of the orchids (Hypoxidaceae). Within this subfamily there are many tribes, including the Cymbidieae which contains all the sympodial vandoid orchids with similar seeds and two pollinia. By this criterion the subgenus *Jensoa*, with four pollinia, should not be included, but this single character seems insufficient to require its removal to another tribe. All these plants have pseudobulbs of several internodes, articulated leaves, and usually a lateral inflorescence. The most similar is the African genus *Ansellia*, but that genus has a particulate inflorescence borne at the apex, not the base, of the greatly elongated pseudobulb. The orchid subfamily relationships can be easily understood Fig.2.1

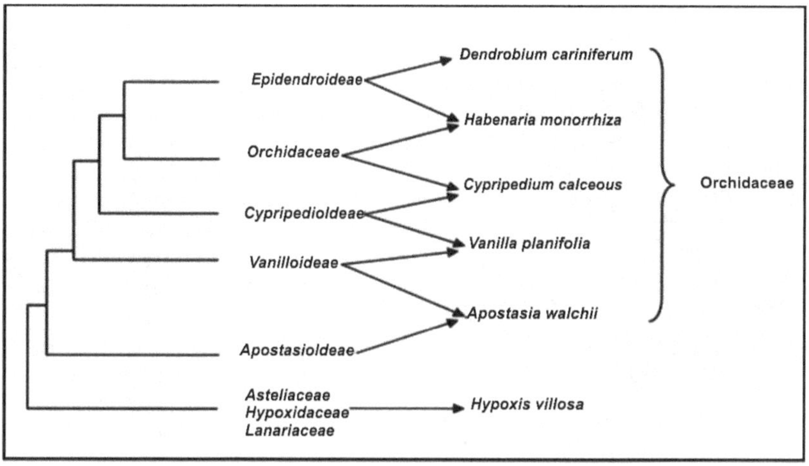

Fig. 2.1 : The diagram of Orchid subfamily Relationships

Botanical Classification

Habit

Orchids are perennial herbs, highly adaptable to most terrestrial ecosystems. Orchids can be grown on rocks (lithophytes), on the ground (terrestrials), or on trees (epiphytes). Terrestrial orchids are rooted in soil; ephiphyte orchids grow on other plants. The epiphyte orchid has an *aerial* root, and is predominately found in tropical habitats. It grows on parasitically on trees; their roots only attach to trees and do not derive any nutrition from them. Orchid species have adapted to varying ecological habitats worldwide; this may be due in part to the root system of the orchid, a complex system capable of capturing and storing water, vital in some landscapes such as desert. It has complex botanical ranges in size from tiny to gigantic, with species growing as tall as 8m. (viz. *Grammatophyllum speciosum*).

Stem

All orchids are perennial herbs and lack any permanent woody structure. According to growth habit they may be classified as sympodial and monopodial.

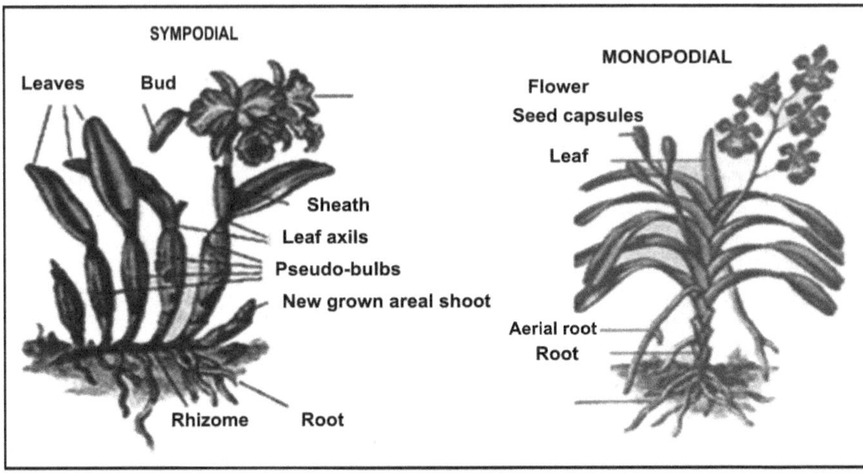

Fig. 2.2 : Sympodial and monopodial orchids.

The base of the stem of sympodial epiphytes may be thickened to form what is called a pseudobulb, a storage organ derived from the part of a stem between two leaf nodes. The pseudobulbs contain nutrients and water for drier periods. Pseudobulbs themselves are relatively short lived (1-5 years), but are continually produced from the growing tip of the rhizome. The pseudobulb has a smooth surface with lengthwise grooves and can have different shapes, often conical or oblong. Its size is very variable; in *Bulbophyllum* it is no longer than two millimeters, while in the largest orchid in the world, *Grammatophyllum speciosum* (giant orchid), it can reach three meters. With aging the pseudobulb sheds its leaves and becomes dormant. At this stage it is often called a backbulb. In warm and humid climates, many terrestrial orchids do not need pseudobulbs.

The word Monopodial is derived from Greek "mono-", one and "podial", "foot", in reference to the fact that monopodial plants have a single trunk or stem. This type of orchids has upward growth habits from a single point. They add leaves to the apex each year and the stem grows longer accordingly. Orchids with monopodial growth often produce copious aerial roots that often hang down in long drapes and have green chlorophyll underneath the grey root coverings which are used as additional photosynthetic organs. They do not have a rhizome or pseudobulbs so these species have adapted to dry periods by fleshy succulent leaves. Flowers generally come from the stem between the leaves.

Leaves

Orchids have simple leaves with parallel veins like most monocot, although some Vanilloideae have a reticulate venation. Leaves may be ovate, lanceolate, or orbiculate and very variable in size. They are normally alternate on the stem, often plicate, and have no stipules.

Fig. 2.3 : Typical leaves and roots of *Cymbidium* Orchid

Orchid's leaves often have siliceous bodies called stegmata in the vascular bundle sheaths (not present in the Orchidoideae) and are fibrous. The structure of the leaves corresponds to the specific habitat of the plant. Species that typically bask in sunlight, or grow on sites which can be occasionally very dry, have thick, leathery leaves and the laminas are covered by a waxy cuticle to retain their necessary water supply. On the other hand shade species, have long, thin leaves. The leaves of most orchids are perennial, while others, especially those with plicate leaves, shed them annually and develop new leaves together with new pseudobulbs, as in *Catasetum*. The leaves of some orchids are considered ornamental. The leaves of the *Macodes sandEriana*, show a sparkling silver and gold veining on a light green background. The cordate leaves of *Psychopsiella limminghei* are light brownish green with maroon-puce markings, created by flower pigments. The attractive mottle of the leaves of Lady's Slippers from temperate zones (*Paphiopedilum*) is caused by uneven distribution of chlorophyll. Also *Phalaenopsis schillEriana* is a lovely pastel pink orchid with leaves spotted dark green and light green. The Jewel Orchid (*Ludisia discolor*) is grown more for its colorful leaves than its fairly inconspicuous white flowers. Some orchids, as *Dendrophylax lindenii* (Ghost Orchid), *AphyllOrchis* and *Taeniophyllum* depend on their green roots for photosynthesis and lack normally developed leaves, as of the heterotrophic species.

Roots

Orchid roots are more complex than roots of other plants (Fig 2.3). Throughout the process of evolution they have adapted to living conditions high in tree canopies. Despite the regular precipitation and abundance of decaying organic material, their living environment is quite poor since water and nutrients are washed away very quickly from the tree. The plant replaces this shortage in many ways (e.g. fleshy leaves and pesudobulbs).

Orchids exhibit the root structure of monocotyledonous plants and have no primary roots, but only secondary roots, which start directly from the stem or from other secondary roots. Often the roots store nutrients and water, helping to retain nutritional substances that deposit on their bases. In some cases, the roots have chlorophyll organs that can sustain photosynthesis when the plants lose their leaves. The roots vary in thickness, from very thin to very thick. Terrestrial orchids may be rhizomatous or form corms or tubers. These may be spherical or elongated cylinders, and they store water and nutrients, replacing the role of pseudobulbs in epiphytic species. Occasionally, these tubers split from the mother plant, giving rise to new plants. Some Australian species of the subtribus Caladeniinae have almost no roots, only a small ovoid tuber. At the other extreme, the tubers of the Brazilian *Cleistes* are thin and delicate, measuring more than one meter, spreading in several directions. As in the *Cypripedium* species, terrestrial orchids can have extensive underground rhizomes that vernalize and produce new growth in the following year.

The root caps of terrestrials are smooth and white. Sympodial terrestrials, such as *Orchis* and *Ophrys,* have two subterranean tuberous roots. One is used as a food reserve for wintery periods, and provides for the development of the other one, from which visible growth develops. The velamen of an orchid is the white or gray covering of aerial roots (when dry, and usually more green when wet as a result of the appearance of underlying photosynthetic structures). It is many cell layers thick and capable of absorbing atmospheric moisture and nutrients. Often, the roots of orchids are associated with symbiotic fungi or bact*Eria,* the latter may fix nutrients from the air. This functionality allows the orchid to exist in locations that provide a reproductive or vegetative advantage such as improved exposure or reduced competition from other plant species. The velamen serves a mechanical function, protecting the vascular tissues in the root cortex, shielding root from transpirational loss and adhering the plant to the substrate.

There are quite a few orchid genera which do not grow leaves at all as their function is performed solely by roots (*Dendrophylax, Microcoelia* and *Polyradicon etc.*). These orchids are slightly more delicate to grow since they have to be grown on tree bark where the roots are exposed to daylight. *Phalaenopsis* is among the most widely known orchid types where roots are capable of performing photosynthesis.

When an orchid starts to prepare for a rest period, due to low temperature and short days, the root growth will slow down and the green tips will be covered in a grey velamen layer. In spring, when the temperature increases, the roots will start to grow again, followed by the growth of vegetative offshoots - new leaves. Some orchids develop root hair on the surface called rhizoids, which is grown to ensure better fixation of the plant in some genera and in others it is active in absorbing humidity and nutrients. Rhizoids are even more sensitive to physical damage and will die away if the plant is not watered in time. New roots usually shoot during or at the end of each vegetative growth period. Although not the primary nutritional source of orchids, they usually benefit from a kind of symbiosis with a fungus (Mycorrhiza) that is lodged on the vellamen exterior cells of their roots, and which excretes nutrients that are absorbed directly by the roots.

Some species, particularly those of the subtribus Catasetiinae, behave as if they were pneumatophorous, with many thin roots growing up forming a sort of 'wig' that catches leaves and sediments that are shed from above during the rains. It is made of dead cells and can have a silvery-grey, white or brown appearance. The cells of the root epidermis grow at a right angle to the axis of the root to allow them to get a firm grasp on their support. The root life varies according to environmental conditions, and generally is less than the stem life.

Inflorescence and Flowers

Inflorescence is indeterminate, sometimes reduced to a single flower. Flowers usually bisexual, bilateral, usually resupinate (twisted 180° during development). Orchid flowers have an array of varying colours and fragrances. The fragrance of an orchid can be sweet or foul. Orchid flowers are shaped differently (slippers, buckets and helmets). There flowers differ from the majority of flower species in both their pollination systems and structural make-up; the male and female parts of the orchid are actually fused together and located to one side of the flower.

Since, the chance of pollinated is minimum in nature, it have long periods of pollination. Orchids rely on insects for pollination, although some species of orchids are capable of self-pollination. Even with such diversity, the flower on every orchid expresses bilateral symmetry, containing three petals and three petal-like, outlying sepals. Outer tepals 3, distinct to connate, usually petaloid, imbricate; inner tepals 3, distinct, sometimes spotted or variously colored, the median one clearly differentiated from the 2 laterals, forming a lip (labellum), often with fleshy bumps or ridges and of an unusual shape or color pattern. This 'lip' contains specialized cells that secrete nectar to attract insects during reproductive pollination. Botanists theorize that many of the physical characteristics peculiar to specific orchid species have selectively evolved to attract specific pollinating insects or animals, and even include structures that so resemble female insects that the male insects try to mate with them. At the centre of the flower is a structure formed by the fusion of stamens and pistil called the column. The various parts of orchid flowers are given in Fig. 2.4 (A-D) Stamens 3 or fewer (usually 1 or 2), adnate to style and stigma, forming a column. Pollen usually grouped into soft or hard masses (pollinia). Carpels 3, connate, ovary inferior, usually with parietal placentation, style and sigma highly modified, with portion of the latter usually nonreceptive (rostellum), a portion of which may form a sticky pad (viscidum) attached to the pollinia.

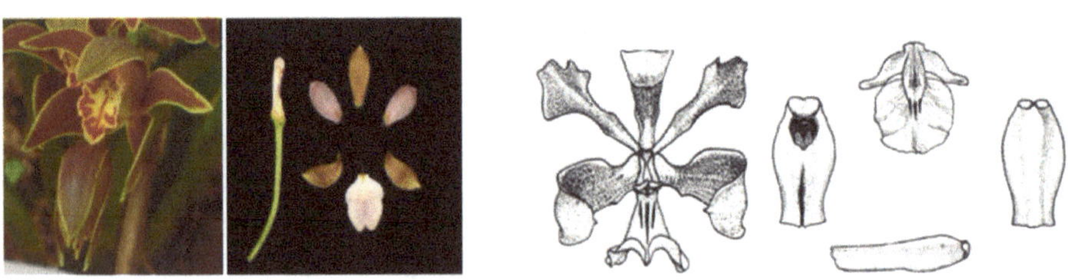

A B C D

Fig. 2.4 : Various parts of Orchid flower(A-D)

Floral Formula

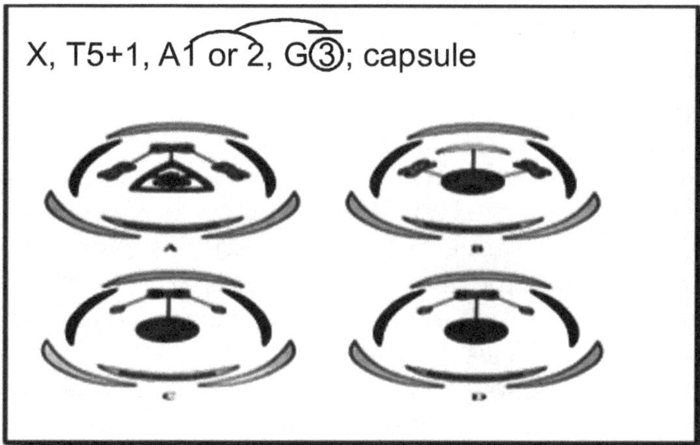

Fig. 2.5 : Showed floral formula

Pollination

Orchids reproduce by cross-pollination with birds, insects, and other animals carrying pollen from the stamen of one orchid to the pistil of another. Cross-pollination between different plants, rather than self-pollination, has helped orchids evolve and adapt to their environment. Orchids have evolved some of the most complicated pollination systems of all plants. Many orchids look like or smell like bees, which is what attracts bees to them. *Ophrys insectifera* emits a chemical mixture that attracts male hymenopteras of the genus *Argogorytes* . The smell of this orchid resembles female insects and hence male insect tries to copulate with the plant. During this time particles of pollen attach themselves to the male insect to be deposited at its next visit on a different orchid plant. Some parts of orchid plants resemble female insects encouraging male insects to try to mate with them thereby spreading pollen. The mirror orchids of the western Mediterranean also emit a pheromone very similar to that of female bees, which sends the male bees into an excited frenzy. Some orchid, the dead horse arum, emits a smell similar to that of rotting flesh in order to attract flies to assist in its pollination. Another orchid species, the *Oncidiums* of South America, attract bees by a process called as pseudoantagonism. That is, parts of the plant resemble an enemy insect and the bee attacks it and try to drive it away. A large number of natural hybrids both intergeneric and interspecific have been reported in different genera like *Odontoglossum, Phalaenopsis, Cattleya, Laelia, Miltonia and Oncidium. Cattleya guatemalensis* collected from Guatemala in 1861 by Skinner was found to be a natural hybrid between *Epidendrum aurantiacum* and *Cattleya skinneri*. It was later named as *EpiCattleya*. Similarly, *Phalaenopsis intermedia* is a natural hybrid between *P. aphrodite* and *P. rosea*.

Fruits and Seeds

Orchid seed contains very few nutrients, thus they require food from fungi for their embryo germination. The seeds are generally almost microscopic and very numerous, in some species over a million per capsule. After ripening they blow off like dust particles or spores. They lack endosperm and must enter symbiotic relationship with various mycorrhizal basidiomyceteous fungi that provide them the necessary nutrients to germinate, so that all orchid species are mycoheterotrophic during germination and reliant upon fungi to complete their lifecycle.

Seed Structure

The most interesting and adaptive feature of the family Orchid is the physiology of its seed germination. Orchid seeds are unique in several respects. They are exceedingly small and dust like and are produced in very large numbers. As many as 1,300 to 4,00,000 seeds per capsules are produced. Their colour may be white, cream, pale green, reddish orange or dark brown and have very diverse shapes. Orchid embryo consists of relatively undifferentiated, most isodiametric cell with dense granulated cytoplasm and loses their viability very fast.

Symbiotic Seed Germination

Under natural conditions, the orchid seeds germinate after being infected by fungus, the orchid mycorrhiza which term was coined by Frank in 1885. Most of the Mycorrhizal fungi of orchids fall into a non-sporing group known as *Rhizoctonia*, the major species being *R. repens*, *R. mucoroides* and *R. languinosa*. Subsequent isolates were known as *Mycelium radicis*. For quite some time these fungi were called as *Orcheomyces*. Later on, however, other fungi were also isolated from orchid cultures. In Japan 54 different fungi were isolated from 20 orchid species. Orchid seeds cannot utilize their own reserve or do so very slowly. They can also not hydrolyze large molecules like starch or cellulose. As a result asymbiotic germination in the absence of sugar proceeds only to the early protocorm stage, after which they wait for external supply of simple sugars through the help of mycorrhizal fungus. This technique was carefully exploited by commercial growers for the rare plants and hybrids which while could not be propagated by any other means, responded very well to tissue culture at a much faster speed. Vacherot and Lecoulfe firm in France has developed a new technique of virus-free plants of *Cymbidium*. Today tissue culture is preferred for commercial propagation of orchids.

Cytogenetics of Cymbidioid Orchids

The Cymbidioid phylad consists mainly of pantropical epiphytic species, with approximately 275 genera and 4300 species. The phylad is formed basically by the ancient subfamily Vandoideae, excluding the tribes Polystachieae and Vandeae and is characterized by having two polinia whose texture varies from firm to hard. It is a morphologically variable group, including ornamental species, mainly in the subtribes Cyrtopodiinae (*Cymbidium*) and Oncidiinae (*Odontoglossum*, *Miltonia* and *Oncidium*), which have been more widely studied cytologically. The Cymbidioid phylad presents the widest chromosome number variation among orchids, with records varying from 2n = 10 in *PsygmOrchis pusilla* to 2n = 168 in two species of *Oncidium*. Chromosome numbers observed were: 2n = 54 (subtribe Eulophiinae), 2n = 44, 46, 92 (subtribe Cyrtopodiinae), 2n = 54, 108 (subtribe Catasetinae), 2n = 52, 96 (subtribe Zygopetalinae), 2n = 40, 80 (subtribe Lycastinae), 2n = 40, 42 (subtribe Maxillariinae), 2n = 40 (subtribe Stanhopeinae), 2n = 56 (subtribe Ornithocephalinae), and 2n = 12, 20, 30, 36, 42, 44,56, 112, 168 (subtribe Oncidiinae). Interphase nuclei varied widely from simple chromocenter to complex chromocenter types,with no apparent cytotaxonomic value. In the genera *Catasetum* and *Oncidium*, the terrestrial and lithophytic species presented higher ploidy levels than the epiphytic species, suggesting a higher adaptability of the polyploids to those habitats. The primary base number x = 7 seems to be associated to the haploid chromosome numbers of most Cymbidioid groups, although n = 7 was observed only in two extant genera of Oncidiinae. For each tribe, subtribe and genus the probable base numbers were discussed along with the possible relationships to the primary base number x = 7 admitted for the whole phylad.

Cultivars and Hybrids

Orchids are very complex flowers due to the many different characteristics that have been developed by orchid cultivators. This has a very complex nomenclature consequence. Interpreting orchid named is thus quite important for becoming a serious orchid collector or cultivator or even an orchid grower. Within an orchid species, any one orchid plant may exhibit features that make it unique within that species. It can be either in size of colour. These differences within a species are known as varieties or forms and are usually denoted as follows. The word "Cultivar" is a combination of cultivated and variety and means that the plant has been vegetative propagated through dividing or cloning, rather than from seed to produce a genetically unique orchid plant. Within a cultivar all the offspring are genetically identical. When a cultivar is named, the customary Botanical name is used and the cultivar name is added to it. A Hybrid on the other hand is a cross between genetically different orchid plants. Orchid hybrids are identified by a "grex". Orchid hybrids, when produced from seed and regardless of variations or form or colour share a common lineage and thus are classified as members of the same grex. This in essence means hybrids that have the same parentage. Cultivars may be given individual names, but the grex name will precede them when classified. For classification purposes and to understand the names of orchids around the world it is best to remember the following points:

- The genus name is started with a capital letter. It can also be abbreviated as in *Cymbidium gigantium* or *C. gagantium*.
- Species and variety names are always in lower case.
- Genus, species, and variety names are always in italics.
- Variety or form is written as 'var.' or 'forma' in Roman typeface.
- Grex names are written in Roman typeface with initial capitalization.
- Cultivar names are also written in Roman typeface, with initial capitalization, and enclosed in single quotation marks.
- Awards are abbreviated and appear in capital letters.

Table 2.1 : *Cymbidium* (Cymbidieae: Orchidaceae)

Kingdom:	**Plantae**
(unranked):	Angiosperms
(unranked):	Monocots
Order:	Asparagales
Family:	Orchidaceae
Subfamily:	Epidendroideae
Tribe:	Cymbidieae
Subtribe:	Cyrtopodiinae
Alliance:	Cymbidium
Genus:	Cymbidium
	Swartz, 1799
Synonyms	
• Cyperochis Blume	
• IridOrchis Blume	
• Jensoa fRaf.	
• Pachyrhizanthe (Schltr.) Nakai	

Natural Hybrids

Most of the orchid plants in commerce are hybrids; however, they have been derived, after more than a hundred years of artificial breeding. The search for novelty, which is ever-present in the orchid industry, has stimulated the use of some of the smaller flowered species in new hybridising programmes in the last 30 years with, in some cases, conspicuous success.

Asian *Cymbidium*

Asian *Cymbidiums* or Chinese *Cymbidiums* refer to mainly five species of *Cymbidiums* orchids that are found throughout East Asia in areas of China, Korea, Japan, India, and in parts of Thailand and Vietnam. These species are usually grown for their variegated leaves. But plants are also grown for their fragrant flowers and peloric flower structure. Plants are usually grown in long and thin vase like pots. The five species are:

- *Cymbidium sinense*
- *Cymbidium ensifolium*

- *Cymbidium kanran*
- *Cymbidium goeringii* (all the varieties like v. *longibracteatum* , v. *lianpan*, subsp. *gracillimum* are treated as different groups in the culture traditionally)
- *Cymbidium faberi*

Common hybrids of above five species are:-

Cymbidium × *ballianum* (= *C. eburneum* × *C. mastersii*) (Myanmar)

Cymbidium × *baoshanense* (= *C. lowianum* × *C. tigrinum*) (SC. Yunnan)

Cymbidium × *chiu-lih* (= *C. lancifolium* × *C. ensifolium*) (China)

Cymbidium × *nishiuchianum* (= *C. goeringii* × *C. kanran*) (Taiwan)

Cymbidium × *nishiuchianum* (= *C. goeringii* subsp. *goeringii* var. *formosanum* × *C. kanran*) (Taiwan)

Cymbidium × *florinda* (= *C. erythrostylum* × *C. iridioides. CyperOrchis* × *florinda*) (Vietnam)

Cymbidium × *gammieanum* (= *C. elegans* × *C. erythraeum. CyperOrchis* × *gammieana*) (Nepal to Sikkim)

Cymbidium × *glebelandensis* (= *C. insigne* × *C. schroederi*) (Vietnam)

Cymbidium × *jy-shiang* (= *C. lancifolium* × *C. sinense*) (China)

Cymbidium × *rosefieldense* (= *C. hookErianum* × C. tracyanum. Cyper*Orchis* × rosefieldensis*)* (Vietnam)

Cymbidium × *woodlandense* (= *C. mastersii* × *C. tracyanum. CyperOrchis* × *woodlandensis*) (Myanmar)

The database storing biological information of Thai orchid was designed by Kaewtrakulpong *et al.,* 2011 based on Thai Darwin's core. Gerlach (1992) studied the flowering periodicity of several orchid species in a screenhouse under approximatetly 50% shade. The time interval between a heavy rainfall and opening of the flowers was 7 days for Flickeringia comata, 8 days for *Dendrobium vaupelianum* and *Dendrobium* sp. (1), 9 days for *Dendrobium biflorum, Dendrobium sladei* and *Dendrobium* sp. (2), and 12 days for *Diplocaulobium fililobum*. On dry, non-inductive days, when flowering was not induced, the difference between the daily minimum and maximum temperatures was 9°C to10°C, the minimum being approximately 21°C. On days when flowering was apparently induced, the temperature range was only 4-5°C , with a minimum of

23°C, and daily precipitation was 0-150 mm. Apang and Rao (2008) studied the flowering under Arunachal Pradesh condition and reported that the maximum number of species flower during May and June and the least in December and January. Flowering calender of the North-East Indian orchids with a special reference to ornamental species was developed by Kataki and Abbas (1992) showed the majority of the orchids flower during April-May, but that most *Paphiopedilum* and *Cymbidium* species flower during the winter months. The duration of flowering varies considerably between species. The growth and flowering behaviour of *Liparis cordifolia* species under Darjeeling conditions were studied by Pal *et al.*, 2010 and observed that the shoot bud arised in July and August from basal part of the pseudobulb and inflorescence emerged from the top of recently formed stem (pseudobulb) days after initiation of shoot bud. The first flower bud opened in the last week of September to October, after emergence of flower buds. The raceme remained in flower up to first week of December. The growth phase marked by emergence of roots from the basal part of stem and swelling of pseudobulb in last week of November to first week of December. The plants proceeded to dormancy after senescence and drying of leaf in April and May.

Phalaenopsis orchids require a day temperature of 26°C or less to initiate inflorescences, whereas the night temperature has little or no effect on inflorescence initiation. Newton and Runkle (2009) determined the duration of high temperature required each day to prevent inflorescence initiation of four *Phalaenopsis* and *Doritaenopsis* clones. Exposure to 29°C for 8 hours or longer inhibited inflorescence initiation of *Phalaenopsis* Miva Smartissimo x Canberra 'Mosella' and *Phalaenopsis* Brother Pink Mask x Brother Success 'Explosion', but *Phalaenopsis* Baldan's Kaleidoscope 'Golden Treasure' and *Doritaenopsis* 'Newberry Parfait' required exposure to 29°C for 12 hours or longer to inhibit inflorescence initiation. Flowering was completely suppressed only when high-temperature exposure time was continual for *Doritaenopsis* 'Newberry Parfait' and *Phalaenopsis* Baldan's Kaleidoscope 'Golden Treasure' and 12 h for *Phalaenopsis* 'Mosella'. Plant life span generally increased as duration of exposure to 29°C increased, but high-temperature exposure had few or no significant effects on flowering characteristics of flowering plants. These studies indicate that as few as 8 hours of high temperature can prevent flowering of some *Phalaenopsis* hybrids, whereas others require greater than 12 h of high-temperature exposure. Effects of $KClO_3$ on *Phalaenopsis* orchid flowering were investigated by Duan *et al.* (2006). Increased concentrations of 2, 4, 8 and 16 mmol/L $KClO_3$ delayed spike emergence by 5, 6, 18 and 26 days, respectively. Moreover, they reduced final spike length by 2.1%, 4.0%, 16.2% and 46.1%, respectively. Nonetheless, application of $KClO_3$ at 4 and 8 mmol/L advanced the time to appearance of the first open flower by 13 and 24 days, respectively. Use of 8 mmol/L $KClO_3$ also increased the number of floral buds by 16%. Treatments with $KClO_3$ tended to reduce flower size. Overall, the data suggest that application of $KClO_3$ at an appropriate concentration (e.g. 8 mmol/L) can increase the number of floral buds and advance the time to *Phalaenopsis* orchid flowering, but may reduce flower size. Paradiso and Pascale (2012) investigated the effects of supplemental lighting, applied during the 60-day phase of flower induction, on growth and flowering of *Phalaenopsis* 'Premium', grown in pot under Naples condition. Supplemental lighting slightly anticipated flowering compared to natural lighting (144 vs. 150 days from the beginning of treatment) and positively influenced plant architecture, promoting the emission of the second flower stem and the flower stem branching.

In south India just after monsoon (August and September) most of the terrestrial tropical orchids like *Habenaria* and its allies, *Liparis, Malaxis etc.*, come into bloom. Aerial shoots arises with the first rain, grow up rapidly during the two months of heaviest rainfall, produce flowers during August and September and by the time rain abates the aerial portions die down leaving the tuber to remain underground till next year. Epiphytic orchids burst into bloom during this season are *Rhynchostylis retusa, Aerides ringens, Eria pauciflora, Bulbophyllym fischeri, Bulbophyllum kaitense*, few of the *Oberonias, Ceologyne mossiae, C. nervo, Aerides crispum, A. maculosum* etc. *Phalaenopsis, Vuylstekeara, Odontoglossum, Miltonia* also start flowering in this season.

Temperate Orchid like *Cymbidium, Dendrobium, Paphiopedilum, Phalaenopsis* start flowering during October till January. *Dendrobium heterocarpum, Bulbophyllum aureum, Chiloschista pusilla* are a few species, bloom in response to the cool nights of January and February. Most of *Neottieae*, like *Zeuxine, Anoectochilus, Goodyera, Cheirostylis etc.*, come into flower during this time.

March and April are the flowering time for most of the epiphytic species like *Acampe praemorsa, Bulbophyllum firmbriatum, Bulbophyllym tremulum, Dendrobium herbaceum D. crepidatum, Luisia* species, *Vanda testacea, Coelogyne odoratissima C. glandulosa, Trichoglottis tenera etc.* The terrestrial species *Eulophia macrostachya, Acanthephippium bicolor, Eulophia nuda* also flowered at this time.

Dendrobium haemoglossum, Thrixspermum arachnites, Ephemerantha macraei etc., In the case of the Malaysian species, *Dendrobium crumenatum,* it has been reported that a sudden drop in temperature of -12°C will induce flower bud initiation.

There is no particular time of the year which can be quoted as the flowering time for orchids in general. The majority of orchid flowered only once in a year. The flowering time may vary according to species and the climatic conditions of the particular area. As far Indian orchids are concerned, the peak flowering periods are as under. Flowering season of common orchids are mentioned in following table.

Table 2.2 : Flowering season of genera of orchids

GENUS	JAN	FEB	MAR	APR	MAY	JUN	JUL	AUG	SEP	OCT	NOV	DEC
Cymbidium	X	X	O				O	O	O	X	X	X
Dendrobium	X					O	O	O	O	X	X	X
Miltonia	O	O	X	X	X	O	O	O	X	X	X	O
Miltassia	O	O	X	X	X	X	X	O	O	O	O	O
Odontoglossum		X	X	X	X	O	O	O	X	X	O	O
Oncidium	O	O	O	X	X	X	X	O			O	O
Paphiopedilum	X	X	X	X						X	X	X
Phalaenopsis	X	X	X	X	X	X	X	X	X	X	X	X
Vuylstekeara		X	X	X	X	O	O	O	X	X	O	O
Zygopetalum	X									X	X	

X = Basic Flowering Period O = Sporadic Flowering

Pollinator and Pollination

Pollination is the process of transfer of pollen from anther to stigma. In orchids pollination process may briefly be summerized as when the pollinator enters into the flower, it touches a viscidium, which promptly sticks to its body, generally on the head or abdomen. While leaving the flower, it pulls the pollinium out of the anther, as it is connected to the viscidium by the caudicle or stipe. The caudicle then bends and the pollinium is moved forwards and downwards. When pollinator enters another flower of the same species, the pollinium has taken such position that it will stick to the stigma of the second flower, just below the rostellum, pollinating it.

There are two kinds of pollination, cross-pollination and self-pollination. In cross-pollination the pollen is transferred from the anther of a flower of one plant to the stigma of a flower of another plant. In self-pollination the pollen is transferred from the anther to the stigma of the same flower or to another flower of the same plant. In orchids self-pollination is discouraged by nature. Even in the few orchids capable of it, the process seems only to be used when insects fail to appear. In these species the caudicles may dry up if the flower hasn't been visited by any pollinator and the pollina fall directly on the stigma. Otherwise the anther may rotate and then enter the stigma cavity of the flower (*Holcoglossum amesianum*). However, some orchids rely on self-pollination, especially in colder regions where pollinators are absent. In orchid breeding pollination is artificially controlled by humans. Orchid growers may pollinate flower with a pencil, small paintbrush or other similar device.

Fertilization is the fusion of two germ cells after pollen felt on stigma under favourable conditions. The pollen develops a tube which grows downward through the style and into the ovule. It is in the ovule that the male and female germ cells unite. Once fertilization has occurred development of the ovule begins and eventually seeds are formed. The parent flower that furnishes the pollen is called the pollen parent; the one that bears the seed is called the seed parent.

The flowers of the Mediterranean orchid genus *Ophrys* imitate important specific sexual releasing factors of aculeate hymenopteran (*Andrena, Anthophora, Blithopertha, Bombus, Chalicodoma, Chlorandrena, Chrysandrena, Colletes, Euandrena, Eucera, Hoplandrena, Hyperandrena, Melandrena, Melecta, Rufandrena, Osmia, Tetralonia, Zonandrena*) females. They attract female-seeking males in a species-specific way. The males attempt to copulate with the labellum. During the pseudocopulation the males remove the pollinia. If sexual mimicry is again successful, the next flower of the same *Ophrys* species can be pollinated. As the flowers use the premating isolation factors of their pollinator species for their own genetic isolation from similar *Ophrys* species, the relationship is species-specific. Each *Ophrys* species has its own pollinator species. This is illustrated by numerous examples in many experimental choice tests in the field under natural conditions. Evolution of new species is always a consequence of a change in pollinator. The new pollinator-species selects those individual flowers for pollination which come nearest to its own female's releasing stimuli. If the systematic status of an *Ophrys* type is unclear, discovering its own pollinator is proof for its biospecies status. Identical pollinator species in different *Ophrys* species are possible only in those cases where the questionable plants are distributed allopatrically or, if syntopically distributed, if they are visited by bees pollinating both by means of head and abdomen. Examples for both types are given. When different *Ophrys* species have identical pollinator species, then the appearance of the flowers can become very similar. This convergence or parallelism occurs as a result of identical selection (Paulus 1997, Vereecken *et al.*, 2010, Schluter *et al.*, 2011).

Mutualism, or a mutually beneficial interaction between two organisms, are ubiquitous in ecological systems. However, some empty flowers, which offer pollinators no any kinds of rewards, design different strategies to attract pollinators without providing rewards to the pollinators. These pollination mechanisms are called deceptive pollination. The family Orchidaceae, representing one of the largest groups in angiosperms, is distinguished by high floral diversity and intricate adaptations to pollinators. Darwin described and identified most of the functional floral morphology and biomechanics in orchid pollination. However, he never recognized that many of the flowers that he examined lacked food rewards for pollinators. Floral evolution in the Orchidaceae appears to be dominated by modes of deceptive pollination, and more than one third of orchid species are thought to be pollinated by deceit. Deceptive pollination is thought to be one of key roles which have lead to relatively high species diversity within the Orchidaceae. Deceptive orchids frequently exploit the food foraging, sexual, oviposition and sleep/warmth behaviors of insects. The most common deception mechanisms include generalized food deception, Batesian floral mimicry, sexual deception, brood-site imitation and shelter imitation. Additionally, floral color, morphology and fragrance play key roles to cheat target pollinators. Relationships between deceptive orchids and their pollinators possibly involve asynchronous evolution; therefore orchids track the diversification of their pollinators. However, deception has negative impacts on the pollinators, which may exert selection on the pollinators. Because duped pollinators tend to avoid rewardless flowers, deceptive orchids suffer low visitation rates and fruit set; various environment factors can affect the reproductive success of these orchids. Deceptive orchids depend largely on insect pollinators for reproduction, and the proclivity of these species to use deceptive pollination strategies puts many of these species at a relatively higher rate of extinction. Therefore, pollination biology studies are needed to provide a scientific basis for proper conservation of orchids. At the community level, co-occurring mimic, non-mimic, and mimic signal providing plant species affect the reproductive fitness of orchids. Therefore, it is necessary to further study the co-evolution webs of deceptive orchids and pollinators along with other related biological and ecological factors (Ren *et al.*, 2012).

Most Mediterranean orchids are deceptive, offering no reward to insect pollinators. In the absence of a reward, inflorescence size and flower position within inflorescence may be associated with the timing and the numbers of pollinators attracted, thus influencing the final reproductive success. Pollination biology and breeding system were investigated in three deceptive orchids, *Orchis italica, O. anthropophora,* and *Anacamptis papilionacea.* Although all examined orchids were self-compatible, bagged inflorescences produced no fruits. Artificial pollination resulted in a 76-79% fruit set by induced autogamy, a 70-76% one by geitonogamous pollination, and a 79-87% one by allogamous pollination. The natural fruit set in the open-pollinated control was 14-16%. Fruit production was neither related to the inflorescence size, nor to the number and position of flowers within the flowering spike, suggesting that variation in these floral traits does not influence pollinator attraction or female reproductive success (Pellegrino *et al.,* 2010).

Pollination of orchid flowers was critically reviewed by Rhodehamel (1994) and general advice was given to growers on hand pollination, the care of plants following pollination, and the harvesting and storage of seeds. He has compiled a comprehensive table showing the interval, in days, between pollination and harvest of seed capsules for green-capsule culture of more than 280 species. Stokl *et al.,* (2011) presented a new case of pheromone mimicry found in the terrestrial orchid *Epipactis veratrifolia.* Flowers are visited and pollinated by several species of aphidophagous hoverflies, the females of which also often lay eggs in the flowers. The oviposition behaviour of these hoverflies is mainly guided by aphid-derived kairomones. They showed that the flowers produce alpha - and beta -pinene, beta -myrcene and beta - phellandrene, and that these compounds attract and induce oviposition behaviour in female hoverflies. This floral odour profile is remarkably similar to the alarm pheromone released by several aphid species, such as *Megoura viciae.* They therefore, suggest that *E. veratrifolia* mimics aphid alarm pheromones to attract hoverflies for pollination.

The extent to which plants depend on pollinators for outcross pollen transportation is a key issue in plant reproductive ecology. Valdivia *et al.,* (2010) evaluated the putative breeding system, floral display, and natural pollination in two Southern Cone of South American orchids, *Gavilea araucana* and *G. venosa,* by performing four hand pollination trials (agamospermy, autogamy, geitonogamy, and xenogamy tests) and by counting each fower and fruit produced by plants in natural conditions. Fruit set differed signifcantly depending on the origin of pollen as well as on the presence of a pollen vector. None of these species produced fruits by the agamospermy tests, therefore indicating that they require the presence of pollen on the stigmas for fruit setting. By contrast, only *G. araucana* was capable of fruit setting following the autogamy test, therefore demonstrating that this orchid needs not depend on pollinators for fruit set. Furthermore, *G. araucana* and *G. venosa* produced 100% fruits by geitonogamy and by xenogamy, thereby indicating that both plants are totally self-compatible. The mean number of flowers per plant was 2.1 times higher in *G. venosa* with respect to *G. araucana* ; however, natural fruiting success was 28.8% and 98.9%, respectively. Furthermore, while the increase in fruiting success signifcantly and positively correlated with an increase in foral display in *G. araucana*, in *G. venosa* these variables were unrelated. Certainly, further studies concerning the reproductive strategies in orchids of southern South America are of great importance.

The white flowers of *Mystacidium venosum* orchid have long spurs (ca. 4.5 cm) and contain dilute sucrose-dominated nectar which is secreted during the afternoon and early evening. Scent, dominated by Jasmine lactone and (E,E)-farnesol, is emitted in the evening. A breeding system experiment established that self pollination results in fruits inferior in quality. Field observations at three sites in Kwa-Zulu Natal, South Africa, in 1998 and 1999 showed that hawkmoths (*Nephele accentifera accentifera*) were frequent visitors to the orchid shortly after dusk, and carried pollinaria along the length of their tongues. No evidence for directional selection on spur length was found at any of the three sites. Results showed that flower longevity is strongly reduced by pollination, while pollinaria removal had only a small effect. Pollinia removed from flowers remained viable for up to 20 days under outdoor conditions Luyt and Johnson(2001). Barrett *et al.*, (2010) investigated ombrophily in the deceptive orchid *Acampe rigida* to determine the mechanism by which this species is able to maintain high fecundity, despite flowering during the rainy season in south-west China when pollinators are scarce. During rainfall, raindrops physically flicked away the anther cap exposing the pollinarium. Raindrops then caused pollinia to be ejected upwards with the strap-like stipe pulling them back and causing them to fall into the stigmatic cavity, resulting in self-pollination. Neither flower nor pollen function were damaged by water. Although *A. rigida* is self-compatible, it is incapable of autonomous self-pollination without the assistance of rain splashes. The results of the rain-sheltering experiment indicated that rain pollination contributed substantially to increasing fruit-set, although there was variation among sites in the intensity of this effect.

In the many ways orchids attract pollinators. Some orchid flowers are shaped similarly to the female of an insect, so the males attempt 'pseudo-copulation', with no other benefit but that the orchids are pollinated in that manner. Each type of orchid places the pollinia on a different body part of a different species of insect, so as to enforce proper cross-pollination. Rare calcicolous *Paphiopedilum thaianum* and *P. niveum* were studied in S Thailand, 2007-2011. Their flowers are rewardless kettle-traps, the pouch-like lips keeping pollinators prisoner for many minutes. To leave, they have to climb a "tunnel", past the stigma to one of two anthers, where they squeeze themselves out of the narrow exit, thereby acquiring an amorphous smear of sticky pollen. Because the flowers were unusual, being fragrant and conspicuously white, they were compared with two typically malodorous and mottled/ yellow species, *P. godefroyae* and *P. concolor*. During more than 450 man-hours of watching, behaviour of all insects near, on and in the flowers was carefully observed to distinguish between legitimate and accidental pollinators, or mere visitors. Unexpectedly, melittophily was discovered in *P. thaianum* and *P. niveum*, the pollinators being the newly described halictid bee Lasioglossum orchidodeceptum and the meliponine bee Tetragonula testaceitarsis, respectively. Pollinators of *P. godefroyae* and *P. concolor* were, as expected, milesiine hoverflies, viz. *Eumerus nicobarensis* and *E. figurans*, respectively, confirming these two orchids as myiophilous like *Paphiopedilum bellatulum*. The five orchid species belong to the subgenus *Brachypetalum*, the second-most basal subgenus of *Paphiopedilum*. The most basal subgenus, *Parvisepalum*, is the closest relative of the mainly holarctic genus *Cypripedium*, in flower structure and in melittophily. Hence *P. thaianum* and *P. niveum* are proposed as the hitherto missing link between *Cypripedium* (via Parvisepalum) and the more derived *Paphiopedilum*, all myiophilous. The very similar *P. thaianum* and *P. niveum* were found to be reproductively isolated by enticing pollinators differing at family rank and by allowing pollen acquisition only by appropriately-sized bees, smaller in *P. thaianum*. This supports them as good species. Oviposition on *Paphiopedilum dianthum* by syrphine hoverflies, supposedly visually misled by dark dots mistaken for aphids, the prey of the flies' progeny, is reassessed because oviposition is known to be olfactorily induced (Banziger *et al.*, 2012). Some attract carrion flies with horrid smell, and others emit a lovely fragrance that lures bees or other insects. Most angraecoid orchids are white, attracting night-flying moths. Insects, such as the highly specialized orchid bees and the long-tongued hawkmoths, pollinate the myriad species of orchids that dominate the tropical regions. However, some orchids relegate their pollination process to other methods and must therefore; make modifications in both their pollinating mechanisms and nectar production.

In *Cypripedium acaule* the bee is attracted to the entrance by white lines, pushes through the softly drawn drapes of the pouch and sips the nectar, but when he seeks to retreat, the drapes are closed tight behind. By squeezing through the open place, the hole at the top, he first scrapes the pollen on to the stigma and then, pushing past the pollen masses at the top, picks up another load and flies to another rosy slipper (*Cypripedium acaule)* to repeats the routine. Many neotropical orchids are pollinated by male orchid bees, which visit the flowers to gather volatile chemicals they require to synthesize pheromonal attractants. Orchids of the genus *Ophrys* are well-known for their flowers that mimic bees and wasps. The Mediterranean orchid *Ophrys fusca* seduces a male bee before the female bees have emerged from their pupal cases. Orchids that are pollinated by birds make the texture of its pollen stickier and larger so that it adheres to the bird's beak, which prevents self-pollination and prevents it from being removed when the bird preens. The path of pollination in *Cymbidium pendulum* (Sw.) Roxb. has been traced by Attri and Kant (2011) . The honey has been identified as *Apis mellifera* was found to act as main pollinator and this bee is the only insects, among others, who succeeded in performing pollination because probably due to its structural compatibility with the plant species. Pollination by *Apis mellifera* bees was suggested to occur in a number of families but rare phenomenon in orchids and it is first time that the species was observed to pollinate the *Cymbidium pendulum* flowers. Bee moved around the flowers for some times, entered the flower and carried on pollinia along with on the back during its journey. It revisited the different flower and deposited its pollinia on to it and the act of pollination was accomplished.

General visual bee mimicry and specific chemical mimicry by flowers to solitary female bees or wasps are well known in several orchid genera, for example, the Mediterranean genus *Ophrys*, the Australian genera *Cryptostylis* and *Chiloglottis*, and the South-African Disa. This mimicry has been shown to attract solitary male bees or wasps, which are their species-specific pollinators. The visual and chemical signals are considered to be a type of deceptive pollination mechanism based on mimicry for the exploitation of perceptual biases of animals. We propose that in addition to this unique pollination mechanism, these plants exhibit another, rarely mentioned and practically forgotten, non-exclusive function of bee or wasp mimicry (Batesian mimicry). This mimicry may deter large mammalian herbivores, and possibly also insects from the plants and especially from their flowers by a type of visual and olfactory deceptive aposematism. While visiting the flowers, bees and wasps may add a Mull*Eria*n effect to this defense. We extend this hypothesis to many other rewarding flowers that are bee or wasp pollinated and propose that abundance of pollinating bees or wasps may deter herbivorous mammals and insects from the plants during their peak flowering season (Lev and Neeman, 2012).

The mimicry of malpighiaceous oil-flowers appears to be a recurrent pollination strategy among many orchids of the subtribe Oncidiinae. These two plant groups are mainly pollinated by oil-gathering bees, which also specialize in pollen collection by buzzing. In the present study, the floral ecology of the rewardless orchid *Tolumnia guibertiana* (Oncidiinae) was studied for the first time. The orchid was self-incompatible and completely dependent on oil-gathering female bees (*Centris poecila*) for fruit production. This bee species was also the pollinator of two other yellow-flowered plants in the area: the pollen and oil producing *Stigmaphyllon diversifolium* (Malpighiaceae) and the polliniferous and buzzing-pollinated *Ouratea agrophylla* (Ochnaceae). To evaluate whether this system is a case of mimetism, we observed pollinator visits to flowers of the three plant species and compared the floral morphometrics of these flowers. The behavior, preferences and movement patterns of Centris bees among these plants, as well as the morphological data, suggest that, as previously thought, flowers of *T. guibertiana* mimic the Malpighiaceae *S. diversifolium*. However, orchid pollination in one of the studied populations appears to depend also on the presence of *O. agrophylla*. Moreover, at the two studied populations, male and female pollination successes of T. guibertiana were not affected by its own floral display, and did not differ between populations (Vale, 2011).

Shape and colours of the labellum usually attract pollinators. For attracting insects, flowers may produce odours, although absent in most of the species. Asparagales produces nectar in a spur of the labellum, on the point of the sepals or in the septa of the ovary. Genus *Catasetum* launches viscid pollinia with explosive force when an insect touches a seta. Pollen is released as single grains in the subfamilies Apostasioideae, Cypripedioideae and Vanilloideae. At the upper edge of the stigma of single-anthered orchids, in front of the anther cap, there is the rostellum, a slender extension involved in the complex pollination mechanism. In the other subfamilies, that comprise the majority of cut flowers, the anther carries into two pollinia. A pollinium is a waxy mass of pollen grains held together by the glue-like alkaloid viscin, containing both cellulosic stands and mucopolysaccharides. Each pollinium is connected to a filament which can take the form of a caudicle, as in *Dactylorhiza* or *Habenaria* or a stipe, like in *Vanda*. Caudicles or stipes holds the pollinia to the viscidium, a sticky pad which sticks the pollinia to the body of pollinators. The ovary is always inferior (located behind the flower). It is three-carpelate and one or, more rarely, three-partitioned, with parietal placentation (axile in the Apostasioideae). Catling and Kostiuk (2011), collected 17, 2010, data on pollination in a population of *Galearis rotundifolia* at the confluence of the Maligne and Athabasca Rivers north of Jasper, Alberta. The primary pollinator was the bee *Osmia proxima* and this species was also the most frequent visitor. While most flies were visitors, four species, Eriozona (Megasyrphus) laxus, Eristalis (Eoseristalis) hirta, Eristalis (Eoseristalis) rupium, and Eupeodes (Lapposyrphus) lapponicus also served as pollinators. It is estimated that 25-44% or more of flowers were pollinated in the previous year, a relatively high percentage that supports the "advertisement model" for evolution of food deception in orchids. The pollinating bee or fly lands on the lip and probes the spur which is approximately the same length or longer than the tongue. In the process of pushing into the flower and backing out, the sticky contents of a bursicle are either discharged by the backward movement of a flap or by forward pressure. Either of these actions may release adhesive fluid which fixes the viscidia onto the front of the insect's head. The gradual bending forward of the caudicles reduces likelihood of pollination of consecutively visited flowers with pollen from those recently visited flowers on the same plant (geitonogamous pollination) and thus promotes outcrossing.

Cham *Orchis alpina* was found to be predominantly ant pollinated, with some occasional pollination by ichneumonid wasps. Two species of ants, *Formica lemani* and *Leptothorax acervorum* visited the flowers and removed pollinaria. These two pollinator ants were found to be among the most common ant species in all habitats. Floral scent consistently comprised five terpenoid compounds, beta -phellandrene, 1,8-cineole, linalool, alpha -terpineol, and beta -caryophyllene. A synthetic blend of these five compounds emitting from rubber septa, was found to be attractive to one pollinator ant-species, *F. lemani*, in the field. The floral scent of C. *alpina*, through attracting only specific ants, may thus play a role in filtering floral visitors (Schiestl and Glaser, 2012).

Eria coronaria , which has an unusual autumn flowering season in some locations, has a bright yellow spot on its labellum. This spot has been hypothesized to function as a visual attractant to potential pollinators because its color may be attractive to insects. Shangguan *et al.,* (2008) tested this hypothesis using field observations between October and November of 2006 in the Yachang Nature Reserve, Guangxi, southwestern China. Honeybee (*Apis cerana cerana*) was the only pollinator of this orchid. Generally, honeybees landed directly on the yellow spot on labellum, and then adjusted their position and entered the flower. When honeybees retreated from flowers, the pollinaria were adhered on their thorax, or the pollinaria carried by honeybee were stuck on the stigma. The anther cap, however, did not separate from the column when the pollinaria were carried away from the flower. This orchid did not provide any rewards to honeybees. The flowers of a co-blooming plant, *Pittosporum glabratum,* with abundant nectars and pollens, attracted numerous honeybee visitations during this period. The flower color and size of *P. glabratum* were similar to that of the spot on the labellum of *E. coronaria.* Based on the behaviors of honeybees on the flowers of both *P. glabratum* and *E. coronaria,* they speculated that the bright yellow spot on labellum of this orchid functions as an attractant to honeybees. Hand-pollination experiments showed that *E. coronaria* was self-compatible, but that reproductive success depended on pollinators. Under natural conditions, the rate of fruit set of *E. coronaria* was 20.72%, which is very close to the average fruit set of other food deceptive orchids (20.7%).

Looking for clues to explain the low rate and unpredictability of the pollination of allogamous, epiphytic orchids in the context of the success of the orchid family in general, Perez *et al.,* (2011) compared the pollination of two miniature twig epiphytes in Mexico: Notylia barkeri, a 'weedy', rewarding species, and *Erycina crista-galli,* a rare, deceptive species. They measured the effects of the spatial organization of the flowers and various parameters of visibility and height above the ground in relation to the spatial organization of the seed capsules (which retrospectively measures the activity of the pollinators). The spatial presentation of the flowers of *E. crista-galli* explained only 5% of the distribution of the seed capsules. For *N. barkeri,* with compound flowers, the spatial presentation, considering individual flowers or inflorescences, explained 12% and 45% of the distribution of the seed capsules, respectively, and all other interactions between the variables were insignificant. Both the deceptive and reward strategies resulted in the production of large numbers of seeds despite a low pollination percentage (1-5%). Notylia barkeri produced 11.8 (2005-6) and 53.7 (2007-8) times more seeds in total than *E. crista-galli.* Furthermore, unlike *E. crista-galli, N. barkeri* responded to loss of individuals after high winds with an increased production of flowers per plant. They suggest that orchids have evolved to specialize in chance, and instead of maximizing pollinator attraction, they maximize the seed production resulting from every casual encounter.

The process of pollination is comparatively simple in genera with large, open flowers, having fairly wide, easily accessible nectaries as in *Cattleya, Laelia, Phaius, Sophronites, Bletia, Coelogyne* and *Cymbidium. Epipactus latifolia* finds its needs best served by wasps. In *Pterostylus*, after the insect enters, the labellum shuts a little trap door, forcing him to leave by the back through a passageway where the cramped quarters facilitate the performance of his function. *Masdevallia fenestra* never fully opens but has tiny windows that remain open until pollination occurs, when they are drawn shut. Artificial pollination (Hand pollination) and seed setting of *Spathoglottis plicata* was observed by self and cross pollination. Flowers of *S. plicata* were pollinated and subsequent capsule development was carefully monitored. Hand pollination of orchid flowers provides an opportunity to discuss floral morphology and associated reproductive biology. Capsules were harvested 60 days after the pollination prior to capsule dehiscence. *In vitro* asymbiotic germination was 87.3% when seeds were cultured on MS media. The effect of BA and NAA hormones were also observed on seed germination. Seedlings were healthy and stronger when cultured on MS media supplemented with hormones. The cultured seedlings were ready to greenhouse acclimatization after 110 days (Thakur *et al.*, 2012).

Disa fragrans subsp. *fragrans*, a taxon from the Drakensberg Mountains of South Africa, pollinated by insects belonging to at least four orders (flies, beetles, bees, and moths). Pollinaria of *D. fragrans* are attached to the feet of these visitors and pollination thus occurs in a rather haphazard fashion. Nevertheless, its pollination success and pollen transfer efficiency are comparable to those of its close relative, *Disa sankeyi*, which is pollinated by a single genus of wasps. *D. fragrans* has an exceptionally strong floral scent: volatile emission is 19-86 micro g per inflorescence per hour, which is up to 100 fold greater than in *D. sankeyi*. The scent bouquet is comprised of at least 46 compounds, mostly benzenoids and phenylpropanoids, which are known to be general attractants to a wide range of insects. In contrast to *D. sankeyi*, the flowers of *D. fragrans* have a high level of spectral purity (chroma) as is typical of many generalist insect-pollinated plants. At a site where *D. fragrans* co-occurs with *D. sankeyi* they found a plant with intermediate characteristics that may be a hybrid between the two taxa. The novel case of generalist pollination in *D. fragrans* documented here serves as an example of how floral advertising traits might evolve during an evolutionary shift from specialized to generalized pollination (Johnson and Hobbhahn 2010). Two populations of *Coelogyne fimbriata* with fragrant, nectarless flowers are pollinated by females of the same *Vespula* species (Vespidae, Hymenoptera). Experiments on wasps show that they crawl towards the source of the odour. The flowering period appears to coincide with an annual peak in *Vespula* colony expansion when additional workers forage for carbohydrates. Rates of pollinarium removal (0.069-0.918) and pollinium deposition on stigmas (0.025-0.695) are extremely variable. However, fruit set in *C. fimbriata* is always low (0.014-0.069) and appears to be based on self-incompatibility coupled with intraclonal (geitonogamous) deposition of pollinia. Conclusions: *Coelogyne fimbriata* and *Steveniella satyrioides* are now the only orchid species known to have food-deceptive flowers that are pollinated exclusively by eusocial, worker wasps. In *C. fimbriata*, floral odour appears to be the major attractant. Sub-populations may go through flowering seasons when pollinators are abundant or infrequent, but fruit set always remains low because the obligate pollinator does not often appear to transfer pollinaria between intercompatible gemots (Cheng *et al.*, 2009).

Due to frequency-dependent negative selection, a strong relationship between reproductive phenology traits and pollination success is expected in deceit-pollinated species. Parra and Vargas (2007), assessed the effects of floral display size on both female (fruit production) and male (pollen removal) pollination success in a population of the deceit-pollinated tropical orchid Myrmecophila christinae during two consecutive years (1998-1999). Low pollen removal (~9% of total flowers) and fruit production values (~3% of total flowers) were recorded during both years. As expected, binary logistic regressions showed a significant negative effect of floral synchrony, and a positive effect of floral display size on both male and female success, although these effects varied across years. Pollination rates in the field and in hand pollinations suggest a doubling in pollinator abundance between years. Results suggest that floral display size and flowering synchrony are of adaptive value for M. christinae. However, between-year fluctuations might indicate that reproductive phenology traits in deceit-pollinated species undergo fluctuating selection regimes among years and are probably linked to short-term changes in environmental (abiotic and biotic) conditions. The pollination biology of five aerangoid orchid species (*Rangaeris amaniensis, Aerangis brachycarpa, A. confusa, A. thomsonii,* and *A. kotschyana*) was investigated by Martins and Johnson (2007) in Kenya. Four of these have long spurs (>10 cm) and were pollinated by *Agrius convolvuli* and

Coelonia fulvinotata. Aerangis confusa, which has relatively short spurs (ca. 4 cm), was pollinated by the short-tongued hawkmoths *Hippotion celerio* and *Daphnis nerii*. Nectar frequently filled the entire spur in some of the studied species, even at anthesis. Sugar concentration of the nectar of four species was found to vary from ca. 1% at the mouth of the spur to 20% at the tip. Gradients were expressed more strongly in species with long, straight spurs. Species with spirally twisted spurs showed both steep and shallow nectar gradients. These gradients, previously unknown in plants, may function as a "sugar trail," enticing long-tongued hawkmoths to probe deeply into spurs without incurring the cost of filling an entire spur with concentrated nectar. In addition, the most concentrated nectar is kept out of reach of short-tongued pollinators.

It is widely believed that most orchid flowers attract insects by using deception or chemical rewards in the form of nectar. Flowers of *Bulbophyllum vinaceum* produce a large array of phenylpropanoids that lure tephritid fruit fly males and also act as floral reward, which the flies subsequently convert to pheromone components. The major floral volatile components identified are methyl eugenol (ME), trans-coniferyl alcohol (CF), 2-allyl-4,5-dimethoxphenol (DMP), and trans-3,4-dimethoxycinnamyl acetate, whereas the minor components are eugenol, euasarone, trans-3, 4-dimethoxy cinnamyl alcohol, and cis-coniferyl alcohol. Among the various floral parts, the lip (which is held in a closed position up against the sexual organs) has the highest concentration of the major compounds. An attracted male fly normally lands on one of the petals before climbing up onto and forcing the "spring loaded" floral lip into the open position, hence exposing the floral sexual organs. The architecture and location of chemical attractants of the lip compel the fly to align itself along the lip's longitudinal axis in a precise manner. As the fly laps up the compounds and moves towards the base of the lip, it passes the point of imbalance causing the lip to spring back to its normal closed position. The fly is catapulted headfirst into the column cavity, and its dorsum strikes the protruding sticky base of the hamulus and adheres to it. The momentum of the fly and the structural morphology of the long stiff hamulus act to pry out the pollinia from its anther cover. Hence, the pollinarium (pollinia+hamulus) is detached from the flower and adhered to the fly's dorsum. In this unique mutualistic association, both species receive direct reproductive benefits - the flower's pollinarium is transported for cross pollination, and the fly is offered a bouquet of phenylpropanoids (synomone) that it consumes, converts, and/or sequesters as sex pheromonal components, thus enhancing sexual attraction and mating success (Tan *et al.*,2006).

Flowers specialized for pollination by long-proboscid flies was particularly prevalent in the southern African flora. Although many orchids in this region possess flowers with traits that are consistent with long-proboscid fly pollination, evidence from field observations is lacking for most of these species. Flowers of the critically endangered orchid *Disa scullyi Bolus* (Orchidaceae) were observed to be pollinated by the large nemestrinid fly species *Prosoeca ganglbaueri* Lichtwardt at a remote site in the Cape Drakensberg mountains, South Africa. The orchid's floral spur contains copious amounts of dilute sucrose-rich nectar and its length (c. 42 mm) closely matches that of the fly proboscis (c. 40 mm). Flies caught on the orchids carried pollinaria of *D. scullyi* on the basal portion of their proboscides. Although flies were not common at the study site, they showed considerable fidelity to *D. scullyi*, resulting in high levels of pollen removal and deposition in flowers in the population. Habitat loss and trampling by cattle, rather than pollination failure, appear to pose the greatest threat to the few remaining populations of this species (Johnson, 2006).

The long-standing notion of pollination syndromes, which postulates that plants form recognizable groups according to pollinator type, has been challenged by Pauw (2006) on the basis of apparent widespread generalization in pollination systems. As a test of the pollination syndrome concept, he examined the pollination biology of a group of 15 orchids that share a recognizable syndrome of floral features that includes yellow-green coloration, oil secretion, pungent scent, shallow flowers, and a September peak in flowering. According to the pollination syndrome concept, the similar floral features of this group indicate a shared pollinator. To test this prediction, he observed pollinators on *Pterygodium alatum*, *P. caffrum*, *P. catholicum*, *P. volucris*, *Corycium orobanchoides*, and *Disperis Bolus*iana subsp. *Bolus*iana. They shared a single species of pollinator, the oil-collecting bee, *Rediviva peringueyi*. Female bees collected oil from the lip appendage using modified front tarsi. The orchids reduce interspecific reproductive interference through differences in pollinarium length or the use of mutually exclusive pollinarium attachment sites on the body of the bee. The results are contrary to the expectation of generalization in pollination systems and suggest that pollinators play an important role in mediating selection on floral traits.

The rare orchid *C. fasciculatum* offers no reward to the small wasps that pollinate its flowers. Non rewarding species typically suffer from low rates of pollination and fruit set, yet they appear frequently in the Orchidaceae. To understand the dynamics of pollination and fruit set in non rewarding orchids and to evaluate whether a conservation strategy for *C. fasciculatum* should include managing for pollinator service. Lipow *et al.* (2003), evaluated factors influencing reproductive success in three widely separated populations. The percentage of open-pollinated flowers maturing fruit varied greatly among populations. These values are greater than is typical for a non rewarding orchid. Like many non rewarding orchids, the species appears to have a self-compatible breeding system; fruit set did not differ between flowers receiving self pollen and flowers receiving cross pollen. Within-season fruit set may be pollen limited, however, since supplemental hand pollination with cross pollen significantly increased fruit production by ca. 4.5 times in Colorado and 1.5 times in Idaho in comparison with open pollination. Even so, many pollinated flowers failed to mature fruit. Pollen tube growth was studied using fluorescence microscopy, and the percentage of pistils with germinated pollen at the end of the flowering season was found to exceed significantly final fruit set by a rate of approximately two in both Idaho and Colorado.

Pollination and Fertilization

A series of studies on pollination biology of the *Epipactis atroruhens* (Hoffm.) Besser, *E. purpurata* Sm., and *E. palustris* (L.) Crantz populations was conducted by Busse and Kadej (2011). It was found that: (1) the type of pollinators depends on the orchid population size and the surrounding environment, where the temperature and amount of precipitation during the vegetative season are the most vital factors; (2) pollinators and visitors to the examined orchids might differ in successive growing seasons; (3) the studied *Epipactis* species differ in the chemical composition of their nectar and its scent, which can influence their pollination biology; (4) the tendency to autogamy observed in *E. purpurata* might be due to lack of pollinators in its habitats.

The unique floral morphology of the South African orchid *Huttonaea pulchra*, with its twin meranthia, is best explained as an adaptation to pollination by oil-collecting bees. Flowers consisting of meranthia (floral parts that function as single pollination units; commonly observed in garden Iris) are extremely rare among the angiosperms and their significance poorly understood. Unlike all other known examples of meranthia, the novel type described for *H. pulchra* is not bilabiate. All *Huttonaea* species are unique in having twin petal sacs with glandular verrucae that secrete oil and are pollinated by *Rediviva* (Melittidae) oil-collecting bees. But only *Huttonaea pulchra* has long and widely divergent petal claws that place the oil sacs well beyond the reach of a centrally positioned bee. The wide separation of these sacs forces the pollinator, *R. colorata,* to visit each side of the flower independently and effectively divides the flower into two meranthia. Molecular data indicate that the evolution of the *Huttonaea*-type meranthium was dependent on the prior evolution of the oil flower/oil bee relationship. Meranthium evolution was also facilitated by the presence of oil in two separate structures (petal sacs) that were not physically constrained to remain in close proximity (Steiner 2010).

Methods of Orchid Breeding

Artificial selection

Selection is the process by which human use to obtain more desirable types of plants from population. Thousands of years ago people learned about saving seeds from the selected plant to obtain offspring in next generation.

Hybridization

Plants inherit characteristics from their parents. The laws of heredity explain why different traits are inherited to the offspring of the same parents. Moreover, these laws make it possible to predict the number of offspring that will inherit a desire trait. Knowledge of the laws of heredity is essential for effective orchid breeding. All cells contain genes, which are the units that determine the traits of a plant. Except for the germ cells, each cell in a plant contains two genes for each trait for example, for the trait of colour a plant could have one gene for yellow and one for red. The germ cells have only one gene for each trait. Of the thousands of germ cells about half have one of the genes for each trait and about half have the other gene. Sexual reproduction involves the union of a male and a female germ cell. When the male and female germ cells unite in the ovule each contributes one gene for each trait so that the new seed then has two genes for each trait. The various combinations of the many genes inherited from the pollen parent and from the seed parent determine the traits of the offspring. Because of this creation of variation in population, sexual reproduction of orchids is the method by which new strain and varieties may be developed by orchid breeders. The reproductive processes of the orchid are such that, with few exceptions, the genera do not intermingle readily, so the generic lines remain uniquely pure. Hybrid means the offspring of two plants of different genotype. Cross-pollination is usually necessary to develop new varieties. Once the desired trait is fixed, measures have been applied to obtain that trait in its offspring.

Due to the complexity of flower structure in orchids, the method of pollination was not understood by breeders for a very long time. It is recorded that even before the time of John Dominy, Rev. William Hubert, succeeded in producing pods through cross made between *Orchis* and *Ophrys*, but the seedlings did not survive. *Calanthe dominyi* (furcata x masuca) was the first orchid hybrid to flower in 1856, developed by John Dominy. The purposes for developing orchid hybrid have been to increase the stock, size, number of flowers, change the colour of flowers, improve the growth habits and throw strength into a weak line. Constant experimentation in the production of hybrids has been going on for many years. Some genera have been found to be definitely incompatible, but many others are cross fertile and can be made to produce interesting flowers. The important genera which have lent themselves for improvement in the hands of man are *Cattleya, Cymbidium, Paphiopedilum, Odonotoglossum, Vanda, Dendrobium* and to a lesser degree *Phalaenopsis*. The breeding programmes are usually drawn out with the intention of achieving the following standards in the qualities of the flowers:

- Large size
- Heavy texture so that the flower may last longer
- Round full form of sepals and petals with a minimum degree of fenestration and twists.
- The pure quality of the colours and wherever there are more than one colour, the correct blending of them in sepals, petals and lip.
- The correct mode of display of the flowers by the plant.

In recent years the emphasis has shifted from large size and brilliant colours to the element of the unusual for which orchid flowers are famous. Because of this many of the botanical species which have long been neglected, have come to be much sought after. The tempo of production of new and varied combinations increased to such an extent that the number of hybrid produced in orchids alone today exceeds those from all the angiosperm families put together.

Cattleya was one of the first genera upon which hybridisation were initiated for the purposes of improvement. Vast improvements on the original *Cattleya* flower are the result of years of carefully planned breeding. There are three groups of *Cattleyas*: mauve and purple, yellow and bronze types and the white with coloured lips. *Cattleya dowiana, C. trianae* and *C. mossiae* were the popular parents used in breeding programmes. *C. dowiana* was valued greatly for its variety aurea, which played a significant role in the attempts to breed pure yellow forms. *Cattleya mossiae* var. *alba* has figured prominently in the breeding of the pure white forms of *Cattleya*. *Cattleya trianae* has helped a great deal in imparting heavy texture to the hybrids. *Cattleya dowiana* crossed with *Cattleya gigas* resulting in a intrageneric hybrid, *Cattleya hardyana,* a fine round flower of *C.gigas* mauve enriched with the iridescence of the Dowiana (yellow with dark red lip). The two other main lines of breeding in the *Cattleya* group were towards evolving pure white and yellow forms. The dream of a pure alba variety has long since been equal success. Alba varieties of *C. mossiae, C. trianae* and *C. gaskelliana* were utilised in the breeding of the pure white form x Suzanne Hye (Pauwel, 1908), x C. Edithae (Pauwel, 1914), x C. Bow Bells (Black and Flory, 1945), x C. Bob Betts (Mc Dade, 1950) are among the best white *Cattleyas.*

The most famous bi-generic cross is important for purely historic and horticultural reasons, having no commercial value. This cross was *Epiphronitis veitchii* (*Epidettdrum radicans,* scarlet species of Mexico) with *Sophronitis grandieora*. *Cattleya mossiae x Laelia purpurata* became *Laelio, Cattleya canhamiana*. When a *LaelioCattleya* was crossed with a species (or hybrid) of *Brassavola* the hybrid genus name became *BrassoLaelioCattleya*. The practice of creating generic names ending in -ara was adopted for hybrids with genes from multiple genera. Examples are *Vuylstekeara* (*Cochlioda x Miltonia x Odontoglossum*) and *Potinara* (*Brassavola x Laelia x Cattleya x Sophronitis*). One way of distinguishing a natural genus from a hybrid genus is to insert a multiplication symbol before the hybrid name, such as *X Potinara*.

But the most popular hybrids from *Cattleya* were bred with the help of other genera related to *Cattleya* like *Laelia, Brassavola* and *Sophronitis*. In 1889 *Brassavolva digbyana* had been mixed with *Cattleya* by Maron and also by Veitch, to produce *BrassoCattleya*. The beautifully fimbriated lip of *Brassavola* had been an added attraction to the hybrid flower. Charlesworth in 1922 combined all the four genera in one hybrid form:*BrassoCattleya Edna x Sophrolaeliocattlonitis* Marathon = Pontinara. The tetrageneric hybrid Pontinara did not possess many desirable traits and hence was not so popular.

The securing of yellow colour is another problem that has been tackled through hybridization. Yellow is desirable in an orchid flower but difficult to obtain. The pollen of yellow-flowering orchids is frequently infertile and evens many of the yellow hybrid orchids lacking vital element, producing crippled growths. Yellow colour derives chiefly from *Cattleya dowiana aurea,* which is of clear golden tone, with a rich velvety maroon or crimson lip and splashy gold stripes in the throat. *Cattleya bicolor* (greenish-bronze species of Brazil) and *Laelia tenebrosa* (purplish-brown sepals and petals and a purple lip) are also used. The latter in some instances gives a bronzy sheen or coppery overlay to a dominantly purple flower, *C. luminosa aurea* (*Cattleya Dowiana aurea x Laelia tenebrosa*).

The *Cattleya* had been found to cross most readily with *Brassavola digbyana,* insignificant except for an extra-large frilled lip, to produce well-shaped flowers with a full, fringed lip. The *Brasso Cattleya,* in spite of size and beauty of lip and color, frequently has the disadvantage of starry, narrow sepals and petals as well as poor texture, so that it does not keep well. It rarely blooms with more than one flower. By adding *Laelia* to the *Brasso Cattleya* perfection was reached. The *Laelia* stands firmly erect, and its influence was to strengthen the weak dorsal of the simpler cross. The size of the resulting *BrassoLaelioCattleya* is enormous and the color range wide. Many of the modern *BrassoLaelioCattleyas* are so full and round that the hand can be hidden by a single flower.

The path of pollination in *Cymbidium pendulum* (Sw.) Roxb. had been traced by Attri and Kant (2011). The honey has been identified as *Apis mellifera* was found to act as main pollinator and this bee is the only insects, among others, who succeeded in performing pollination because probably due to its structural compatibility with the plant species. Pollination by *Apis mellifera* bees was suggested to occur in a number of families but rare phenomenon in orchids and it is first time that the species was observed to pollinate the *Cymbidium pendulum* flowers. Bee moved around the flowers for some times, entered the flower and carried on pollinia along with on the back during its journey. It revisited the different flower and deposited its pollinia on to it and the act of pollination was accomplished. *Cymbidium goeringii* is one of the most important and popular species in the orchid family in north-east Asia. Lee *et al.* (2012) prepared multiplex PCR system, and used it for the genotyping of eight simple sequence repeats (SSR) markers in subject with 40 samples of cultivated varieties. All the analyzed samples showed different combined genotypes. The average combined power of discrimination was very high value of 7.14 x10-10, and observed heterozygosity was similar with two wild populations of *C. goeringii,* which may indicate no or little occurrence of genetic change after collection from wild habitats. They also developed a two-dimensional barcode to express information of genotype results of eight SSR markers (SSR DNA ID). The discrimination power of DNA ID between two individuals will be statistically more than 99.999999%. The SSR DNA ID and two-dimensional barcode may be very usefully applied for the discrimination and maintenance of cultivars of *C. goeringii.*

Cymbidiums do not combine with other genera, but crosses within the genus have come in for their share of glory, the progeny being much improved in form, shape, texture, and clearness of colour. The two unforgettable names in *Cymbidium* breeding are that of H. G. Alexander of Westonbirt and Lionel de Rothschild. Earlier, these two workers developed the finest *Cymbidium* hybrids. The greatest contribution of Alexander was the legendary hybrid, *xC. Alexanderi* 'Westonbirt' (*Cymbidium eburneum* x *C. insigne* ' Westonbirt'). This hybrid had lovely large white flowers characters, all of which were easily transmissible. Due to the partial dominance of its white colour, it produced beautiful pastel shades, when crossed with coloureds *x C. Alexanderi* ' Westonbirt' gave rise to several coloured flowering plants which are now in use widely, like *x C. Rosanna* Pinkie and *x C. Balkis*. Apart from these pastel shades, highly coloured varieties were also greatly appreciated. To meet this demand dark reds like x *C. Carisbrook,* greens like *x C. Baltic* and amber yellows like *x C. Marcia,* were produced. Most of the *Cymbidium* species in breeding hybrid come from the Indo-Malayan region (*C. eburneum, C. giganteum, C. lowianum, C. erythrostylum, C. devonianum*).

Cymbidium 'Purple Princess' ('Show Girl' x 'Eiko') is a cultivar having dark purple sepal and petal (RHS, RP59A) with purple lip (RHS, RP59A). A cross was made between pink colored C. 'Show Girl' as maternal line and dark purple colored C. 'Eiko' as paternal line in 1994. It was preliminarily selected as Line No. 94019267 in 1999 considering the characters of flower shape and color, leaf growing shape and growth habit in normal culture condition. Multiplication, cultivation, and characteristic trials were conducted from 2000 to 2003. The line was named as Wongyo F1-7 and phenotype was characterized in 2004 as a new 'Purple Princess'. 'Purple Princess' has about 13.2 flowers per flower stalk and flower size of 5.2 cm. General appearance of petals and sepals is slightly incurved shape. The plant size is intermediate having erect peduncle. Blooming starts from early December under optimal culture condition. Leaf attitude and twisting is erect and very weak respectively (Seon *et al.,* 2010).

Miniature *Cymbidiums* have an appealing charm and are more manageable in size than ordinary *Cymbidiums*, which take up a great deal of space in the orchidhouse. The species from which the present day miniatures are derived are mostly from Japan and China, with *C. pumilum* from Japan heading the list. Hybrid miniatures were first developed by England in 1944 (*x C. Lois Sander x C. pumilum*) and later in 1955 by the United States (*C. pumilum x C. Zebra*).

Lady's slipper orchids (*Paphiopedilum*) are of high value in floriculture, and interspecific hybridization has long been used for breeding improved cultivars; however, information regarding the genome affinities of species and chromosome pairing behaviour of the hybrids remains almost unknown. Lee *et al.*(2011) analyses the meiotic behaviour of interspecific hybrids by genomic *in situ* hybridization and cytologically evaluates the genomic relationships among parental species. Genomic *in situ* hybridization was demonstrated as an efficient method to differentiate between *Paphiopedilum* genomes and to visualize the chromosome pairing affinities in interspecific F1 hybrids, clarifying the phylogenetic distances among these species. Comparatively regular chromosome pairing observed in the hybrids of *P. delenatii x P. bellatulum, P. delenatii x P. rothschildianum* and *P. rothschildianum x P. bellatulum* suggested high genomic affinities and close relationships between parents of each hybrid. In contrast, irregular chromosome associations, such as univalents, trivalents and quadrivalents occurred frequently in the hybrids derived from distant parents with divergent karyotypes, such as *P. delenatii x P. callosum, P. delenatii x P. glaucophyllum, P. rothschildianum x P. micranthum and P. rothschildianum x P. moquetteanum*. The existence of multivalents and autosyndesis demonstrated by genomic *in situ* hybridization in this study indicates that some micro-rearrangements and other structural alterations may also play a part in differentiating *Paphiopedilum* species at chromosomal level, demonstrated as different chromosome pairing affinities in interspecific hybrids. It is also the most difficult of all orchid species to breed because of the high amount of sterility in hybrids and the special requirements of the seeds for proper germination. The sterility in *Paphiopedilum* hybrids is due to the variations in the chromosome number from 2n = 26 , 28, 30, 32, 34, 36, 40 and 42. However, *Paphiopedilum insigne, P. villosum, P. spicErianum* and *P. bellatulum* are the most common species useful in hybridisation. *P .spicErianum* has 30 chromosomes, while others have 26 . Yet *P. spicErianum* is quite compatible with the other species.

Paphiopedilum delenatii is an endangered slipper orchid with delicate flowers that is native to Vietnam. An understanding of this plant species' reproductive biology is required for its effective propagation and conservation. *P. delenatii* has a bisporic pattern of embryo sac development with a six-celled number in a mature embryo sac. Prior to pollination, the nucellar filaments are present within the ovary. The archesporial cell at the tip of the nucellar filament differentiates into a megasporocyte fifteen days after pollination. As the megasporocyte matures, the nucleus becomes located at the micropylar end of the cell, resulting in an uneven distribution of the cytoplasm with the starch granules becoming more abundant at the chalazal end. After the first meiotic division, a distinct callose layer separates the micropylar dyad from the chalazal one. Soon after, a callose wall completely encloses the functional dyad at the chalazal end. As the functional dyad increases its size, the callose remains in the common wall separating the two dyads as well as in the chalazal end wall of the functional dyad. One notable finding of this study is the presence of cuticular material in the lateral walls of the mature embryo sac. The deposition of the cuticular material may serve to strengthen the fairly large embryo sac and helps to conserve water for the developing embryo after fertilization (Lee and Yeung 2012).

An experiment was carried out using 20 native *Dendrobium* orchids by Moniruzzaman *et al. (*2012) to estimate genetic variability, genetic parameters and correlation coefficient among different flower components. Highly significant variations were observed among the genotypes for all the characters. High broad sense heritability with high genetic advance was observed in spike length , rachis length and), flower durability and floret number indicating additive gene action, suggesting the possibility of improvement of these traits through selection. Other characters exhibited moderate broad sense heritability with low genetic advance. The number of florets per spike recorded a highly significant association with flower durability, rachis length and spike length. Durability of flower had significant positive correlation with rachis length, flower size and spike length at both genotypic and phenotypic level. Highly significant positive correlation was also found between rachis length and spike length. However, the correlation study revealed that selection of parents should be done based on characters such as floret number, rachis length and flower durability which would be useful in a breeding program. The path analysis revealed that rachis length, durability of flower and number of flowers per spike had contributed maximum direct effects on flower yield, indicating the importance of these three characters as selection criteria for native *Dendrobium*

Odontoglossum and its allies *Miltonia, Oncidium* from tropical America are especially liked by orchid growers, because they occupy the minimum of space and put forth the most gorgeous of sprays among orchids. *Odontoglossum* has the added advantage of not being seasonal, with the result that a spray may appear at any time of the year. The various members of the group interbreed freely. The most popular hybrids are *x Odontioda* (*Odontoglossum x Cochlioda*), x *Odontonia* (*Odontoglossum x Miltonia*) and *Vulylstekearas (Odontoglossum x Cochlioda x Miltonia*). *Odontoglossum* has the largest flowers of the group with a colour range of white, reddish purple, plum as well as yellows and bluishmauve. *Cochlioda*, particularly, *C. noezeliana* has brought to its hybrids for a dominant brilliant red colour. The contributions of *Miltonia* to hybrids are large lip and wide range of flower colours. *M. roezelii* and *M. vexillaria* are the two most common parents for development of *Miltonia* hybrids.

Among the hybrids showing great promise of improvement over the parents are the *Odontonias, Odontoglossum crispum* and its hybrids are spray flowers of outstanding texture, color, and beauty, but in some place they have a blooming problem. Similarly, *Miltonia* is a plant of fragile characteristics and difficult for some growers to treat successfully. *Miltonia* blooms, with the distinctive butterfly mask around the lip and throat of the pansy face, are of delicate beauty and unsurpassed color. The *Odontonias* resulting from the union of these two species retain the beauty of both and in some way have their life and growth habits strengthened so as to grow successfully in the *Cattleya* house.

Brachypetalum hybridizing started in the late 1800s with a few primary hybrids. Another handful of secondary hybrids were created at the turn of the twentieth century. The group was virtually neglected for another half century with no hybrids being registered. Then in the late 1970s and 1980s about five hybrids were again registered. In the last few years, this group has caught on with the *Paphiopedilum* fanciers with many new registrations of hybrids.

In nature, *Vanda* occurs in two vegetative forms -terete- leaved and strap-leaved. The first recognised *Vanda* hybrid x *Vanda* Miss Joaquim, was a cross between two terete -leaved species, *V. hookEriana* and *V. teres*. This is a very floriferous form and is still very popular as the lei orchid in South East Asiatic countries. These hybrids were, however, found to be quite sterile and for a time it appeared as though this would be a dead end to *Vanda* breeding. However, these semiteretes when crossed with natural diploid, strap leaved *Vandas*, produced progeny of outstanding merit and which turned out to be fertile than the hybrid parent. Storey (1956) was the first to clarify the cytological situation in these hybrid plants and explain their unusual behaviour on the basis of his findings. He and later Kamemoto (1959) found out that the sterility of the first generation semi-terete hybrids is due to the incompatibility between the genomes of the strap-leaved and the terete -leaved species. There is no pairing of chromosomes during meiosis and distribution at anaphase is at random, resulting in a high degree of sterility. But polyploid gametes also result from this random distribution and consequent restitution, giving rise to diads, triads and tetrads. These on fusing with haploid gametes from the other parent, produce tripoids and tetraploids. A diploid gamete fusing with a triploid one produces a pentaploid.

The noble type *Dendrobiums* of Eastern Himalayas and *D. Phalaenopsis* of Eastern Asia were frequently used parents. Colour has always been of prime importance in *Dendrobium* breeding, ranging from chalky white to yellowish brown and intense crimson.

It has been noted that certain genera will not cross with others. The failure of a plant to produce fertile seed when crossed with another plant may be the result of a difference in chromosome number, although there are many other possible causes. The failure of a genus to cross with another does not, however, entirely rule out hybridization.

Epidendrums are popularly known as 'Crucifix orchid' and 'Poor man's orchid', has a long flowering period with 2-3 flowerings in a year, and are easy to multiply. An interspecific *Epidendrum* hybrid was developed by Devadas *et al.* (2010) using *Epidendrum radicans* as female parent and *E. xanthinum* known as 'yellow orchid' as male parent. The selected line (NRCO-*Epidendrum* cross/2005-01) was characterized for morphological and floral traits. Flower size of selected line was bigger than both parents, with bright saffron-orange colour (RHS 44A). Dorsal sepal size, lateral sepal size, petal size, lip size and column size were bigger than in parents. Shape and fimbriated side lobes of lip with deep cleft of anterior margins was similar to the male parent (*E. xanthinum*), except colour. The F1 progeny of 'NRCO-*Epidendrum* cross/2005-01' flowered with different red-orange to yellow shades is categorized broadly into three types: Red-orange, Orange-yellow and Yellow

Extensive artificial hybridization among members of some 86 genera of *Vanda*ceous orchids within the subtribe Sarcanthinae in Singapore over the five decades has created many thousands of exotic hybrids. A small number of these are selected for commercial cut-flower production, while others are cultivated as pot-plants. Two classes of intergeneric hybrids are of special commercial importance as export cut-flowers, the bigeneric Aranda (*Arachnis* x *Vanda*) and the trigeneric Mokara (*Arachnis* x *Vanda* x *Ascocentrum*) hybrids. Cytogenetic studies of these hybrids in recent years have provided information on the significance of ploidy levels and genomic constitutions on the horticultural performance of cut-flower cultivars, and also on the strategy for effective breeding in *Vanda*ceous orchids (Lee 1995, Lee *et al.*, 1996).

Phalaenopsis orchids with regular meiotic behavior can often produce seeds successfully. Most *Phalaenopsis* hybrids are obtained after interspecific hybridization. Unfortunately breeding barriers may occur during the meiotic process thereby affecting fertilization events and reducing seed set. Through pollen tetrad analysis, a correlation between successful seed set and a high tetrad percentage was observed in both *Phalaenopsis* and *Doritaenopsis* orchids. To improve the breeding of hybrid polyploid cultivars, mitotic inhibitors were applied to young flower buds before pollen meiosis in order to monitor changes in the production of tetrads. When compared to the control, a higher number of dyads were observed after treatment with mitotic inhibitors. Treated pollen grains were then attached to the stigma of female parents for self- and cross-pollination, resulting in successful fruit set in several cases (Hsu *et al.*, 2010). Polyploidy played an important role in the development of *Phalaenopsis* varieties. Many popular commercial hybrids are tetraploids. However, analysis of the pedigree of these hybrids indicated that their genetic base was quite narrow. On the other hand, it is difficult to transfer favorable genes of the diploid wild species to the commercial hybrids because of the difference in the ploidy levels. In addition, the chromosome sizes of *Phalaenopsis* species varied tremendously. Hybrids developed from the crosses between these species were usually sterile due to the problem of genome homology. Chromosome doubling is one of the methods to restore the fertility of these hybrids. Due to the above reasons, a research project was initiated in 2004 to develop simple and effective methods to detect the nuclear DNA content by flow cytometry and to double the chromosome of *Phalaenopsis* species and varieties by the use of tissue culture without application of anti-microtubule agents. The nuclear DNA contents of the native species and some of the related hybrids of *Phalaenopsis* in several germplasm banks and orchid nurseries in Taiwan were determined. Based on this information, *Phalaenopsis* species were grouped into low, medium and high nuclear DNA content which reflects the chromosome sizes of the species (Chen *et al.*, 2010).

Chen *et al.* (1990) collected twenty-five wild species and 252 commercial varieties of *Phalaenopsis* for the improvement. Cytogenetic studies and computer analysis of the pedigree of parental varieties were carried out to design breeding programmes and to understand sexual incompatibility. P2 medium was used to germinate the dominant buds of flower stalks and induce development of adventitious shoots. The adventitious shoots multiplied rapidly on P4 medium. The use of P19 medium allowed the formation of protocorm-like-bodies from a leaf explant. Plantlets which rooted on P4ac medium were transplanted into pots and grew well in the greenhouse. The isolation and electrofusion of viable protoplasts of *Phalaenopsis* were successfully undertaken to achieve genetic modification *in vitro*. By using E4 enzyme solution, numerous protoplasts were rapidly released from leaf, petal and root tip of *Phalaenopsis*. A microscopic examination of protoplasts stained with fluorescence diacetate or trypan blue showed that the frequency of the viable protoplasts was about 90%. Plating of protoplasts from leaves on 1.2% agarose KP8 medium resulted in the division of protoplasts and formation of cell clusters. Protoplasts from different species of *Phalaenopsis* fused to form hybrid cells (ca. 10%) when alternating current field of 70 V RMS/cm, 800 KHz and direct current pulse (2 msec) of 2500 V/cm passed through protoplasts suspended in 0.5M mannitol solution.

Breeding Techniques

Equipment

The equipment required for orchid breeding is relatively inexpensive and easy to use. Here are some items that may find useful:

- Magnifying glass (10 or 15 power)
- Tweezers
- Small sharp-pointed scissors
- Camel-hair brush
- Small containers or vials
- Alcohol
- Rubber bands or soft wire
- Paper or cellophane bags
- Paper clips
- Tags
- Notebook

Steps of Prepollination

Pre-pollination steps begin before the flower opens otherwise, it may be pollinated by natural means, which would make the flower useless. Extremely high temperatures or moist conditions are harmful to pollen. For best results, plants should be pollinated on dry days during the cool hours of the morning or as soon as the anthers have split open.

Selection of Parents

The parent orchids should be checked for natural barriers to cross- or self-pollination. Before begin breeding, barriers should overcome however, some orchids cannot be artificially pollinated. An example of a barrier that cannot be overcome is the self-pollination prohibitor of some orchids; the stigmas of certain orchids produce a substance which kills the pollen of flowers of the same plant. Some orchid flowers become pollinated inside the bud before they open, this phenomena is known as cleistogamy. It is not a desirable trait for hybrid development. Most orchids prefer to be fertilized with pollen from a different plant, ensuring genetic versatility.

Several orchids are so far apart in their genetic make-up that they do not cross-breed. *Zygopetalums* are infamous for not accepting any pollen from another orchid. The mechanism that performs this cannot be removed without destroying the pistil. Selected plant for breeding should be sturdy and healthy. A fresh flower is more likely to have healthier pollen than one that has started to wilt or dry out. In choosing a seed parent stigma should properly examined. It should have either a glistening substance on it that is sticky to the touch or a hairy surface. It is this surface that retains the pollen, thus making fertilization possible. After selection of pollen and seed parents pollination should be started.

In the first step pollen parents and seed parents should marked carefully. This can be done with coloured thread, one colour for the male and another colour for the female. Paper label may also be used for identification. For protection of flowers from weather, they should be covered with varnish. The next step is to protect the plant from unwanted pollen. If the plant is to be cross-pollinated, the stamens will have to be removed to prevent the possibility of selfing. The removal of the stamen is called emasculation. It should be performed before the anthers split to release pollen. Emasculation may require opening the flower by hand before it is ready to bloom. Emasculation can be accomplished by: pinching off the stamens or anthers with tweezers, snipping off the stamens or anthers with sharp-pointed scissors and removing the petals to which the stamens are sometimes attached. A magnifying glass will be particularly useful in emasculation. If the plant is grown indoors there is little likelihood of contamination by foreign pollen and therefore no need to cover or close the flower. However, indoor as well as outdoor plants require emasculation to avoid self- pollination.

Steps of Pollination

Crossing

The transfer of pollen from pollen parent to stigma of seed parent can be achieved by suing toothpick. Some pollen on the toothpick should be put and bring onto the sticky surface of the stigma. Sometimes, toothpick should be first dip into the stigmatic cavity to pick up some of the moisture there. That can assist in holding the pollen in place during transfer. It is a good idea to have the mother plant standing on a large surface with a clean white paper underneath. Often the pollen roll away before they are securely placed and they are easier to find on a large piece of white paper. Each time different pollen can used, but be sure to first wash with alcohol the camel-hair brush, tweezers and any other item which might have touched pollen. This step is very important to prevent pollination of the seed parent with unwanted pollen adhered to the equipment. After we wash the instruments are sure that they are dry before using them again.

Postpollination Steps

Pollinated seed parent should be covered and labelled just after pollination. The standard method of labelling is as follows:

1. The tag should be written in the following order for labelling of plant and flowers: (a) the number that has assigned or the variety name of the seed parent (b) the letter X (c) the number or variety name of the pollen parent and (d) the date of crossing was made.
2. Label should be attached to the stem just below the pollinated flower.

Keeping complete and accurate records of breeding operations is very important. The information that recorded should contain all essential facts regarding the cross or self so that it can refer at a later time and even do the job again from the beginning if necessary. A separate form or page should be used for each cross or self that make.

An easy way to keep track of the offspring is to assign consecutive numbers to each generation resulting from each cross or self.

Table 2.3 : Offspring notes

Date of planting	Date of first flowering	Traits			Date of pollination		Harvesting information	
		size	colour	others	first	last	date	no. of seeds
———	————	———	———	———	———	———	———	———
—	—	–	–	–	–	–	—	—

If the pollination has been effective, the column will begin to swell and ovaries enlarge, within 2-4 days. Otherwise these organs will shrivel and die. After pollination the sepals and petals fade and wilt, but they usually remain attached to the ovary. Once the pollination has successfully completed, a period must elapse during which fertilization and subsequent development of the fertilized ovule take place. The ovary typically develops into a capsule that is dehiscent by 3 or 6 longitudinal slits, while remaining closed at both ends. After fertilization the ovule develops into the ovoid embryo, and the integument develops into the water impermeable seed coat. In most species embryogenesis lasts about 2 weeks. The ripening of a capsule can take 2 to 18 months. The success rate for artificially pollinated orchid can often be improved when this is done during the morning, afternoon, evening, night or dark time. Artificial pollination results about 20 per cent viable seeds. Seeds are dispersed into the air like minute dust particles or single-celled spores. The seeds of some species are no larger than fungal spores and occur in a loose cellular sheath.

Inheritance

In spite of the great enthusiast evinced in hybrids and hybridisation of orchids, the knowledge of the mode of inheritance of characters in this group of plants is still very inadequate. In this field, Hurst's (1925) pioneering work on *Cattleya* flower colour still remains very much of reference record. The apparently erratic expression of flower colours noticed during the breeding experiments with white and coloured *Cattleyas* piqued the curiosity of Hurst and led him on to the first experiments in orchid genetics and formulation of his theories on colour inheritance. Coloured piqued the curiosity of Hurst and led him on to the first experiments in orchid generics and formulation of his theories on colour inheritance. Coloured *Cattleyas* bred to coloured forms sometimes produced coloured progeny. Hurst suggested that two genes C and R are responsible for the expression of colours in *Cattleya* flower. The colour is expressed only when these two factors are present in dominant condition either heterozygous(CcRr) or homozygous(CCRR). When either of both the genes are in the recessive state, no colour is expressed. The recessive state (ccrr) represents the pure alba form of all outward appearances, but on crossing with each other give all coloured progeny as they have CcRr genotype. When the parent plants are alba forms but heterozygous for the colour factors, ie. Ccrr or cc Rr, one fourth of the progeny is coloured and the remaining are white. The genotype Ccrr produces gametes of the constitution Cr and cr. Similarly ccRr produces gametes of cR and cr. When these two forms are brought together, the progeny will be of the following constitution:

	Cr	cr
cR	CcRr(coloured)	ccRr(white)
cr	Ccrr (white)	ccrr (white)

Similarly when two coloured forms with CcRr of genetic constitution are crossed with each other, some of the progeny would be white and others coloured in the ratio of 9: 7 (coloured : white).

	CR	Cr	cR	cr
CR	CCRR(coloured)	CCRr(Coloured)	CcRR(coloured)	CcRr (Coloured)
Cr	CCRr (coloured)	Ccrr (white)	CcRr(Coloured)	Ccrr (white)
cR	CcRR(coloured)	CcRr (Coloured)	ccRR(White)	ccRr(white)
cr	CcRr(coloured)	Ccrr(white)	ccRr(White)	ccrr(white)

Besides the pure white and coloured forms, there is also a third form of flower in *Cattleya* that has white flower with a coloured lip, popularly known as semialbas. Mehlquist (1958) proposed a genetic theory for their occurrence. He suggested a gene P is responsible for the method of distribution of colours in orchids. This gene, when present in homozygous dominant (PP) or heterozygous (Pp) condition along with the genes C and R, makes the whole flower coloured. On the other hand, when this gene is present in the recessive state (pp), with the genes C and R in dominant condition, the colour is expressed in the lip alone. When genes C and R are in the recessive state, the gene P has no action and the result is a pure white form. Thus CCRRPP or CcRrPp are both fully coloured; CCRRpp and CcRrpp are both white with coloured lips or semi albas and ccrrPP, ccrrPp and ccrrpp are pure white or albas. Storey (1958) conducted studies on the breeding behaviour of *Spathoglottis plicata,S. plicata var. alba* and *S. grandifolia*. He found out that the purple colour in both *S. plicata* and *S. grandifolia* is controlled by two genes -P, which is the dominant gene for pink colour and red spotting on the calli, its recessive allele being p, which produces neither colour nor spotting and T, which is the dominant gene for pale pink colour, producing tinging of sepals and petals, its recessive allele being t, which produces no colour. When these two genes are present together, they produce deep purple colour. When either one is the dominant state, other being recessive, then the respective pink or tinged condition results.

A triploid form of *Phalaenopsis* was created by Amado and George Vasquez at Zuma Canyon Orchids in California.The tetraploid form of *P. equestris* which was used to create the triploid hybrid was produced through chemical treatment. Robert (1981) developed procedures for using colchicine to double the chromosome number of *Phalaenopsis* seedlings. The colchicine-induced tetraploid form of *P. equestris* was seed propagated.

The complete chloroplast genome of the economically important *Oncidium* variety Onc. Gower Ramsey was by Hui *et al.* (2010) determined using a polymerase chain reaction (PCR) and Sanger based ABI sequencing. The length of the *Oncidium* chloroplast genome is 146,484 bp. Genome structure, gene order and orientation are similar to *Phalaenopsis*, but differ from typical Poaceae, other monocots for which there are several published chloroplast (cp) genome. The Onc. Gower Ramsey chloroplast-encoded NADH dehydrogenase (ndh) genes, except ndhE, lack apparent functions. Deletion and other types of mutations were also found in the ndh genes of 15 other economically important Oncidiinae varieties, except ndhE in some species. The positions of some species in the evolution and taxonomy of Oncidiinae are difficult to identify. To identify the relationships between the 15 Oncidiinae hybrids, eight regions of the Onc. Gower Ramsey chloroplast genome were amplified by PCR for phylogenetic analysis. A total of 7042 bp derived from the eight regions could identify the relationships at the species level, which were supported by high bootstrap values. One particular 1846 bp region, derived from two PCR products (trnHGUG -psbA and trnFGAA-ndhJ) was adequate for correct phylogenetic placement of 13 of the 15 varieties (with the exception of Degarmoara Flying High and *Odontoglossum* Violetta von Holm). Thus the chloroplast genome provides a useful molecular marker for species identifications.

Conclusion - In this report, we used *Phalaenopsis*. aphrodite as a prototype for primer design to complete the Onc. Gower Ramsey genome sequence. Gene annotation showed that most of the ndh genes in Oncidiinae, with the exception of ndhE, are non-functional. This phenomenon was observed in all of the Oncidiinae species tested. The genes and chloroplast DNA regions that would be the most useful for phylogenetic analysis were determined to be the trnHGUG-psbA and the trnFGAA-ndhJ regions. We conclude that complete chloroplast genome information is useful for plant phylogenetic and evolutionary studies in *Oncidium* with applications for breeding and variety identification.

The general genetic principles governing the inheritance of white flower colour in orchids are outlined and their application to the genus *Paphiopedilum* is discussed, with examples of hybridization. The species of greatest importance are *P. niveum*, *P. insigne* var. sanderae, *P. bellatulum* and *P. godefroyae* (Rogerson, 1991).

Breeding barriers in producing hybrids by intergeneric, inter- and intra-specific hybridization in *Phalaenopsis* have been observed by Huei *et al.*, 2010. Four cultivars of red *Phalaenopsis* which are known to be difficult in hybridization, including *Doritaenopsis* Ever Spring Prince (ESP), Dtps. Queen Beer 'KHM 159' (QB), Dtps. Taida Salu 'KHM 101' (TS), and *Phalaenopsis* Brother Girl 'B' (BG) were investigated for the mechanisms underlying the crossing barriers. Root tip squash observation showed that all of these selected cultivars are triploids with 2n=3x=57 chromosomes. To test their crossing behaviors, these four cultivars were either self- crossed, or used as the pollen and pistillate parents in reciprocal crossing with three cultivars with normal behavior in crossing, including Phal. Brother Irene 'Feng Fong', Phal. Sogo Yukidian 'V3', and Phal. Tai Lin Redangel 'V31'. These normal cultivars are tetraploids with 2n=4x=76 chromosomes. The crossing results showed that the closure of stigma cavities occurred in TS and BG, while such stigma closure was never observed when QB or ESP was used as the pollen parent in crosses. However, hybrid progenies were obtained only from the crosses between ESP or TS (pistillate parent) with the normal cultivars. Taken together, these results suggest that crossing difficulties in these four selected cultivars may be caused by odd chromosome set and some other factors of male or female parents.

Breeding barriers of producing hybrids by intergeneric, intra- and interspecific hybridization in *Phalaenopsis* were found. *Phalaenopsis* Sunrise Goldamour KHM637 was found to be tetraploid and *Phalaenopsis* Taipei Gold STM was a triploid by using the root tip squash method.

Neither closure of the stigma cavities nor elongation of pollen tubes occurred when KHM637 or STM was self-crossed or was used as the pistillate parent during intraspecfic crossing. Sealing the stigma cavity with a wax film or applying mechanical injury did not correct the problem, but auxin treatment was helpful. It was suggested that the major breeding barrier for Taipei Gold STM is the triploidy and both STM and KHM637 had another barrier, i.e. difficulty in stigma cavity closure (Ting *et al.*, 2008).

Endopolyploidy was observed in the protocorms of diploid *Phalaenopsis aphrodite* subsp. formosana with ploidy doubling achieved by *in vitro* regeneration of excised protocorms, or protocorm-like bodies (PLBs). Thirty-four per cent of the PLBs regenerated from the first cycle of sectioned protocorms were found to be polyploids with ploidy doubled once or twice as determined by flow-cytometry. The frequency of ploidy doubling increased as the sectioning cycles increased and was highest in diploid followed by the triploid and tetraploid. Regeneration of the endopolyploid cells in the tissue of the protocorms or PLBs is proposed as the source of the development of ploidy doubled plantlets. The frequency of ploidy doubling was similar in seven other *Phalaenopsis* species, although the rate of increase within cycles was genotype specific. In two species, a comparison of five parameters between 5-month-old diploid and tetraploid potted plants showed only the stomata density differed significantly. The flowers of the tetraploid plant were larger and heavier than those of the diploids. This ploidy doubling method is a simple and effective means to produce large number of polyploid *Phalaenopsis* species plants as well as their hybrids. The method will be beneficial to orchid breeding programs especially for the interspecific hybridization between varieties having different chromosome sizes and ploidy levels (Huei *et al.*,2009).

Natural hybrids among the specialized terrestrial oil-secreting orchids of South Africa are extremely rare even where multiple closely related species co-occur. Steiner and Cruz (2009) putative hybrids between *Pterygodium catholicum* Sw. and *P. acutifolium* Lindl., two closely related oil-secreting orchids that lack morphological barriers to inter-breeding. The purpose of this study was to confirm the parentage of the putative hybrids using molecular data from one nuclear (ITS) and two plastid (matK-trnK and trnL/F introns) DNA regions. Phylogenetic analyses of sequences as well as nucleotide substitution patterns of the putative hybrids, putative parents, and their closest relatives were consistent with a hypothesis of hybridization. They suggest that the hybrids were the result of visits to both orchids by a single species of oil-collecting bee during a brief period of flowering overlap. These results suggest that the rarity of hybrids between these orchids is due to factors other than genetic incompatibility.

Many researchers have postulated that introgression increases genetic variability through the production of recombinant genotypes, which provides populations with a means of coping with environmental change or evolving novel adaptations. Forty-four morphological characters and 21 allozyme loci were analysed from 17 populations of *C. pubescens* [*C. calceolus var. pubescens*] 9 populations of *C. candidum,* and 5 sites that appeared to contain hybrid swarms, to assess the direction and magnitude of interspecific exchange of genes. The data demonstrated the existence of bidirectional gene flow wherever the 2 species were sympatric. Hybrid populations consisted largely of individuals that were later-generation backcrosses or recombinants, with few pure parents or F1 hybrids noted. Some individuals that appeared to be of one morphological species contained marker alleles from the other species. In Iowa, *C. pubescens* is generally a woodland species but occasionally occurs on prairies. Allozyme and morphological data and ecological considerations suggest the prairie ecotype of *C. pubescens* may have arisen as a direct consequence of its acquisition of genetic information from C. candidum. Implications for conservation of *C. candidum* were discussed by Klier *et al.,* 1991.

Vanda Mimi Palmer (VMP), an orchid hybrid of *Vanda* tesselata and *Vanda* Tan Chay Yan is a highly scented tropical orchid which blooms all year round. Isoprenoids are well known to contribute significantly to the scent of most fragrant plants. They are a large group of secondary metabolites which may possess valuable characteristics such as flavor, fragrance and toxicity and are produced via two pathways, the mevalonate (MVA) pathway or/and the 2-C-methyl-D-erythritol-4-phosphate (MEP) pathway. Song *et al.* (2012) studied, a sesquiterpene synthase gene denoted VMPSTS, previously isolated from a floral cDNA library of VMP was cloned and expressed in *Lactococcus lactis* to characterize the functionality of the protein. *L. lactis,* a food grade bacterium which utilizes the mevalonate pathway for isoprenoid production was found to be a suitable host for the characterization of plant terpene synthases. Through recombinant expression of VMPSTS, it was revealed that VMPSTS produced multiple sesquiterpenes and germacrene D dominates its profile.

Propagation of Orchids

Orchids may be propagated either sexually or asexually. There are six main techniques currently in use to propagate orchids like division, back bulbs, keiki, aerial cuttings, meristem or tissue culture and seed. Since most of the commercial orchids are highly heterozygous they are not raised through seed and are propagated through vegetative means to get true-to-type plants.

Yong *et al.* (2011) discussed the propagation of orchids through tissue culture, sexual propagation through seeds, and vegetative means using backbulbs (pseudobulbs that are severed from the parent plant and planted in pots or trays) or offshoots (plantlets separated from the parent plant and transferred to pots), or through division (splitting an existing plant into two or more parts so that each new section contains at least one new shoot or vegetative bud).

Division

Division means splitting the plant in to two or more parts with at least one new shoot and each will produce a fully grown flowering size plant, capable of flowering in the following season. This is one of the simplest methods of producing more plants of the same variety or species. Splitting a plant will encourage the plant to produce more shoots of a better quality and with much more vigour than would be the case if left to it's own devises. The reason for this new vigour is the plants response to being placed under threat and introduction of new fresh potting medium which the plant will take full of advantage. Division of plants is best in the early spring just as new growth starts; this would ensure that each division has a full growing season to establish itself so it can flower in the following

Fig. 2.6 : Division of *Cymbidium*

season. Care should be taken not to divide the plants unless there are 5-7 pseudo-bulbs.

Genera having sympodial growth like *Aerides, Arachnis, Coelogyne, Cattleya, Cymbidium, Epidendrum, Renanthera, Phalaenopsis* , *Vanda* and *Dendrobium* can be propagated by division. *Dendrobiums* which are very fast growing can be divided every year. Division of *Dendrobium* and *Phalaenopsis* need special care to root and should be potted in propagation beds thereafter.Whereas, those of very hardy genera like *Aerides, Arachnis, Vanda etc.* should directly potted in the pots after treating the cut ends with fungicides to prevent rotting. Plants of monopodial growth like *Vanda, Renanthera* and *Angraecum* with the new growth appearing continuously from the top or crown, would not multiplied by division.

Multiplication by Backbulbs

It is a method of producing new plant from previously flowered or unflowered leafless pseudobulbs. This is another simple method of propagation often used with *Cymbidiums* but this may take 3 to 4 years to obtain a flowering plant. The process involves the removal of older back bulbs at the time of re-potting and placing them under ideal growing conditions to induce rooting. Backbulbs can be used to obtain extra divisions of *Cymbidiums.* First of all backbulbs should properly be cleaned along with root and old leaf bases. Then it should be placed in a cool dark humid location for several weeks till eyes at the base of the bulb appear. Excess moisture should be removed to avoid rotting of backbulb. When an eye has started to grow, the bulb should be potted in a 7.5 cm to 10 cm pot with the new growth toward the centre. Proper watering should be done until the growth is about 5 cm or till it starts to form roots. At this point, the plant should be treated as small seedling.

Fig. 2.7 : Backbulb growing along with mother plant

Seedling should be grown until the first bulb forms on the new growth and a second growth has started. After that, the seedling should be potted in a larger pot.

The advantage of the backbulb type of propagation over the growing of seedlings is that the flower will exactly resemble that of the original plant, while in the seedling there is no way to tell whether it will resemble one parent plant or the other or be something entirely different.

Patil and Nevkar (2002) studied the backbulbs of 7 sympodial species of orchids (*Dendrobium* pierardii, *Oncidium* sphacelatum, Phatius takervilleae [*Phaius* tankervilleae], *D. crysotoxum, D. nobile, Vandopsis parishii* and *Epidendrum radicans* [*E. ibaguense*]) and were planted in earthen pots containing coal and *Sphagnum* moss to develop a standardized vegetative propagation of the exotic orchids. *D. pierardii* recorded the highest propagation ratio (1:9) followed by *O. sphacelatum* (1:8), *P. takervilleae* (1:7) and *D. crysotoxum* (1:6). *D. nobile, V. parishii* and *E. radicans* all recorded 1:1 propagation ratio.

Keikis

Keiki is the Hawaiian word derived from Kay-key (baby plant), refers to a plant produced asexually by an orchid of monopodial genera *Ascocenda, Phalaenopsis* and *Vanda*. They produced through the accumulation of growth hormones at a specific point. There are two types of keiki: regular and basal. The regular keiki is a small plant growing from one node along the flower stem, instead of a branch. This is induced by the accumulation of cytokinin at that point, either naturally or artificially. The basal keiki is a baby plant growing from the base of the mother plant. This usually happens when the apex has lost its effectiveness in suppressing axillary buds. Keikis formation may be induced by topping of the stem. Estimates were provided by Zotz 1999 for the *in situ* occurrence of keikis (additional offshoots which develop on vegetative stems or flower stalks of several orchid genera) in the epiphytic orchid Dimerandra emarginata, common in moist lowland forests in central Panama. Their growth and development over several years are described and a possible adaptive function of these structures is discussed.

Barman and Rajni (2003) determined the effects of IBA (0, 250, 500 and 1000 mg/litre) on the rooting of young, 1- and 2-year-old keikis bamboo orchid (*A. graminifolia*). Alongside this, the effects of different potting media viz., sand, soil, farmyard manure, saw dust, leaf mould and rice husk in different combinations were also determined. The highest number of primary and secondary roots were recorded in the 1-year-old keikis bamboo orchid. Rooting percentage, root number and, root length and diameter increased with increasing concentrations of IBA. Among the different potting media, rooting was highest in both soil and soil + sand + farmyard manure + rice husk. The number of secondary roots and the longest roots were recorded in soil.

Two species of *Dendrobium* with horticultural merit were discussed by Webb and Webb(1993), with details of their natural habitats and advice on culture. *Dendrobium moorei* is endemic to Lord Howe Island. Two forms are in cultivation, one being less productive of keikis (offshoots) but better flowering than the other. The white flowers are small, tubular and drooping. Once established in containers, plants will flower during most of the year. The northern Queensland species *D. agrostophyllum* bears fragrant yellow flowers along the pseudobulbs. It requires a cool environment and grows well under temperate conditions. Pseudobulbs do not mature in one season, but once mature they continue to produce blooms over a long period. So far, these species have been little used in hybridization.The effect of injection of high concentrations of 6-benzyladenine (BA) into pseudobulbs (stems) of *Dendrobium* Jaquelyn Thomas 'Uniwai Princess' was evaluated. The flowering cycle for mature plants of *Dendrobium* Jaquelyn Thomas is production of one new pseudobulb per year, with autumn and early winter flowering on leaved pseudobulbs in their first year (1-yr) of growth, and summer and autumn flowering on almost leafless pseudobulbs in their second year (2-yr) and subsequent years. Pseudobulb injections made in November caused the growth of dormant lateral buds into inflorescences (sprays) that flowered in February in the spring off-season. Injection of 100 mM BA into 1-yr stems resulted in a significant increase in flower spray production over control non-treated stems, 8.92 vs. 0.52, respectively. Injection of 100 and 10 mM BA solutions into 2-yr stems also resulted in a significant increase in flower spray production over controls, 6.32 and 4.00, vs. 0.24, respectively. The high concentrations of BA also significantly reduced inflorescence length and caused development of abnormally formed flowers. Addition of gibberellic acid (GA) to the injection solution significantly increased inflorescence length and reduced the percentage of abnormally formed flowers. Timing of injection for year around production of Hawaiian *Dendrobium* cut flower sprays is discussed by Sakai *et al.* (2010).

The history and cultivation of *Lepanthes*, one of the oldest and largest orchid genera, was discussed by Hermans and Hermans (1995).

The genus has a huge variety of flower colour and shape, and although the flowers are often small their abundance makes *Lepanthes* worth cultivating. *Lepanthes* species grow in a number of different habitats and locations, and so cultural requirements vary between species. They had grown *Lepanthes*, along with other orchids, in a cool greenhouse. Keiki-like plantlets are produced from the leaf petiole by a number of species, providing a convenient method of propagation.

Aerial Cutting

It is usually done when the orchids are placed under conditions that will develop small plants instead of producing flowers. Aerial cutting is the easiest to achieve because they can be removed once they are fully grown. The big and healthy new shoots that seem to be surviving on their own should cut out from the plant. Aerial cuttings are the common way to propagate *Dendrobium*. It produces aerial shoot or bulbs on older growth. They take three to four months to develop roots and then should be detached and re-potted. Aerial cuttings are usually a successful method of orchid propagation, as the plant is nearly full grown before being removed. In *Goodyera*, rhizomes produce special lateral branches which turn up and produce aerial shoots. When they are properly rooted they get detached from the mother plant and establish separately.

Tissue Culture

Hossain *et al.* (2013) reviewed and provides an informative broad overview of orchid biotechnology, addressing several important aspects such as molecular systematics, modern breeding, *in vitro* morphogenesis, protoplast culture, flowering control, flower color, somaclonal variation, orchid mycorrhiza, pathogen resistance, virus diagnosis and production of virus-free plants, functional genomics, genetic transformation, conservation biotechnology and pharmaceutical biotechnology.

Orchid is the first horticultural plant cloned by tissue culture methods on a commercial scale (Goh, 1990). Rapid multiplication using tissue culture techniques is an important strategy for preserving the valuable orchid wealth in India. Vegetative propagation through division of rhizome and separation of offshoots and keikis is very slow. Both these problems can now be successfully alleviated through micropropagation. The technique has revolutionized the orchid industry and now it is possible to multiply unlimited number of desirable clones. Rotor (1949) was the first to have employed tissue culture techniques to multiply *Phalaenopsis* , and since then micropropagation potential of a variety of monopodial taxa has been successfully tested (Sheela *et al.*, 2006).

The clonal propagation of *Cymbidium* apical meristem shoot were first applied in 1960 when Dr G. Morel, the distinguished French botanist, while trying to find out a way to prevent virus in potatoes hit upon the idea that the technique of tissue culture can be highly successful in cloning of orchids and to get virus-free plants. It had been the most sensational development which had revolutionized the orchid industry and triggered global expansion of tissue culture for rapid propagation of orchid. The technique has been especially important for orchids as their genotypes are highly heterozygous and sexual reproduction is an extremely slow process. Among different methods available, shoot tip and axillary bud cultures are widely used. In sympodial orchids like *Cattleya, Cymbidium* and *Dendrobium*, a young shoot arising from back bulb provides suitable material for tissue culture. On longer shoots, there may be several axillary buds as well as the apical bud while the monopodial orchids like *Vanda, Aerides, Phalaenopsis* , nodal sections, shoot apex. Keikis or offshoots or even flower stalk cutting can be used. Regeneration potential of alternative plant parts other than shoot tip and axillary buds like stem, leaf, root and inflorescence have been worked out in some species of orchids. More than 43 orchid genera have been mericloned successfully using different plant parts including leaves, roots, flower stalks, axillary buds and apical meristem. Both shoot tips and axillary buds produce protocorm-like bodies (PLBs) which subsequently develop into plantlets. Now a day these techniques are commonly exploited by commercial growers for the rare Orchids and hybrids which could not be propagated by any other means. The most common methods of micropropagation involve the proliferation of shoots via a semi solid system. Micropropagation by conventional techniques is typically, labour intensive and time taking means of clonal propagation. Bioreactor provides a rapid and efficient clonal propagation system utilizing liquid medium to avoid intensive manual handling. Automation of micropropagation in bioreactors has been advocated by several researchers as a possible way of reducing cost of micropropagation. Micropropagation for optimal plant production depends upon better understanding of physiological and biochemical responses of plant to the signals of culture microenvironment and an optimization of specific physical and chemical culture conditions to control the morphogenesis of plants in liquid culture systems. The advances made in other applications of biotechnology in orchid production and improvement are also reviewed by Mudalige and Kuehnle (2004).

A competent protocol for accelerated plant regeneration system via direct induction of protocorm-like bodies (PLBs) from leaf of orchid hybrid Aranda Wan Chark Kuan 'Blue' x *Vanda coerulea* Grifft. ex. Lindl. was developed for the first time to establish a basis for mass production. Using tissue culture technique, the conditions for PLB induction from leaf explants and conversion of PLBs into plantlets were investigated. Leaves were transferred to MS medium containing different types and concentrations of cytokinins (namely, N6-benzyladenine, 6-furfurylaminopurine, N-phenyl-N'-(1,2,3-thidiazol-5-yl)urea/TDZ and zeatin) for PLB induction. By means of exploring the effects of cytokinins, it was determined that the optimum PLB induction occurred on MS media supplemented with 1.5 mg l-1 TDZ; whereby accordingly, PLB induction (with a frequency of 94.8%) was observed as early as 8 days of culture and an average of 25 PLBs was obtained from 1 cm 2 leaf segment after 40 days of culture. Variable pressure scanning electron microscopy indicated the different developmental stages of PLBs in detail. Light/stereo microscopic observation showed the maturation of PLBs and gradual formation of shoot and leaf primordia. More than 96% conversion (with well-developed shoots and roots) was achieved within the next 30 days of culture, when well developed PLBs were transferred in MS medium supplemented with 1 mg l-1 BA, 0.5 mg l-1 IBA plus 60 mg l-1 adenine sulphate. After 60 days of transfer in plastic pots filled with sand and perlite (2:1; v/v) and with charcoal and coconut fibre (1:1; v/v), subsequently, 90% well-acclimatized plantlets were recovered.

Paphiopedilum orchids are generally propagated through the division of axillary buds from mother plant, which limits commercial production due to its unproductive proliferation and time consuming. Udomdee *et al.* (2012) studied, the effect of *in vitro* cutting methods and medium composition on efficient shoot multiplication of *Paphiopedilum hsinying* Rubyweb. Among three different *in vitro* stem cutting methods used, vertical cutting was able to produce more new shoots than horizontal and cross cutting when cultured on Hyponex based medium. After 12 weeks of culture, plantlets regenerated from vertical cutting were able to produce new healthy and well rooted shoots higher than without cutting on the same medium. Moreover, the newly-formed shoots which were divided into single plantlets and subcultured onto half-strength Murashige and Skoog (MS) medium without growth regulators could remain higher shoot multiplication than in other media.

The micropropagation procedure provides a simple means to *in vitro* propagate *Paphiopedilum* plantlets which are able to produce large numbers of uniform plantlets in a shorter time compared to the conventional propagation method.

An *in vitro* propagation and tuberization method was successfully established for *Eulophia nuda* Lindl., an endangered orchid of Aravalli hills. Shoot buds were induced from tuber explants (1 cm cut pieces) on Murashige and Skoog (MS) medium containing 44.4 micro M 6-benzyladenine (BA) and additives (283.8 micro M ascorbic acid, 67.8 micro M adenine sulphate, 143.5 micro M arginine and 118.9 micro M citric acid). After excision of newly formed shoots, explants were successively transferred to MS medium with 22.2 micro M BA and additives for shoot multiplication up to three passages. Further shoot multiplication and tuberization was achieved, by subculturing of *in vitro*-formed shoots, on MS medium containing 8.88 micro M BA, 4.68 micro M Kinetin (Kn) and additives. Shoot multiplication was influenced by cytokinin type, concentration and combinations. Shoots with tubers were successfully rooted *in vitro* and *ex vitro*. Shoots were treated with 2.46 mM indole-3-butyric acid (IBA) for 4 min for *ex vitro* rooting. Plantlets were acclimatized in the greenhouse conditions in bottles containing soilrite. Acclimatized plantlets were transferred to the earthen pots containing soil as a mixture of clay, sand and vermicompost (1:1:1). *In vitro* tuber formation of *E. nuda* was high compared to the ex vitro/ natural conditions(Panwar *et al.*, 2012).

Advantages of Micropropagation of Orchids

1. Micro propagated plants are more or less genetically homogeneous.
2. Shoot tip culture allows the plants to become virus free.
3. Since the culture condition such as light, temperature and humidity are specifically controlled during the procedure, so the plants can be produced irrespective of weather, season and other environment conditions.
4. Propagation through somatic embryogenesis allows higher efficiency of production, therefore a large number of stock plants can be grown by this process in a short time span.
5. Under controlled culture conditions and proper management, a large number of elite stock plants can be maintained in a small area and for a very long time.

Medium

One of the most important factors governing the growth and morphogenesis of plant tissues in culture is the composition of the culture medium. The basic nutrient requirements of cultured plant cells are very similar to those of whole plants. Both liquid and solid media are used for the orchid tissue culture. In most of the cases the simple media used for orchid seed-germination are also used for tissue culture. Plant tissue and cell culture media are generally made up of macronutrients, micronutrients, vitamins, amino acids sugars, other undefined organic supplements, solidifying agents and growth regulators. The same nutrients are used for both liquid and solid media, with the omission of agar on the former. A schematic outline of mass propagation of economically important Orchids involves the following steps:

1. Establishment of aseptic cultures.
2. Rapid growth and multiplication of apical and auxiliary buds.

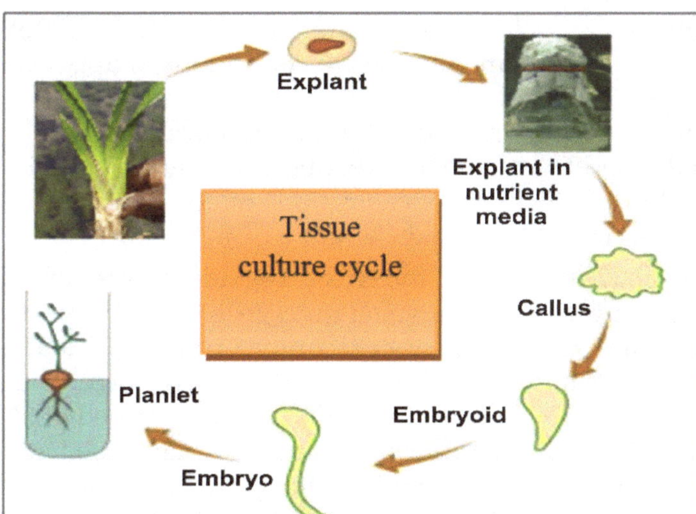

Fig. 2.8 : Steps of plant regeneration

3. Rapid growth and multiplication of shoot buds in culture.
4. Root formation and hardening of the *in vitro* raised plants.
5. Acclimatization of the *in vitro* raised plants under field conditions.

The effects of culture media, i.e. Knudson C, Viacin and Went, Murashige and Skoog (MS), and Rosa and Laneri, on the production of protocorm-like bodies (PLBs) from shoot tip explants of *Dendrobium* cv. Sonia, *Oncidium* cv. Gower Ramsay, and *Cattleya leopoldii* were studied by Saiprasad and Raghuveer (2002). MS medium produced the greatest production of PLBs, multiple shoots, growth, and differentiation of explants in the three orchid genera.

Macronutrient

Nitrogen (N), phosphorus (P), potassium (K), calcium (Ca), magnesium (Mg), and sulfur (S) are macronutrients essential for plant cell or tissue growth. The optimum concentration of each nutrient for achieving maximum growth rates varies considerably among species.

Culture media should contain at least 25-60 mM of inorganic nitrogen for adequate plant cell growth. Plant cells may grow on nitrates alone, but considerably better results are obtained when the medium contains both a nitrate and ammonium nitrogen source. Nitrates are usually supplied within the range of 25-20 mM. Typical ammonium concentrations may vary from 2 to 20 mM. Cells can grow on a culture medium containing ammonium as the sole nitrogen source if one or more of the TCA cycle acids (citrate, succinate, or malate) are also included in the culture medium at concentrations of approximately 10 mM. When nitrate and ammonium sources of nitrogen are utilized together in the culture medium, the ammonium ions will be utilized more rapidly and before the nitrate ions. Most media contain K, in the nitrate or chloride form at concentrations of 20-30 mM. The optimum concentrations of P, Mg, S, and Ca range from 1-3 mM.

Min *et al.* (2012) studied the effects of adding nitrogen (N) (0, 105, 210, and 420 mg L^{-1}) on the vegetative growth and reproduction of *Paphiopedilum armeniacum*. N enrichment improved leaf area and lengthened the leaf lifespan during the vegetative stage. The effects of N application on flower size were minor. The intermediate N level of 210 mg L^{-1} (MN) increased the seed capsule weight, seed germination rate, and improved the growth of seedlings that developed from seeds of MN-treated plants. N fertilizer exerted little influence on ramet emergence and ramet number per plant, but a low N concentration of 105 mg.L^{-1} promoted the leaf number and leaf area of ramets. Appropriate N levels for P. armeniacum in production and cultivation should be determined according to different production objectives.

Micronutrients

The essential micronutrients include iron (Fe), manganese (Mn), zinc (Zn), boron (B), copper (Cu) and molybdenum (Mo). Chelated forms of iron and zinc are commonly used in preparing culture media. Iron is the most critical element among all the micronutrients. Iron citrate and tartrate may be used in culture media but these compounds are difficult to dissolve and frequently precipitate after media are prepared. Murashige and Skoog used an ethylene diaminetetraacetic acid (EDTA), an iron chelate to bypass this problem. Copper and Cobalt are normally added to culture media at concentrations of 0.1 μM, Fe and Mo at 1 μM, I at 5 μM, Zn at 5-30 μM Mn at 20-90 μM, and B at 25-100 μM.

Carbon and Energy Source

Sucrose is the most preferred carbohydrate used in plant cell culture. Glucose and fructose may be substituted in some cases. Glucose is as effective as sucrose but fructose is somewhat less effective. Sucrose concentrations of culture media normally vary from 2 to 3 percent.

Vitamins

Vitamins act as catalysts in various metabolic processes of plants. Thiamin (B_1), nicotinic acid, pyridoxine (B_6) and myo-inositol are the common vitamins frequently used in cell and tissue culture. Thiamin is basically required by all cells for growth. It is normally used at concentrations ranging from 0.1 to 10.0 mg/liter. Nicotinic acid and pyridoxine are often added in culture media but are not essential for cell growth in many species. Nicotinic acid is normally used at concentrations of 0.1-5.0 mg/liter, pyridoxine is used at 0.1-10.0 mg/liter. Myo-inositol is commonly included in many vitamin stock solutions, though it is a carbohydrate, it has been shown to stimulate growth of certain cell cultures. Its presence in the culture medium is not essential, but in small quantities myo-inositol stimulates cell growth in most species. Myo-inositol is generally used in plant cell and tissue culture media at concentrations of 50-5000 mg/litre.

Rodrigues *et al.* (2012) evaluated the *in vitro* growth of orchid (*Cattleya walkEriana* Gardner) seedlings subjected to different rates of a NPK fertilizer added to the growth medium. The fertilizer Peters Reg. was used in the formulation NPK 10-30-20+Mg+micronutrients in the rates of 0.0; 1.0; 2.0; 3.0; 4.0; 5.0; and 10.0 g L^{-1} medium, with coconut milk (150 mL L^{-1}), sucrose (20 g L^{-1}) and activated charcoal (2 g L^{-1}) and pH adjusted to 5.6. Six-month-old seedlings, previously germinated in Knudson C medium were transferred to 320 mL glass jars containing 35 mL of medium, using 10 plants per jar. The maximum production of dry matter from the aerial part at six months of age was estimated at 0.255 g/

jar, with the medium added of 5.22 g L^{-1} of fertilizer. Considering the possibility of high salt concentration in the medium for some orchid species, the fertilizer rate needed to obtain 90% of maximum production was estimate at 3.55 g L^{-1}. There was a drastic growth reduction in root growth when using high fertilizer concentrations, producing fewer roots and shorter.

Amino Acids and other Nitrogen Supplements

Amino acids provide plant cells with an immediately available source of nitrogen, which generally can be taken up by the cells more rapidly than inorganic nitrogen. The most common sources of organic nitrogen used in culture media are amino acid mixtures (e.g. casein hydrolysate), L-glutamine, L-asparagine and adenine. Casein hydrolysate is generally used at concentrations between 0.05 and 0.1 per cent. Tyrosine has been used to stimulate morphogenesis in cell cultures but should only be used in an agar medium. Supplementation of the culture medium with adenine sulfate can stimulate cell growth and greatly enhance shoot formation.

The effects of glutamine, asparagine and phenylalanine, each applied at 1.0, 2.5 and 5.0 mg/litre in the MS basal medium, on the production of protocorm-like bodies (PLB) from shoot tip explants were investigated for three orchid genera, namely *Dendrobium* (Sonia), *Oncidium* (Gower Ramsay) and *Cattleya leopoldii*. The number of days taken for initial differentiation increased or decreased by 1-2 days over their controls among the various amino acid treatments. The production of PLB varied with the type and concentration of amino acid and with orchid genera. For *Dendrobium*, glutamine at 1.0 mg/l resulted in 7- and 3-fold increase in PLB and multiple shoot production, respectively, over the control. On the other hand, *Oncidium* responded only to glutamine at 1.0 mg/litre. Among the amino acids, glutamine was ranked highest and phenylalanine was ranked lowest in terms of PLB production. Among the orchids, *C. leopoldii* responded quite differently to all amino acid treatments, as PLB production in this species was generally lower with amino acid treatment than the control (Saiprasad and Raghuveer 2002).

Polyamines, putrescine, spermidine and spermine all at three concentrations *viz.*, 0.2, 0.4 and 1.0 mM were tested, by supplementing to MS basal plus 4.44 micro m BAP medium. The explant used was fractionated plb. Ethylene and methane was measured at 20 and 40 days after inoculation (DAI). Among various polyamines tested, maximum number of plb's (protocorm-like bodies) were produced in putrescine 0.4 mM treatment. Increase (1.0 mM) or decrease (0.2 mM) in concentration caused a decrease in the production of plb's. All spermidine and spermine treatments resulted in the production of less number of plb's than control. No ethylene evolution was observed in any of the polyamine treatments. However, methane evolution was observed in all the polyamine treatments. The absolute amounts, of methane evolved could not be related to the observed plb's production response. However, when the evolution of methane was more than 1 nmol per gram FW h-1, poor plb's production was observed (Saiprasad *et al.,* 2004).

Solidifying Agents and Support Systems

Agar is the most commonly used gelling agent for preparing semisolid and solid plant tissue culture media. When it is mixed with water, it forms a gel that melts at approximately 60°-100°C and solidifies at approximately 45°C; thus, agar gels are stable at all feasible incubation temperatures. Agar gels do not react with media constituents and are not digested by plant enzymes. The firmness of an agar gel is controlled by the concentration and brand of agar used in the culture medium and the pH of the medium. The agar concentrations commonly used in plant cell culture media range between 0.5 and 1.0 per cent, these concentrations give a firm gel at typical pH_S of plant cell culture media.

Growth Regulators

Four broad classes of growth regulators are important in plant tissue culture; the auxins, cytokinins, gibberellins and abscisic acid. Both auxin and cytokinin are usually added to culture media in order to obtain morphogenesis, although the ratio of hormones required for root and shoot induction is not universally same. Considerable variability exists among genera, species, and even cultivars in the type and amount of auxin and cytokinin required for induction of morphogenesis. The various auxins differ in their physiological activity and in the extent to which they move through tissue, are bound to the cells, or metabolized. The auxins commonly used in plant tissue culture media are 1-indole-3-acetic acid (IAA), 1-indole-3-butyric acid (IBA), 2,4-dichlorophenoxy acetic acid (2,4-D) and 1-napthalene acetic acid (NAA). Auxins are generally included in a culture medium to stimulate callus production and cell growth, to initiate shoots, particularly roots, and to induce somatic embryogenesis and stimulate growth from shoot apices and shoot tip cultures.

The cytokinins are generally added to a culture medium to stimulate cell division, shoot formation, axillary shoot proliferation, and inhibit root formation. Many plant tissues have an absolute requirement for a specific cytokinin for morphogenesis, whereas some tissues are considered to be cytokinin independent, i.e., no cytokinin or a specific cytokinin may be required for organogenesis. The type of morphogenesis that occurs in a plant tissue culture largely depends upon the ratio and concentrations of auxins and cytokinins present in the medium. Root initiation of plantlets, embryogenesis, and callus initiation all generally occur when the ratio of auxin to cytokinin is high, whereas adventitious and axillary shoot proliferation occurs when the ratio is low. The concentrations of auxins and cytokinins are equally as important as their ratio. The cytokinins commonly used in the culture media include 6-benzylaminopurine or 6-benzyladenine (BAP, BA), 6-γ-γ-dimethylaminopurine (2iP), N-(2-furanylmethyl)-1H-puring-6-amine (kinetin), and 6-(4-hydroxy-3-methyl-trans-2-butenylamino)purine (zeatin).

Generally GA_3 is added in culture media to promote the growth of low-density cell cultures, enhance callus growth and elongate stunted plantlets. Inhibition or stimulation of the callus growth (depending upon the species), enhancement of shoot or bud proliferation and inhibition of latter stages of embryo development are the main function of abcissic acid. Morel (1960), first used Knudson C medium for meristem Culture of *Cattleya, Cymbidium, Dendrobium, Miltonia* and *Phaius*.

The effect of various plant growth regulators on protocorm-like bodies (plbs) production was assessed in 3 orchid genera, i.e. *Dendrobium* cv. Sonia, *Oncidium* cv. Gower Ramsay and *Cattleya leopoldii*. BAP [benzyladenine], kinetin, NAA, IAA, 2,4-D and GA3 were supplemented to MS medium at concentrations of 0.25, 0.5 and 1.0 mg/litre. The number of days taken for initial differentiation was observed to be a function of growth regulator supplemented as well as orchid genera. Maximum number of plbs were produced in 1.0 mg BAP/litre. The relative ranking among plant growth regulators for obtaining maximum number of plbs was BAP > kinetin > NAA > IAA > 2,4-D > GA3, and for the 3 orchid genera, *Dendrobium* > *Oncidium* > C. leopoldii (Saiprasad *et al.*, 2002).

The effect of plant growth regulators as a function of ethylene and methane on the production of protocorm-like bodies and multiple shoots was assessed by Saiprasad *et al.*(2003) in *Dendrobium* cv. Sonia. The growth regulators tested were 6-benzyladenine (BA), kinetin (KIN), alpha -naphthaleneacetic acid (NAA), indolyl-3-acetic acid (IAA), 2,4-dichlorophenoxyacetic acid (2,4-D) and gibberellic acid (GA), supplemented to MS basal medium each at 0.25, 0.5 and 1.0 mg litre concentrations. The explant used were fractionated protocorm-like bodies. Ethylene and methane were determined at 20 and 40 days after inoculation (DAI) using Perkin-Elmer (Sigma 2000) gas chromatogram along with data on protocorm-like bodies and multiple shoot production. Maximum number of protocorm-like bodies was recorded with BA (1.0 mg litre) at 20 and 40 DAI. Auxin (NAA and IAA) treatment at all concentrations produced significantly highest number of multiple shoots at 20 and 40 DAI, and no protocorm-like bodies at 20 DAI but a few protocorm-like bodies at 40 DAI. GA and 2,4-D treatment did not produced any protocorm-like bodies and multiple shoots. Significant differences in methane evolution were observed among the different growth regulator treatments at 20 and 40 DAI. More methane was observed to evolve 20 DAI than 40 DAI. The absolute amount of methane evolved did not appear to be related to the production of protocorm-like bodies at both DAI. Methane data at 40 DAI indicated that when methane values were more than 1.0 nmoles g FW- h-1 the possibility of production of callus or lanky, elongated and etiolated type shoots was indicated, as observed in 2,4-D and GA treatments, respectively. Ethylene could not be detected in almost any of the treatments tested at both DAI.

To increase the efficiency of production of protocorm like bodies, ethylene inhibitors like silver nitrate, salicylic acid and cobalt chloride at three concentrations, viz., 2, 5 and 20 micro M and ethrel at 0.7, 1.4 and 14.0 micro M were tested by Saiprasad and Raghuveer(2007), by supplementing MS basal with BAP 4.44 micro M. The explant used was fractionated protocorm like bodies (plb). Ethylene and methane gases evolved by the explant in the culture container were measured at 20 and 40 days after inoculation (DAI). Among various ethylene inhibitors tested, the maximum number of plb's were produced in media containing silver nitrate (5 micro M), cobalt chloride (2 micro M) and salicylic acid (20 micro M). All ethrel treatments produced succulent, vitrified and deformed plb's. No ethylene evolution was observed in any of the ethylene inhibitor treatments. Only in ethrel treatments was evolution of ethylene observed. Methane evolution was observed in all the ethylene inhibitor treatments. Absolute amounts of methane evolved could not be related to the observed plb production. However, when the evolution of methane was more than 1 nano mole g-1 FW h-1, poor plb production was observed.

Stock Solutions

The use of stock solutions reduces the number of repetitive operations involved in media preparation and hence, the chance of human or experimental error. Moreover direct weighing of media components (micronutrients and hormones) that are required only in milligram or microgram in the final formulation cannot be performed with sufficient accuracy in tissue culture. Preparation of concentrated stock solutions and subsequent dilution to makes the final media is standard procedure. Concentrated solutions of some materials are more stable and can be stored for longer period than more dilute solutions. To prepare a stock solution, weigh out the required amount of the compound and place it in a clean flask. It is common practice to make a stock solution 10X or 100X, depending upon the solubility of the compound. Once the chemical is in the flask, dissolve it in a small amount of water, ethyl alcohol, 1 N NaOH, or 1 N HCL. Next, slowly double-distilled water should be added to the flask, together with gentle agitation, it should be continue till the proper volume is reached. Flask should be labelled with the name of the solution, preparation and expiration dates and the name of the person who prepared the solution. Certain such as IAA, must be prepared and stored in amber bottles to prevent photodecomposition. Stock solutions of macronutrients can be prepared at 10 times the concentration of the final medium. A separate stock solution for calcium salts may be required to prevent precipitation. Stock solution of macronutrients can be stored safely for several weeks in a refrigerator at 2°-4°C.

Micronutrient stock solutions are generally made up at 100 times their final strength. The stock solutions of it could be stored in a refrigerator up to 1 year without deterioration. Iron stock solutions should be prepared and stored separately from other micronutrients in an amber storage bottle.

Vitamins and Auxin stock solutions are generally prepared at 100-1000 times the final desired concentrations. Vitamin stock solutions should be made up each time media is prepared if a refrigerator or freezer is not available. Vitamin and IAA stock solutions should be stored safely in a refrigerator at -2°C for 2-3 months but should be discarded thereafter. The auxin NAA and 2,4-D are considered to be stable and can be stored at 4°C for several months. Generally IAA and 2,4-D are dissolved in a small volume of 95% ethyl alcohol or KOH and then brought to volume with double-distilled water; NAA can be dissolved in a small amount of 1 N NaOH or KOH, which also can be used to dissolve 2,4-D and IAA.

The cytokinins are considered to be stable and can be stored at -20°C. Cytokinin stock solutions are generally prepared at 100X to 1000X concentrations. Many of the cytokinins are difficult to dissolve and a few drops of 1 N HCL, 1 N NaOH, KOH or DMSO, are required to bring them into solution.

High plantlet mortality rates have been experienced with some orchid hybrids. The development of new approaches such as the photo-autotrophic culture system with CO_2 enrichment represents a significant contribution to improve the growth and acclimatisation of orchid plantlets *in- vitro* and during transplanting. In the tropics, it may take more than three years for the orchid plantlets to reach the flowering stage. Orchids, particularly those with an epiphytic origin are notoriously slow-growing plants. The slow growth of epiphytic orchids may be attributed to its mode of carbon acquisition.

Paphiopedilum orchids are generally propagated through the division of axillary buds from mother plant, which limits commercial production due to its unproductive proliferation and time consuming. Udomdee *et al.* (2012) studied the effect of *in vitro* cutting methods and medium composition on efficient shoot multiplication of *Paphiopedilum hsinying* Rubyweb. Among three different *in vitro* stem cutting methods used, vertical cutting was able to produce more new shoots than horizontal and cross cutting when cultured on Hyponex based medium. After 12 weeks of culture, plantlets regenerated from vertical cutting were able to produce new healthy and well rooted shoots higher than without cutting on the same medium. Moreover, the newly-formed shoots which were divided into single plantlets and subcultured onto half-strength Murashige and Skoog (MS) medium without growth regulators could remain higher shoot multiplication than in other media. The micropropagation procedure developed in this study provides a simple means to *in vitro* propagate *Paphiopedilum* plantlets which are able to produce large numbers of uniform plantlets in a shorter time compared to the conventional propagation method.

An *in vitro* method was developed by Dhyani *et al.* (2012) for the production of phenolic compounds from callus suspension cultures of *Habenaria* edgeworthii Hook. f. ex. Collett, a rare Himalayan medicinal orchid. Friable and light yellow coloured callus was induced within eight weeks when seeds were cultured on Murashige and Skoog (MS) or 1/2 MS medium supplemented with 1 micro M alpha - naphthaleneacetic acid (NAA). In order to increase the biomass, following another 2 weeks these calli were repeatedly subcultured (4 week interval) onto 1/2 MS liquid medium supplemented with 0.1 micro

M 6-benzyladenine (BA). The effect of BA (0.0-3.0 micro M) and methyl jasmonate (MeJA; 10-1000 micro M) treatment on callus growth, total phenolic content and corresponding antioxidant activity was also determined following 4 weeks of culture on 1/2 MS medium (liquid). Maximum callus biomass (2.61 g fresh weight) was obtained in medium containing 1.0 micro MBA. The total phenolic content varied significantly in callus grown on BA, and ranged between 10.33 and 14.30 mg gallic acid equivalent (GAE) per g dry weight (DW) (compared to control 10.17 mg GAE per g DW). Antioxidant activity, measured by three *in vitro* assays, i.e., 2,2'-azinobis (3-ethylbenzothiazoline-6-sulphonic acid) (ABTS) radical scavenging, 1,1-diphenyl-2 picrylhydrazyl (DPPH) radical scavenging and ferric reducing antioxidant power (FRAP) assays showed maximum activity in callus grown on 3.0 micro MBA containing medium. High performance liquid chromatography (HPLC) analysis showed the presence of high gallic acid content (143.63 mg/100 g DW) when compared to wild tuber (5.5 mg/100 g DW). The result indicate that callus grown on medium supplemented with 3.0 micro MBA showed high phenolic content along with increased antioxidant activity, however, callus growth was higher with 1.0 micro MBA. The result of this investigation has commercial applications as increased biomass production will yield active compounds, much needed by the pharmaceutical and neutraceutical.

Seed Propagation

Low fruit set values in most orchids (especially epiphytic and tropical species) are normally thought to be the consequence of pollination constraints and limited resources. In particular, pollination constraints are modulated by pollinator visitation rates, pollinator visitation behaviour (promoting crossing or selfing), the type and number of pollinia deposited on stigmas (in the case of orchids with subequal pollinia) and the amount of pollen loaded per inflorescence. In order to assess to what extent these factors can affect fruit set in specific orchid-pollinator systems, the repercussions of some of these aspects on reproduction of *Broughtonia* lindenii were examined in a coastal population in western Cuba by Vale *et al.,* (2011). The study focused on plant breeding system, importance of pollen load and type of pollinia on subsequent fruit and seed, limiting factors of seed production and interaction with pollinators. This species presents long-lasting flowers that senesce after all forms of effective visit. Pollinator dependence for fruit production was demonstrated, while hand-pollination experiments revealed self-compatibility and inbreeding depression at seed level.

More pollinia on stigmas enhance the proportion of well-developed seeds. In contrast, the pollinia type used in pollination is not important for seed quality of fruits, suggesting that small pollinia are not rudimentary. Natural fruit set in two consecutive years was substantially affected by pollinator activity, and also by systematic depredatory activity of ants and a caterpillar. Considering that this orchid completely lacks nectar and that the local assemblage of pollinators and predators influenced its reproduction, a minor importance of resource constraints in this epiphyte (with long-lasting reserve structures) .

Deceptive orchids are generally characterized by low levels of fruit set; however, there may be substantial variations in fruit set between sites and years. Within a single population, individual plants may also differ greatly in their reproductive output as a result of differences in inflorescence size or local density. In this study, we determined flower and fruit production over 5 years in two populations of the food-deceptive orchid, *Orchis purpurea*. Jacquemyn and Brys(2010) monitored plants annually for survival and flowering at each site to determine whether flowering and fruiting induced costs. The number of flowers per inflorescence varied considerably from year to year (min: 36.6, max: 49.5). Average fruit set was low (7%) and varied considerably among years and populations. They observed a considerable proportion of plants failed to set any fruit. However, the probability of producing at least one fruit was not affected by inflorescence size or local density. The number of fruits was significantly related to inflorescence size, but proportional fruit set was not. Local density also did not affect the number of fruits, nor proportional fruit set. There was also no evidence that plants with large inflorescence size or high fruiting success had a larger probability of remaining vegetative the year after flowering than plants with small inflorescence size or low fruiting success. Our results suggest that pollinator-mediated selective forces on inflorescence size through female reproductive success alone are weak, most likely because of the low overall level of visitation and the resulting uncertainty of pollination at the individual level. Their results further demonstrate that investigation of patterns of fruit set over several years is needed to better understand the variability in female reproductive success that is typical of most plant-pollinator interactions.

Artificial seeds were produced from encapsulated protocorm-like bodies (PLBs) obtained from six-month-old axenic leaf explants of *Vanda coerulea* Griff. ex. Lindl by Sarmah *et al.*(2010). The percentage of germination of encapsulated PLBs was influenced by the concentrations of sodium alginate and calcium chloride ($CaCl_2.2H_2O$) used. It was found that among the different concentrations tested, 3% sodium alginate and exposure to 100 mM $CaCl_2.2H_2O$ solution for 30 min produced firm, clear, round and uniform optimal beads which were suitable for handling. It was also observed that PLBs obtained from optimization of encapsulation matrix showed the highest percentage of germination (94.9%) when beads were inoculated immediately after formation. Encapsulated PLBs stored at 4°C retain their viability up to 100 days. The findings suggest that the encapsulation method for PLB obtained from leaf explants of *V. coerulea* can be useful as an alternative tool for conservation of this endangered species.

Orchid seedlings grown *in vitro* respond differently to the different culture media. Rodrigues *et al.* (2012) compares the traditional Knudson (KC) medium used in orchid *in vitro* cultivation with a new medium formulation based on the KC medium with modifications, which was named Novais medium (NM), and two other media prepared with NPK fertilizer in the formulations 10-30-20 plus micronutrients and 30-10-10+micronutrients. The media were tested for the *in vitro* growth of six-month-old orchid seedlings derived from the cross *Cattleya walkEriana* x *Cattleya loddigesii* 'Etibaia' and germinated in Knudson C medium. The KC medium was tested in the concentration of 2.0 g L^{-1}, and the other three media were tested in the concentrations of 0.25; 0.50; 1.00; 1.75; and 2.25 g L^{-1}. All the media were added with 20 g L^{-1} of sucrose, solidified with 10 g L^{-1} agar and pH adjusted to 5.7. NM and PetersReg. media were added with 2 g L^{-1} of activated charcoal and 200 ml L-1 of coconut water. The seedlings were cultured in 320 mL glass jars containing 35 mL of medium. The experiment was maintained in a growth room at 27+or-2°C, 16/8 h photoperiod with 48 i mol m-2 s-1 irradiance. The best responses for production of fresh matter were obtained using NPK fertilizers, in comparison with KC and NM medium. The fresh matter production increased linearly with increasing nutrients doses in both MN and KC media. The orchid seedlings grown in Knudson medium had the lowest means for the variables evaluated.

The seed capsule should be harvested while the seed is viable but before the capsule splits. Optimal harvesting time is specific, not only to genera but to individual clones, harvesting time will vary as a result of extremely complex inter-relations between generic and environmental factors. Time intervals listed below are meant as a general guide line for understanding the approximate maturity.

Table 2.4 : Duration of maturity of orchid seeds

Genus/Hybrid	Duration (months)
Aerides spp. and hybrids	5 - 6
Ansellia spp. and hybrids	4 - 5
Brassavola spp. and hybrids.	5 - 6
BrassoCattleya hybrids	4 - 6
Brassolaeliocatteya hybrids	4 -6
Broughtonia hybrids	2
Cattleya spp. and hybrids	5 - 6
Chochlioda spp.	9 - 11
Chysis spp. and hybrids	5 - 6
Cirrhopetalum spp. and hybrids	4 - 6
Cymbidium spp. and hybrids	9 - 12
Dendrobium spp. and hybrids	5 - 9
Doritaenopsis spp. and hybrids	3
Doritis pulcherrima spp. and hybrids	2 - 3
Encyclia spp. and hybrids	4 - 6
EpiCattleya	5
Laelia spp. ,hybrids and *Schomburgkia* hybrids	4 -5
Oncidium spp. and hybrids	3 - 5
Lemboglossum spp. and hybrids	12 -15
Lilia spp.	4 - 6
LaelioCattleya hybrids	4 -6
Maxillaria spp.	4 - 6
Mltonia spp. and *Odontoglossum hybrids*	4
Paphiopedilum spp. and hybrids	8 - 10
Phaius spp. and hybrids	4 -5
Phalaenopsis spp. *and* hybrids	4
Renanthera spp.	5 - 6
RhynchoLaelia spp.	4 - 6
Rhynchostylis hybrids	5 - 8
Rodriguezia spp.	3 -5
Rodriquezia spp.	3 - 5
Sarcochilus spp.	6 - 9
Schomburgkia spp.	4 - 5
SophroCattleya spp.	3 - 5

Contd.

Sophronitis spp.	2 - 3
Stanhopea spp.	4 - 7
Trichopilla spp.	4 - 5
Vanda, Ascocenda and *Ascocentrum* hybrids	5 - 7
Vandopsis spp.	5 -6
Vanilla spp.	2 - 3
Zygopetalum spp.	7 - 9

Harvesting of Seedpod

Optimal germination of seed is largely depends on the time of harvesting. There are two options to harvest the seed capsule. The difference in harvesting time of dry seed and the green pod may be as much as six to eight months in some genera.

In the first option seed capsule may still attached to the plant until full ripening. The ripe pod is yellow/brown in colour. The seed capsule is removed from the plant at the first sign of dehiscence. Then it should be stored in a cool and dry place using coffee filter to let dry them. When the pod is completely dry, it can be opened and the seeds must be stored in the refrigerator. This can be done by placing the filters having dry seeds in an airtight container. Ideally, seeds are stored in a glass vial in an airtight container containing calcium chloride. It is advisable to put the calcium chloride on the bottom of the container and the vials on a table in the container. The container with the seeds can be stored in the refrigerator. Seeds of *Cypripedium parviflorum* and *C.pubescens* can store with this way for 4 years still germinated profusely. Put labels on the vials mentioning at least the parentage and date of storing the seeds in the refrigerator. When sending the seeds to other people, put them in a small paper bag such as the ones used by stamp collectors. Make sure the seals do not leak by putting cello tape on them. Put a label on the baggie with parentage and date of cold storage and put it in a padded envelope to prevent the seeds from being crushed by the machinery handling the mail. The padded envelope is not a luxury but really essential.

The seeds are separated from the seed capsule, treated with a sterilizing agent and may be sown using aseptic procedure. A large number of seeds may be lost by contamination as a result of incomplete seed sterilization. Overexposure to the sterilizing agent may also cause excessive seed loss due to burning. The time required by the sterilizing agent to decontaminate the seeds without burning the proto-embryo is sometimes difficult to estimate. The effect of the sterilizing agent may vary depending on the genus and the age of the seeds being sown. Another advancement in increasing the successful germination of orchid seeds and reducing the duration of plant to be mature to attain flowering stage was the development of a green pod culture. In the green pod culture harvest the capsule at a stage that the fertilization has just occurred. The surface of the seed capsule is decontaminated with a sterilizing agent, the seed capsule is opened and the seeds removed and sown using aseptic procedure. During this process the seeds are not contaminated by exposure to air, eliminating the need for contact with a sterilizing agent and the possibility of burning the seed. This result increased germination of orchid seeds. It is advisable to perform the green pod method on plants carrying multiple seed capsules. When the green pod is harvested in right time, the percentage germination is often higher as compared to the harvesting of mature seeds. The pods may be carefully wrapped in clean tissue paper and kept in a jar lined at the bottom with calcium chloride packets or any other dehydrant. This prevents contamination from pathogen spores which need a humid atmosphere to thrive. When the pods split, the seeds are collected and stored in refrigerator in paper packets over a dessicator. The stored seeds will remain viable for a long period, in some cases as long as six years.

Orchid Seeds

The most interesting and adaptive feature of the family Orchidaceae is the physiology of its seed germination. Orchid seeds are unique in several respects. The seeds are generally microscopic and populated in millions per capsule. After ripening they blow-off like dust particles or spores. Their colour may be white, cream, pale green, reddish orange or dark brown and have very diverse shapes. A spore embryo consists of relatively undifferentiated, most isodiametric cells with dense granulated cytoplasm. Germination of spores typically takes weeks, while there is a report of one *Paphiopedilum* that took fifteen years. The minute amount of nutrients in the seeds is stored in the embryo cells in form of lipid and protein bodies, rarely in starch grains or glucoprotein bodies. Orchids have thus no external nutrient tissues such as an endosperm. The vast majority of orchids don't have an endosperm (except *Bletilla striata* and *Disa uniflora*) but they do grow through suspensor cells.They must enter symbiotic relationship with various mycorrhizal basidiomyceteous fungi that provide them the necessary nutrients to germinate, so that all orchid species show mycoheterotrophic nature during germination and reliant upon fungi to complete their life cycle. The chance of mycorrhizal association between seeds and fungi is very limited, only a minute fraction of all the seeds can grow into an adult plant.

During early stages of development, the fungus supplies critical nutrients to the orchid seedling. Later the orchid may become fully independent or it may retain its mycorrizal relationship throughout its life.The availability of large numbers of orchid seeds, with little or no food reserve and relatively uniform cultural and genetic characteristics have prompted researchers to use orchid seeds for *in-vitro* nutritional and developmental studies.

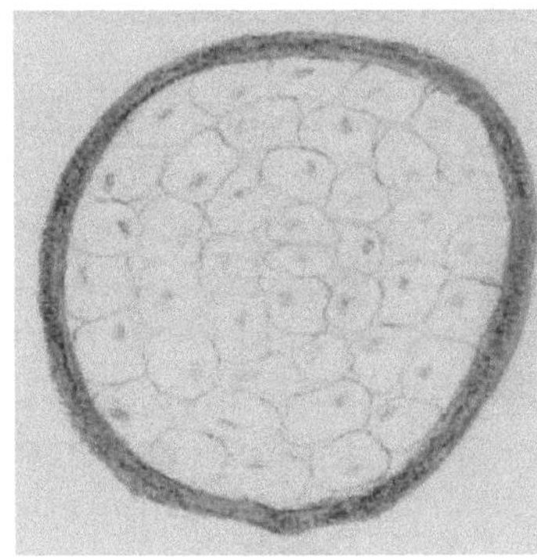

Fig. 2.9 : The cellular structure of the ovoid embryo.

Knudson (1922, 1924 and 1925), demonstrated that orchid embryos could be grown *in-vitro* without the intricacies of a host-fungal relationship, has played an essential role in advancing orchidology. These advances and developments have contributed to a substantial reduction in flowering time. *Phalaenopsis* Blume, *Oncidium* Swartz and *Dendrobium* Swartz, usually requiring four to five years to flower from seed sowing, can now be regularly flowered in less than two years.

Symbiotic Seed Germination

Frank in 1885 coined the term mycorrhiza. All terrestrial orchids are characterised for having a relationship with soil born fungi. After germination, seedling immediately needs infection from a compatible fungus to aid in nutrient mobilization and to transfer mycotrophic nutrition for further development. The hyphae make contact with roots and there's an exchange of organic building blocks - mainly carbon and nitrogen. This is initiated in seed germination and continues into adulthood. When the fungus enter into the testa it penetrates the embryo via the (large) suspensor cell. The fungus hyphae grow inside the inner embryonic cells as little coils called peletons. Orchid digests peletons and thus receives nutrition. Orchid embryo increases in mass to form the protocorm. Meanwhile the number of hyphae enters into the seeds and when a tipping point is reached the protocorm develops shoot roots and leaves. At this point the fungus invades the outer (cortical) cells and starts producing peletons there. Mature roots initiate the emergence of roots hair on them. Root hair grow in length and make more surface contact area with fungi in soil.

Most of the Mycorrhyzal fungi of orchids fall into a non-sporing group known as *Rhizoctonia* (basidiomycota), the major species being *R. repens, R. mucoroides* and *R. languinosa*. They play a large role in decomposition of cellulose and organic forest soil. Subsequent isolates are known as mycelium radicis. For quite some time these fungi were called as Orcheomyces. Later on, however, other fungi were also isolated from orchid cultures. It's difficult to determine whether the fungus is actually benefited from orchid, for this reason it is often said that the orchid is a parasite on the fungus. The orchid digests the peletons as a carbon source until the chlorophyll in the leaves have developed, this can take from months to an eternity as some orchids never build chlorophyll. Dependency on the fungus can decimate once the orchid reaches at its maturity. Especially epiphytic orchids are known for losing the need for peletons, terrestrial orchids typically keeps the relationship going until death does them part. Initially, nearly all orchids are myco-heterotrophic, but some never develop leaves and lose the photosynthetic function. A few species of *Disa* from South Africa on the other hand can be germinated on pure *Sphagnum* moss without a fungus.

Orchids use their symbionts to gain access to organic and mineral nutrients by increasing nutrient absorption and translocation to plants under natural conditions. The benefit of orchid mycorrhizal symbiosis using *Phalaenopsis* as model plants was conducted. Hung *et al.* (2011) inoculated *in vitro* grown plantlets of *Doritaenopsis* Taisuco Wonder 'King Car Butterfly KC1111' and *Phalaenopsis* Tai Lin Redangel 'V31' with two Orchid micorrhizal fungi (OMF) isolates, Ceratobasidium sp. AG-A (R02) and Rizoctonia solani AG-6 (R04). The effects of OMFs on orchid plant growth and Erwinia soft rot progression were examined after two months of *ex vitro* growth. The results showed that the presence of OMFs in *Phalaenopsis* roots significantly increased the growth and soft rot resistance of plants. Selectivity of cultivar type to different OMF was also observed.

Nontachaiyapoom *et al.* (2011) compared effectiveness of eight putative orchid mycorrhizal fungi obtained from mature orchids in the genera *Paphiopedilum, Cymbidium* and *Dendrobium*, in promoting *in vitro* seed germination and protocorm development of Grammatophyllum speciosum Blume and *Dendrobium* draconis Rchb. f., native Thai orchids. The developmental stages of seeds and protocorms cultured on Murashige and Skoog (MS) medium, oat meal agar (OMA), or OMA inoculated with one of the eight fungal isolates were evaluated weekly. Two isolates of *Epulorhiza repens* (Bernard) Moore (=anamorphic species of *Tulasnella calospora* (Boud.) Juel), Da-KP-0-1 and Pv-PC-1-1, were found to be the most effective fungi in promoting protocorm development of *G. speciosum*. At week 13, protocorms co-cultured with either one of these two fungal isolates, on the average, were significantly more advanced than those sown on OMA. Protocorms co-cultured with isolate Pv-PC-1-1 were also significantly more advanced than those cultured on MS medium. For *D. draconis* seed germination, three fungal isolates of different anamorphic species of Tulasnella, C1-DT-TC-1, Pv-PC-1-1, and C3-DT-TC-2, were found to be the most effective fungi in promoting protocorm development. However, none of these fungal isolates outperformed MS medium.

Cultivation of Mycorrhiza

Isolation of the fungus

Orchid roots are the best place for isolation of fungus. Amputating a root will normally not lead to a weaker plant and can be done without disturbing the plant and soil. A few tips regarding this are as under:

- Successfully isolation of fungi from orchid is not possible during dormancy period. Sympodial plants grow new roots every year and they become impregnated with the fungus as they contact the soil. Hyphae activity is at its peak when the plant is at its vegetative peak.
- Rhizomatous genera such as *Bletilla* and *Cypripedium* use the roots during vegetative growth for storing and extracting starch, it's only after the leaves have withered that the hyphae re-enter the cells to produce peletons.
- It is better to choose the largest, fastest growing plants for fungal isolation. They yield the best results for mycorrhizal germination.

Root cutting

Root tips of selected plant should be cut at 1-1.5 cm length. The cleaning and sterilization of the root cuttings is done by washing them with normal water to get rid of soil particles. Then 8 petri dishes or test tubes are prepared accordingly. The first test tube contain30% Hydrogenperoxid and second one has 20% Hydogenperoxid, the rest six contains sterile water. Root tips should be taken with a tweezers and kept in the first petri dish for 5 seconds, 20 seconds in the second, and 10 seconds in each of the other six. These steps sterilise the surface and then the root without killing the mycorrhiza. Mocan and Bularda (2012) study the relationship between fungi and the roots of orchids under greenhouse conditions. They recorded the presence of the mycorrhizal fungi in the zone of the root cells and did not see permanent infection, especially in the vegetative period of the plants.

Seed baiting

Baiting is a technique which is used to isolate the mycorrhizal fungus from a site or soil sample, being be sure that the soil actually contains the mycorrhiza. The idea behind this is pretty simple: instead of putting the soil on a medium and isolating/testing all the fungi that grow out of it, seeds should be sown on the soil and look whether there is germination. These germinated seeds are then put on a culture medium and the mycorrhiza, which is now present in the seed - is allowed to take over the medium.

Culture of Fungus

A Few Tips for Fungus Culture

1. For various stages of sterilisation it is better to use heat-treatment from pressure-cooking or flaming the utensils.
2. The medium should be transparent that makes easy to observe all about of germinating seeds and associate.
3. Everything should mix normally to result in 1 litre of medium - when a recipe tells us to add 3 grams, it means 3 grams for every litre of finished medium. One litre (1000ml) of water weighs 1 kilogram (1000g) so when adding 10 grams of something else we could try to think of it as 1% of the resulting mix.
4. The liquid medium (without agar) sometimes we need to filter liquid from a broth - use a handkerchief or some thick clean fabric, not cheesecloth.
5. We often need a gel to solidify the medium; the most commonly used is agar. Use just the amount needed to keep the medium solid - normally around 8g for soft and 14g for hard agar.
6. The acidity of a medium is denoted as pH (power of Hydrogen) which should be maintained by pH meter at the final stage of media preparation.

7. Reheating the medium should be avoided after the pH has been lowered as this can lead to the agar not gelling when cooling.
8. Cultures should not be smell, spores of unfriendly fungi can get into our lungs.

The media requires three components for the fungi to grow on:

- Nitrogen: peptone, yeast extract, malt extract, amino acids, ammonium or nitrate compounds are the major source of nitrogen supply into medium.
- Carbon: can be supplemented with glucose, fructose or mannose for carbon sources.
- Vitamins: fungi have a natural deficiency for vitamins, but these are provided by the organic ingredients.

Type of Media

Potato Dextrose Agar (PDA)

Potato Dextrose Agar is a common medium for isolation of fungi and bact*Eria*. PDA is a very rich medium which causes excessive mycelia growth and inhibits the formation of spores on most fungi, use it for rapid multiplication. The medium is commercially available from a large number of suppliers. However, it can be prepared in the laboratory using method giving below:-

1. Boil 200g of slice of peeled potatoes in 1 litre of water until the potatoes are soft.
2. Filter the extract in a container and add distilled water to make volume one litre.
3. Add 10 -20g of dextrose and 12-17g agar.
4. Bring this mixture to a gentle boil with constant stirring. Then dispense in cleaned and rinsed baby food jars or jam-jars that can stand the heat of a pressure cooker.
5. Place the punctured lids on the jars and wrap/cover them with aluminium foil.
6. Sterilize the jar in the pressure cooker at 15 PSI for 20 - 30 minutes and let everything cool before removing the lid of the cooker.

Germination of Seeds

Orchid seeds can be germinated on the pure strains of mycorrhiza without mixing special media or it can be inoculated in the extreme sterile working conditions as in asymbiotic *in- vitro* culture.

Stock should be maintained in jars with the following content:

- Five sterilized jars having agar of low strength (10g/L) are inoculated in anticipation of new mycorrhizae or for emergency rescue of contaminated cultures. The advantage is that if these jars remain free of contamination for more than a few weeks it will be absolutely sure about contamination and they can be stored pretty much indefinitely.
- The pure culture of fungus is maintained into 3-5 jars having the medium of 15 g agar/litre. They are known as mother cultures and store separately for further use. The strength of the medium is rather high but this is for two reasons:
- Makes contamination visible as soon as possible.
- If samples are taken out to inoculate new jars the damaged culture will recover very fast.
- Three cultures at reduced strength (5g/L) for each of most-used strain should be maintained. The strength is low to reduce the speed at which contaminants on the seed grow. Culture should allow covering the entire agar surface before sowing on it.

Until the middle of 19th century growers had no method for germination of orchid seeds and no hybrid could be produced, the only method followed was sowing of seeds at the base of the mother plant but in most of the cases the seed either did not germinate, got mixed up or took long time in germination. This was later followed by Bernard's method in which culture tubes were inoculated both with seed and fungi but the method was very elaborate and left much to be desired.

Asymbiotic Seed Germination

Lewis Knudson in 1916 conducted an experiment on germination of orchid seeds on the basis of the analysis of orchid salep which contained starch, protein, sugars and minerals. He formulated a medium and got successfully germination of *Cattleya, Laelia, Epidendrum* seeds and concluded that fungus was not necessary for orchid seed germination. This was the breakthrough in orchid seed germination and was followed by other who tried to improve the original media of

Knudson in order to germinate the seeds of other orchid species. When the seeds are planted on a nutrient medium *in-vitro* not only the percentage of germination is improved but it also takes less time for further development. Sometimes cent per cent germination was also achieved.

A successful attempt was made for asymbiotic immature embryo culture and *in vitro* mass scale production of plantlets by Deb and Pongener (2011). For successful culture initiation seed pods of various developmental ages, various nutrient media, sucrose concentrations, different quality and quantity of plant growth regulators were surveyed. Immature embryos of 9 months after pollination was successfully germinated on MS medium containing sucrose (2%) (w/v) and alpha - naphthalene acetic acid (NAA) and benzyl adenine (BA) (3 and 6 micro M respectively in combination) within 45 days of culture where 90% germination was recorded. The germinated seeds formed PLBs on the optimum germination medium within two passages. The protocorm like bodies (PLBs) differentiated into rooted plantlets within 3 weeks on regeneration medium containing sucrose (3%), casein-hydrolysate (0.1 gl-1) and BA 3 micro M. Amongst the three media studied, optimum regeneration was registered on MS medium where as many as 12 shoot buds developed per explants per subculture of 4 weeks duration. The well rooted plantlets of 6-7 cm long with 3-4 roots were hardened *in vitro* 3-4 weeks before they were transferred to potting mix. The potted plants were exposed to full sunlight periodically and watered at regular interval. About 70-80% transplants survived after 2 months of potting. Mercado (2012) studied the asymbiotic germination and seedling development of seeds of *Cattleya mendelii,* in several *in vitro* culture media. It was found that seed viability was 93%, the highest percentage of germination was found in the culture medium Murashige-Skoog plus coconut water (MS+AC) with significant differences compared to other culture media ($P<0.05$, Tukey). He showed that MS medium supplemented with coconut water and pineapple juice, were more efficient in asymbiotic germination and seedling development of *C. mendelii* orchids compared with other culture media. It could be an option to reduce the costs generated by using phytohormones.

Breaking Seed Dormancy *in-vitro* and Sterilization

Germination is impeded or prevented by illumination in several orchid species and thus darkness is often required for germination *in-vitro*. Scarification and sterilization *in vitro* by using hypochlorites, is used to break dormancy by rupturing the water-repellent testa in orchid seeds. Calcium hypochlorite is usually preferred, since it often yields better germination results after sterilization than sodium hypochlorite (NaOCl). Hypochlorites are strongly alkaline in solution and oxidizing agents. Other oxidants, such as alcohols, do not promote germination. However, other strong bases, such as sodium hydroxide (NaOH), promote germination, suggesting that it is the alkalinity of the sterilant that improves germination. NaOCl is used in concentrations ranging from 0.2 to 5 percent. Addition of a detergent to the sterilizing solution will improve the contact between water and lipoid surface (seed coat), since it contains both hydrophilous and lipophilous groups. Tween 80 has been widely used with good results in sterilizing solution for orchid seeds. The best germination percentage for many species is usually achieved in between 22 and 25°C.

Culture of Immature Seeds

Immature seeds have been found to germinate better *in-vitro* than ripe seeds. Embryo culture is important for a number of horticulturally important temperate and terrestrial species. If orchid seeds are excised before the water-repellent testa has fully developed, imbibition will be rapid. If the embryo cells are still hydrated and mitotically active, nutrient reserves will be mobilized.

Cymbidium lowianum Rchb.f., orchid was cultured *in vitro* on Mitra (M) and Murashige and Skoog (MS) medium using immature seeds derived from mature capsules. Plant growth regulators such as BAP (6-benzylaminopurine), NAA (alpha-naphthaleneacetic acid), and IBA (indole-3-butyric acid) were used either singly or in combination with or without activated charcoal (AC). The best seed germination response (85%) was observed in M medium supplemented with AC. M medium with BAP and AC induced highest leaf multiplication (6.6) in the culture while maximum rooting (6.2) was obtained in NAA supplemented M medium without AC. Complete seedlings with 2-3 leaves and 1-2 roots were successfully established in the green house after routine hardening procedures. This *in vitro* regeneration protocol may permit rapid propagation and conservation of this floriculturally important *Cymbidium* species (Nongdam *et al.*, 2006).

Different Media

Since the advent of Knudson medium, a large number of modified media have been standardized and are available in market even with pre -adjusted pH. The most common media used are Knudson *C. medium* and Vacin and Went medium. Many medium recipes for temperate terrestrial orchids have been developed through modifications of salt compositions for tropical orchids. The mineral salt concentration has then been reduced and organic compounds increased. Additions of complex organic compounds containing vitamins, growth regulators and amino acids are suggested to some extent mimic the conditions provided by the fungus. The media developed are convenient for horticultural plant production purposes since they are composed to promote both germination and subsequent seedling development.

Sowing of Seeds and Transplanting

In the beginning the pod are cleaned thoroughly using tap water followed by washing with 5 per cent Tween 20 for 10 minutes and finally with distilled water. Surface of pod should be sterilized with an aqueous solution of 0.1 per cent mercuric chloride for 15 minutes and are subsequently rinsed 3 times in autoclaved distilled water. The segments are then washed in 70 percent ethanol for 2 minutes. Then the explants are removed from the sterilizing solution and rinsed thoroughly for two times with sterile double distilled water. After surface sterilization, the capsules are taken in a sterile petridish containing filter paper to absorb the surface water of the pod. Sterilized pods are transferred aseptically to sterilized glass plate under the laminar flow hood. The capsules are cut longitudinally to find powdery mass of yellowish seeds and are inoculated on the slant surface of 0.8 percent agar solidified nutrient medium. pH of the media is adjusted to 5.4 and autoclaved at 15 kpas for 15 min at 121°C. The seeds further aseptically inoculated into test tubes of culture medium. Cultures are then incubated at dark condition at 20±1°C and later cultures should be kept on the same media without any subculture at 20±1°C under 16 hr photo period under the cool white fluorescent light of average 2500 lux (cool white fluorescent tube light 40 W.

Establishment of Cultures and Production of Protocorm Like Bodies (PLB's)

After a few days of inoculation the proliferation of seeds could be seen. When the seeds attain proliferation, these cultures should then transfer to jars containing fresh medium. When seed is cultured, instead of developing into a leafy shoot, they formed a spherule -like body with rhizoids at the base these structures are known as protocorm like bodies. After 21-25 days of incubation with a clean and sterilized forceps under the laminar flow hood, the initiated PLB'S should take out from the test tube without medium. Undesirable/brownish tissues should be removed from the PLB'S and be taken to the culture bottles containing autoclaved semi-solid media having the same combinations as that for the culture initiation.

Various procedures have been developed for protocorm-like bodies (PLBs) formation and proliferation, subsequently plant regeneration from Orchid seeds. These protocorms proliferated and formed a cluster of dozen protocorms. These protocorm mass should be cut into three to six pieces and planted them on fresh Knudson C Medium. Each piece formed additional protocorms by the activity of the superficial cells. This process of protocorm multiplication could be repeated indefinitely by regularly chopping the protocom mass and planting the pieces on fresh medium. This approach can yield over 100 orchid plants in one year from a single shoot-tip. The technique is widely used in orchid industry for micropropagation of both sympodial as well as monopodial orchids. But some orchids are not amenable to this approach.

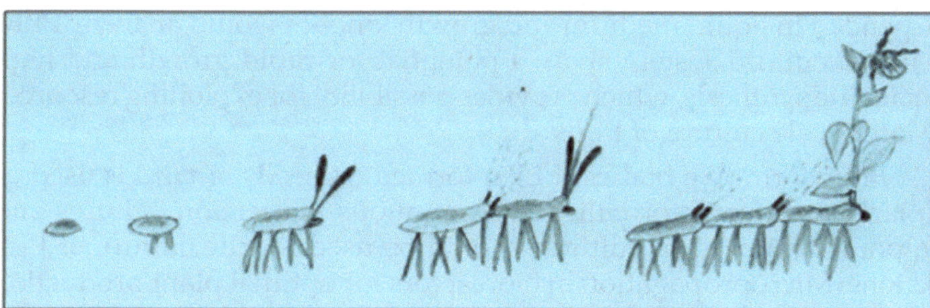

Fig. 2.10 : Development of plants from protocorm

Multiplication of Shoots by Repeated Sub-Culturing in Multiplication Media

Multiple shoots/cluster should transfer from the culture bottle to a sterile glass plate using flamed sterilized forceps. The brown leaves should remove from the primary shoots and sectioned into one shoot piece. These shoots should transfer to the multiplication media. All this work should be done with extreme care and inside the laminar flow hood to avoid any possible chance of contamination. These culture bottles then incubated in the growth room. Rapid clonal propagation would be achieved by subsequent induction of protocorm-like bodies (PLBs) and its conversion to shoots. These steps should be repeated every 25-30 days for the next sub-culturing.

Axillary shoots developed in medium *in-vitro* in the presence of cytokinin generally lack roots. To obtain full plants the shoots must be transferred to a rooting medium, which is different from the shoot multiplication medium, especially in its hormonal composition.

Multiplication of PLB's in Biorector

Bioreactors are special devices developed for the effective *in-vitro* cultivation of cells, tissues and organs, mostly guided by an automatic control unit. Bioreactors have a programme for direction of the optimum growth of vegetative tissues for example buds, shoots, seeds, embryos and roots, by the regulation of the chemical, physical and chemical factors of the inner and outer space. The bioreactors used for culture of plant cells may be of four types as batch bioreactors, continuous bioreactors, and multistage bioreactors and immobilized cell bioreactors; all bioreactors, except the last one are commonly called stirred tank reactors since the vessel has a device for stirring. PLBs should be suspended in liquid medium and growth kinetics would analyze: PLBs in liquid culture system show a potential for rapid growth and high metabolite synthesis, which provides possibility for exploiting resources by large scale culture of PLBs.

Protocorm-like bodies (PLBs) formed on seeds *in-vitro* is used as explants for bioreactor cultures. Continuous immersion cultures and temporary immersion cultures should be used for the culture of PLB sections. Micropropagation in bioreactors for optimal plant production is depend on a better understanding of plant responses to signals from the microenvironment and on specific culture manipulation to control the morphogenesis of plants in liquid cultures. PLBs grown in bioreactors could be cultured on solid Murashige and Skoog, Vacin and Went, Knudson C, Lindemann and Hyponex media. Hyponex medium would be suitable for conversion of PLBs into plantlets and 83per cent of PLBs transformed into plantlets on this medium.

Physical and chemical factors affecting growth include dissolved oxygen of the liquid medium, pH of the medium, inoculum density, culture period, light and temperature conditions and the configuration and type of culture vessel. The dissolved oxygen content indicates the amount of oxygen available to the tissue in the liquid medium. This available oxygen for plant cell is determined by oxygen transfer coefficient (KLa values). As the tissue grows, the need of oxygen increases which can be meet out by airflow and continuous agitation of the medium. Current interest towards photoautotrophic micropropagation revealed that carbon dioxide and light enriched environment during the *in- vitro* culture facilitates and supports the shoot growth. The various bioreactors used for Orchid propagation are mention in the following table:-

Table 2.5 : Types of bioreactor may be used for Orchid propagation

S.No.	Types of bioreactors	Example
1	Mechanically agitatedbioreactors	a) Aeration agitation bioreactors
		b) Rotating drum bioreactors
		c) Spin filter bioreactor
		d) Stirred tank bioreactors
2	Pneumatically agitatedbioreactors	a) Simple aeration bioreactor
		b) Bubble column bioreactor
		c) Airlift bioreactor
		d) Ebb and flow bioreactor
3	Non-agitatedbioreactors	a) Gaseous phase bioreactor
		b) Overlay aerationbioreactor

At four-leaf stage the seedling are taken out from flask and after through washing should be planted in community pots in a 1:1 mixture of shredded tree fern fibre or charcoal.

Advantages and Disadvantages of Bioreactor Culture System

The major advantage of using bioreactor culture system for micro propagation of economically important Orchid is the potential for scaling-up in lesser time, reduction in the production cost as well as an automated control of physical and chemical environment during growth phase of the plant cultures. Different designs and modifications in the agitation systems have made the technique more advantageous. In the case of airlift and column type bioreactors, where simplicity of design and construction, low shear stress, low contamination rate and less power consumption are the positive points towards plant propagation. There are certain demerits also such as foaming induced by large volumes of air leading to impaired gas transfer within the medium. Another problem in airlift type bioreactors is the evaporation of culture medium. To overcome the problems encountered during the use of air lift and bubble column bioreactors, the bottom-type bubble bioreactor (BTBB) was designed, in which foaming was drastically reduced by the use of cell lifting devices or tube(s) at vessel bottom. In addition, there was the provision of gas recycling system which allowed the examination of different gases in the medium. Stirred tank bioreactors (STR) were later designed to ensure the proper mixing of medium and large sized bubbles. The use of mechanically agitated stirrers resulted in an even flow of the medium in different directions of the vessel, which enhanced the proper oxygenation of the cultured tissue. The stirred tank reactors (STR) have some demerits like high shear force, complicated configuration, high contamination rates, difficulty in optimizing culture conditions *etc.* Recently, the use of ebb and flow type of bioreactor for plant propagation has overcome a number of problems faced during the culture in other types of bioreactors. This type of bioreactor consists of a fixed or floating support system inside the vessel, which initially helps to hold the explant apart from the agitator and later supports the growing shoots. In these reactors, medium is pumped into the vessel from storage tanks through a series of channels evenly to the plant material resulting in fast and better growth. The medium remains in the vessel for some time and then drains back to the storage tank. This has eliminated the problems of those culture systems where different growth phases need different compositions of nutrient concentration (as in case of somatic embryo development). Under these circumstances the ebb and flow type systems would be helpful because each step can be separated from the other and can be easily monitored and controlled independently. In such systems each step of plantlet production can be carried out in the same culture vessel simply by changing the culture medium. The nutrient medium outflow and the level regulated by specific pumps and controller, facilitate the tissue growth. Besides, the system has the advantage of short term exposure of growing cultures to plant growth hormones for the induction of different and desirable growth responses i.e., embryogenesis, organogenesis or multiple shoot induction, *etc.* This system completely eliminates the step of regularly performed sub-culturing, which is the most labor intensive and adds to the production cost of micropropagated plants.

Chapter - 3

Production Technology Management

Orchids are perennial ornamental crop, grows well in shady and humid places in wild habitat. The same growing conditions can be achieved by making special structures known as orchidaria or orchid houses.

Basics for Orchid House Cultivation

The following points should be considered during the construction of orchid houses:-

- Topography of land.
- Climate of region.
- Type of soil and water.
- Accessibility of area, transport and marketing facilities.
- Proper drainage facilities.
- Availability of good quality irrigation water.
- A good infrastructure like availability of electricity.

The size and shape of the area determine where the orchid house can be built. An orchid house can be of any size depending on the necessity, but about 18.4 m^2 area (200 square feet) should be smallest that considered because of proportionately more expensive to build and operate than larger ones. Orchid house costs vary widely depending on the type, size of the structure and the covering material.

Orchid house should be built on a horizontal terrain. The direction of orchid house should be east to west, so that the maximum sunshine is available. Future extension of the orchid house should also be considered at the outset. Construction of orchid house should be done in a free-standing separate unit. Orchid house attach to an existing structure are easier to access, have lower costs of construction and heating, however, the plants receive less light. Orchid house characteristics may limit the choices for locating an attached orchid house. A free-standing Orchid house can be located away from existing orchid house. Plants in a free-standing orchid house receive light from all directions, are more expensive to build, heat, extra expense is required to run water and electrical lines to them.

Orchid house cultivation needs more attention than outdoor cultivation (daily climate control). Therefore, orchid house should be at a easy reach at all times. Orchid house must be located in a well-drained spot that is not heavily shaded. The house should not be constructed in shade.

Rainfall, temperature, solar radiation, air humidity and the interrelationship of these climatic factors pay closer attention in designing orchid house and climate regulating equipment inside orchid house.

Oncidium are planted in open protected culture or under shading nets in subtropical regions. Because of the lack of cooling techniques for the inside microclimates, the production period is limited, and the quantity and quality of flowers is unstable. Without ventilating, high humidity and temperature induces severe growth problems. In this study, mechanical ventilation fans with intermittent misting were applied to modify the internal microclimate of the *Oncidium* greenhouse. Nine experimental setting of misting intervals were tested. The misting efficiency is defined and served as the performance index. The misting intervals of 3 min on to 3 min off or 5 min on to 5 min off had the higher efficiency. The cooling technique can keep the interior air temperature lower than that of outside air temperature. The temperature of plants could be maintained close to the greenhouse air temperature. The intermittent misting function with the best evaporative cooling efficiency was controlled with interval operation. This technique can be applied to other protected structures of *Oncidium* and other orchids (Shen and Chen, 2012).

Orchid houses can be constructed by split bamboos, glass, shade nets *etc.* which provides favourable environment and protect the crop from scorching sunshine, heavy rains, hailstorms, thus ensures quality production with full exploitation of genetic potential. However, all types of orchids can not be grown under one roof. Tropical orchid enjoy humid, warm atmosphere and burst into activity during rainy season. However, temperate orchid should be grown in cool houses. Sudden change in temperature is not preferable for orchids and a difference of 10-20°C between day and night temperatures is beneficial. The best suitable range is 18-30°C; proper ventilation is must to provide fresh air and also helps in reducing the temperature. Some orchids favour open places. Various species of *Vanda, Aranda, Arachnis, Renanthera, Kegawara* and *Mokara etc.* grow better in open trenches filled with brick pieces, charcoal.

Fig. 3.1 : Cultivation in orchid house

The concept of polyhouse cultures came from communities of farmers and gardeners concerning simple solutions to protect their crops using plastic film in subtropical and tropical countries as well as in the cooler mountainous regions. Orchid houses are frames of inflated structure covered with a transparent material which reflects back about 35-45 per cent of the net solar radiation incident upon it allowing the transmittance of the "photo synthetically active solar radiation" in the range of 400-700 nm wave length. The sunlight admitted to the orchid house is absorbed by the orchids, floor and other objects. These objects in turn emit long wave thermal radiation in the infra red region for which the glazing material has lower transparency. As a result the solar energy remains trapped in the orchid house, thus raising its temperature and providing heat to overcome extreme cold conditions. However, in the summer season due to the above stated phenomenon the ventilation and cooling are required to maintain the desirable temperature inside the structure. Natural or forced ventilation system can be used for proper ventilation.

The good orchidaria must faciliate with suitable temperature, light, relative humidity, carbon dioxide levels, water, plant nutrients and pest controls system as per the need of orchids. Manually controlled system or semi automatic controlled system of orchidaria requires a lot of attention and care. Microcontroller based control system is the solution to over comes this problem and to maximize returns. An ideal orchid house should have following components:

1) Orchid house G.I. structure.
2) Covering material-UV stabilized polyethylene / polycarbonate / acrylic.
3) Root ventilation & side wall roll up curtains.
4) Cooling pad and fan system.
5) Shading / thermal net manually / motorized.
6) Micro irrigation system.
7) Fertigation system
8) Misting system.
9) Heating system.
10) CO_2 generator.
11) Control System: manual/semi automatic/automatic. Fully computerized weather station.
12) Planting material, soil media.

Types of Orchid Houses

Classification of orchid house can be made on the following basis.

- On basis of cost of construction / technology
- On the basis of type of structure.
- On the basis of type of glazing material.
- On the basis of number of spans.
- On the basis of environmental control

Classification on Basis of Cost of Construction/Technology

a) **Low cost or low tech orchid houses:** Low cost orchid house is a simple structure constructed with locally available materials such as bamboo, timber *etc.* The ultra violet (UV) film is used as cladding materials. Unlike conventional or hi-tech orchid houses, no specific control devices inbuilt in these houses for regulation of temperature and humidity, however, these parameters can be adjusted manually. For instance light intensity can be reduced by incorporating shading materials like nets. The temperature can be reduced during summer by opening the side walls. Such structure is used as rain shelter for crop cultivation. Otherwise, the house temperature increased. This type of orchid house is mainly suitable for cold climatic orchids.

b) **Medium-tech orchid house :**These structures have manually or semi automatic control system owing to minimize investment. This type of Orchid house is constructed using galvanized iron (G.I.) pipes. The canopy is attached with side frame with the help of screws. Whole structure is firmly fixed into the ground. Exhaust fans are provided to maintain temperature, humidity and CO_2 level in the house. Evaporative cooling pads and misting arrangements may also provide in the house to maintain humidity. As these systems are semi-automatic, hence, require a lot of attention and care, and it is difficult and cumbersome to maintain uniform environment throughout the cropping period. These orchid houses are suitable for dry and composite climatic zones.

(c) **Hi-tech Orchid house :** The hi-tech orchid house is fully automatic and coupled with all devices, monitors and controls systems to regulate inside air temperature, relativity humidity, soil pH, electrical conductivity *etc.* The desirable values of all parameters maintain automatically with reference of standardized values. This type of orchid house consists of a sensor, a computer and an operator. The temperature, humidity and light are automatically controlled. These are indicated through sensor or signal-receiver. Sensor measures the variables, compare the measurement to a standard value and finally recommend running the corresponding device. For example, temperature control system consists of temperature sensor, heating / cooling mechanism and thermostat-operated fan. Similarly, relative humidity is sensed by optical tagging devices. Boiler operation, irrigation and misting systems are operated under pressure sensing system. This modern structure is highly expensive, requiring qualified operators, maintenance, care and precautions. In fully high tech orchid house computerized control system provides a faster and precise operation in the orchid house and also stores, displays and prints the orchid house information as needed. Computer can do the following operations as per the pre-scheduled programme:

- Starting and closing of micro irrigation system.
- Application of water soluble fertilizer or liquid fertilizer and other nutrients to the plant.
- Operation of misting system as required.
- Opening and closing of ventilators and side wall roll up curtains as needed.
- Operation of shading net / thermal screen.
- Operation of cooling pad and fan.
- Operation of heating system.
- Operation of CO_2 generator, climate control, temperature, humidity, heat radiation, control of electrical conductivity, pH, parts per million levels in irrigation water *etc.* as required to the orchids.

A low to medium-tech orchid houses could cost between Rs. 125 to Rs. 500 per square meter in India, whereas a high-tech, fully-automated orchid house costs Rs. 2,000 per square meter.

Classifications on the Basis of Type of Structure

a. Quonset type (semicircular / subtropical orchid house)
b. Gothik arc type
c. Gable roof type (slopping roof)
d. Slant type
e. A frame type
f. Dome shape

Classification as per Glazing

a. Glass glazing
b. Fiberglass reinforced plastic glazing
    i. Plain sheet
    ii. Corrugated sheet
c. Plastic film
    i. Ultra violet stabilized low density polyethylene
    ii. Silpaulin
d. Multi span or ridge and furrow or gutter connected

Classification Based on Environmental Control

a. Naturally ventilated
b. Passive ventilation

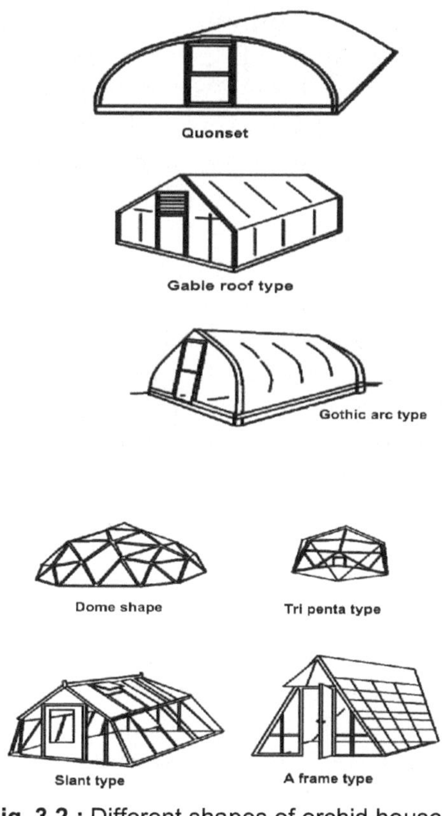

Fig. 3.2 : Different shapes of orchid house

Classification Based on Number of Spans

a. Free standing or single span

Material of Construction of Orchid House

Covering material: An orchid house cover is usually made by glass, fibreglass, acrylic, polycarbonate or polyethylene plastic. Glass-covered, metal-framed Orchid houses are usually most expensive. Each of the coverings has advantages and disadvantages, the main determining factors usually being the trade-off between cost and length of service. Glass is more expensive but will generally have a long lasting life than either polycarbonate or polyethylene. Glass is heavy, requires a strong frame, expensive and also subject to breakage. It is usually considered more attractive and permanent. It is not recommended in areas prone to hail. Fibreglass is widely used because it has good light transmission, little breakage and requires less roof support. Only fibreglass made for polyhouse / greenhouse should be used. Fibreglass is also highly flammable and at higher elevations, it will discolour and degrade at an even faster rate due to high ultraviolet radiation levels. Acrylics are more durable, harder plastics and last in 15 to 20 years. Polycarbonate coverings are very strong and are recommended in areas where hail can be a problem but are not hail proof and can be damaged if hailstones are large enough and Polycarbonate orchid house covers lasts in 12 to 15 years or more. Polythene is an economical cladding material. Two layers of polyethylene, with pressurized air filling the space between the two layers are provided for rigidity of covering. Now a day's long lasting, unbreakable and light roofing panels-UV stabilized clear fibre glass and polycarbonate panels are available. Plastics are used in tropical and sub-tropical areas compared to glass/ fibre glass. Low density polyethylene (LDPE) / Linear low density polyethylene (LLDPE) will last for 3-4 years compared to polythene without UV stabilizers. Ordinary polyethylene is broken down rapidly by ultraviolet light and must be replaced at least every year especially at higher elevations and its substitute is special polyethylene that contains ultraviolet-inhibiting compounds which may last one to two years. Many orchid houses are protected with a double-layer covering to reduce heating costs. The two layers slightly reduce the amount of light that is admitted but the same time cuts the heating costs by about 40 par cent. The two layers should be spaced 2.5 to 7.6 cm apart. Double layers of polyethylene can be separated by air inflation with a small squirrel-cage fan. Orchid houses may be constructed with double polyethylene skins. For insulation plastic bubble wrap can be used in winter inside of an orchid house. Glass or fibre glass orchid houses can be built with a spaced double covering. They may also have a polyethylene liner installed inside the outer layer or a double polyethylene air-inflated cover can be put over the roof. Acrylics and polycarbonates are available in double and triple layers from manufacturers. Selection of covering materials during the construction of orchid houses is very important factor as it allows and regulates energy inside the orchid house during winters. New coverings are being developed which selectively exclude certain wavelengths of light and as a result can help in reducing insect and disease problems.

Galvanized steel

Steel is used for framework on kits. It is strong, heavier in weight and prone to rust. It is not readily available for self construction job. Galvanized steel is rust resistant tends to be used on the lower end of orchid house kits, but it is more expensive than others.

Aluminium

Aluminium is strong, light in weight and has no risk of rust. It is round or square in shape, depending upon the covering material to be used on it. Drilling holes in it is fairly easy, so fibreglass panels or polycarbonate sheets could be fastened to the structure.

Poly Vinyl Chlorite (PVC)

PVC pipe can also be used in the construction of a Quonset type orchid house frame. It is readily available, cheap and easy to work. Up to 2-3 cm diameter can be bent for form hoops over which polyethylene film can be stretched. Ends can be constructed from either plywood or fibreglass sheets. Fibreglass sheets can also be screwed to the hoops to form a cover over the frame.

Wood/Bamboo

Wood is the most versatile material to be used for an orchid house structure. Treated wood can also be used to construct an orchid house. This type of structure would be a bit harder to build. Any type of covering could be installed over a wood frame. It is strong enough for glass, polyethylene films and polycarbonate or fibreglass sheets. These covering materials can be fixed easily on the wood frame. Pressure treated lumber, cedar and redwood can be used for orchid house construction. Cedar and redwood are lightweight and strong. Redwood is the easily available and used in high end kits which are long-lasting. The low-cost orchid house is generally made by polythene sheet and bamboos. At least 3 years old bamboos are suitable for construction of orchid house. Bamboos are selected and fumigated for minimum period of 2 weeks for 24 hours each day. Selected bamboos can be fumigated by smoke. The active ingredients of fumigation are neem cake and tar along with dry wood. Bamboo is a bad conductor of heat and electricity. So even in the scorching sun or in extreme cold, the outside environmental changes do not have an adverse effect on the ecology inside the orchid house. The benefit is also evident at the contact point where the polythene touches the bamboo member elements. At this contact point polythene expansion is less compared to the contemporary steel orchid house due to less heating, thus enabling polythene to last longer.

Fibreglass

Fibreglass tubing is also used as framing material in orchid house construction. An advantage to this material is that it will not conduct heat or cold easily, thus helping keep heat in and cold out from house.

Benches for orchid house

In Orchid house plants are grown in some type of container. These containers can be placed on the floor or on benches. Benches are not necessary, but they make to work easier with the orchids. Benches should be made of sturdy, decay-resistant material such as cedar or redwood or rust-resistant metal. The wood used in bench construction should be treated with a wood preservative containing copper naphthalene for longer life. Benches made of metal work better. Over the bare ground, there should be a layer of landscape fabric down and then an inch or two of gravel. This will allow adequate drainage and also provide aeration under containers.

Design of orchid house

The structure carries the following loads and designed accordingly.

a) **Dead load:** Weight of all permanent construction, cladding, heating and cooling equipment, water pipes and all fixed service equipments to the frame.

b) **Live load:** Weights superimposed by use (include hanging baskets, shelves and persons working on roof). The high tech orchid house has to be designed for a maximum of 15 kg per square meter live load. Each member of roof should be capable of supporting 45 kg of concentrated load when applied at its centre.

c) **Wind load:** The structure should be able to withstand winds of 110 kilometer per hour and at least 50 kg per square meter of wind pressure.

d) **Snow load:** These are to be taken as per the average snowfall of the location

The orchid house should be able to take dead load plus live load or dead load plus wind load plus half the live load.

Wind effects: If the orchid house is naturally ventilated, the advantage of natural wind direction has to be taken to the maximum possible. The maximum dimension (length) of orchid house should be perpendicular to the wind direction especially in summer. For fan and pad orchid house the natural wind direction should be same as the air blown by fan.

Height of orchid house

The maximum height can be up to 5m for 50m x 50m orchid house and this can be reduced as per the reduced size of the orchid house. Higher is the orchid house more is the wind load for structure and glazing. The side ventilation can better achieve with 2 m width and roof ventilation is 1m in width.

Orientation of orchid house

Orientation of the orchid house is a compromise for wind direction, latitude of location and type of temperature control. Single orchid houses with latitude above 40°N should have ridge running east to west to allow low angle light to enter from side rather than ends. Below 40°N the ridge of single orchid houses should be oriented from north to south, since the angle of sun is much higher. This orientation permits the movement of shadow of the gutter across the green house. The location and orientation of the orchid house should avoid falling of shadow on the adjacent orchid houses. To avoid the shading effect from one orchid house to another, these should be oriented east to west. However, the wind direction and latitude are also to be considered.

Pot and Potting Management

Orchids should be potted in small pots according to the size of the plants. As a thumb rule, it should be under potted to get more flowers. Any kind of pot/container which can hold medium and provide aeration is suitable. The most common container is the plastic or clay pots. The big differences between standard garden pots and those used for orchids are the number and size of drainage holes in the pots. Orchid pots have larger holes on the bottom and sides of the pot to ensure better drainage. Some are shallow and shorter than standard garden pots, with a larger base, especially useful for top-heavy orchids. Orchids can also plant in wooden baskets usually constructed of teak or some other rot-resistant wood. Plastic pots should be preferred because it retains moisture longer than mud pots. Plants should not be disturbed frequently and re-potting done only when absolutely necessary. Potting of orchids varies according to genera and species. *Dendrobium* & *Cattleyas* should be potted in clay pots or mount them in drift wood. *Vandaceous* and *Sarcenthene* orchids can be grown in teak-wood baskets. *Vandas* and *Ascocendas*, small seedlings are potted with little osmunda fiber and a little charcoal.

Oyamada *et al.* (2011) have developed a new growing technique called 'The Paper Lite cultivation method' and have now confirmed its efficacy for cultivating *Cypripedium* macranthos var. speciosum. They used Paper Lite medium (PL medium), which is made from a mixture of rolled paper and vermiculite and placed an aseptic air ventilation filter on the top of the growth container. They transplanted young cultured plants on to the PL medium. Immediately after transplantation, we cultured the plants in a dark place keeping the temperature at 4°C for 120 days. Then the plants were moved to a bright places keeping the temperature at 20°C. Just prior to their acclimatization and potting, plants produced by our method were compared with others produced by the standard methods. The former surpassed the latter in total weight, height and number of leaves, number of roots, and number and development of dormant buds. A year after being potted and moved outdoors, plants grown by our method continued to surpass those grown conventionally in height and number of leaves. The plants were maintained outdoors in pots for one year. They then transplanted them to wild forests. A year after this transplantation, their survival rate was 84%, and the flowering rate was 5%.

Potting Media

Different types of growing media may be used for orchid cultivation, but it is crucial that it must retain a certain amount of moisture and be well draining. Different orchids require different substrate in which they would thrive. Conditions of the plant experiences in nature should think before choosing the media. Tropical plants require more water naturally compare to subtropical and temperate ones. For irrigating with sprinkling can or hose, it is better to use materials that drain well and decompose slowly while for drip irrigation heavy potting material should be used. The most important thing is to match watering habits to the potting material. Growing conditions play a most significant role in deciding on a medium. If there is no problem in providing a constantly humid environment due to climate or if growing in a orchid house, the moisture retention on the part of media is less of a concern. Ambient temperatures play a role here, as well, as warmer temperatures tend to increase the evaporation of water and along with the drying rate of the medium. The personal tendencies and how much time might be spending to water plants is another significant factor in choosing a medium. The combination of potting materials that will work best for orchid depends mainly on plant structure and root structures:-

1. **Plant Structure:** Terrestrial orchids should keep slightly damp all the time, while epiphytes or lithophytes want to dry off thoroughly between watering. Orchid with pseudobulbs can generally store water for longer periods of time than those of without pseudo bulb. Those having thin, strap-like leaves, such as *Oncidium* and *Cymbidium*, will lose their stored water faster than those with shorter, thick leaves, such as *Cattleyas*. Slipper orchids lack pseudobulb for water storage, so they should not be allowed to dry out entirely. Plants that don't have pseudobulbs in some cases developed thick, fleshy leaves (*Phalaenopsis*) as their water storage scheme, but that's obviously not as effective as pseudobulbs. Others that have neither pseudobulbs nor thick leaves, such as *Phragmipediiums*, are needed a steady supply of moisture. Large plants usually do best in coarser potting materials and smaller plants do better in finer potting materials. Orchids which grow in the ground in a moderate climate will grow in regular soil (*Bletilla*), while terrestrial orchids from tropical areas (some species in the genus *Cymbidium* and *Paphiopedilum*) require very porous yet rich substrate. Young plants and seedlings need finer substrate than adult plants. Clearly, there is no universally applicable substrate which would suit all orchids. *Paphiopedilum* should not be grown in waterlogged soil because its roots will rot.

2. **Root Structure:** It is another important aspect that decides medium texture. In general, smaller roots grow better in finer water-retentive materials, while larger roots perform best in coarser materials. The factors to be considered are the extent of root growth (long versus short, branched versus single), the thickness of the roots, particularly that of the velamen layer. A very extensive root system suggests that the plant has to work hard for its water and nutrition, so wants to gather as much as it can when it is available. By contrast, a short root system suggests easy and/or frequent availability of them. Velamen is actually dead tissue that acts as a sponge to rapidly absorb water and nutrients, so they can be passed onto the plant. Thick velamen is generally found on plants that like to grab a lot of water, but prefer to have their roots dry rapidly. *Vandaceous* plants are good examples. It should not mean that roots with thin velamen like to stay moist. *Oncidiums*, for example, tend to have thin velamen on thin roots, but their root systems are quite extensive, and they have pseudobulbs for storage, so they also like to dry fairly rapidly. *Phalaenopsis*, with their minimal storage capabilities, tend to have thick roots to add to the "reservoir," so do well in a damp, but not soppy environment. Some plants like *Arachnis* and *Renanthera* are planted directly in a well-drained bed or in pots with soil.

Fig. 3.3 : Types of containers for growing orchids.

Media Ingredient

Commonly available media ingredients and their specific properties are as under.

1. **Fir bark**: Shredded bark is the basic ingredient contained in most substrates for orchid growth. The most common bark found in these substrates is pine while some growers also uses the bark of *Pseudotsuga menziesii*, known as Douglas fir. Shredded bark is sold in various fractions, from finely shredded (up to 25 mm in size), mid-sized (up to 40 mm), and roughly shredded (over 40 mm) particles. For larger orchids (some species in *Cattleyas, Angraecums etc.*) bark with fine shredded particles should be used. While for which demand good aeration of roots (*Vanda* and its hybrids), mid-sized and roughly shredded particle bark should be used which can be turn into smaller pieces if necessary. It is essential that bark is soaked in water for minutes and then washed at least once to get rid off tannins and fine dust particles, which are not good for tender orchid roots. Good quality bark is getting harder and harder to find, but the stuff that is available is typically steamed to extract the resins that could be toxic to plants. The removal of that resin increases the water-holding capacity of the bark, but also shortens the life of it. The irregular shape of the pieces of bark can lead to excess packing, which may reduce the airflow to the root system. Fine grade is often used for small seedlings, and the coarser grades for large plants. Bark should be moist by boiling water to facilitate wetting before use.
2. **Coconut husk chips (CHC):** It is made from the pithy covering outside of the spherical, hard coconut shell, is gaining popularity as a replacement for fir bark. It is readily available, relatively cheap and wets and rewets better than bark. It also tends to be more regular in shape - chunks or cubes as opposed to coarse flakes - so facilitates better air flow throughout the root mass. CHC holds a lot more water than bark, so keep that in mind when using it. It lasts considerably longer than bark as well, often going three years before decomposition becomes an issue, as opposed to a year for bark.
3. **Coconut husk fibre:** It is the wiry result of stripping and shredding the interior of the coconut husk, rather than chipping it. It is springy and flexible, and is often matted to form liners to hanging baskets, but it can also be use straight as a medium for plants that really like to dry out fast. If fairly tightly compressed, water will stay in pockets in the mass for a longer time period. Coir, sometimes called "Coco-Peat," is

the result of grinding the husk into coarse powder. It is often used as a substitute for peat moss in blends, but can be used alone for seedlings that like to stay damp.

4. **Osmunda fiber**: Roots of a number of ferns from the genus *Osmunda,* used to be a staple in the orchid growing community, but it is now difficult to find a high quality fibre. It is tough and springy and requires vertical alignment in the pot to ensure proper drainage. It almost totally ensures that the plant roots get all of the air they need, and as it decomposes, is an excellent source of nutrients, requiring little or no supplementation via fertilizers.

5. **Tree fern fibre:** The trunks of tropical tree ferns have replaced most osmunda in orchid culture. It is very stiff and airy, and can be obtained as shredded fibers, chunks or slabs. It has little moisture holding capacity, so aerates media blends very well. The slabs are often used for mounting plants directly.

6. **Redwood fibre:** It is sometimes known as redwood fibre, is a light, fluffy wood fibre that does not decompose, and is added as a moisture-retention aid and as a means of increasing the acidity of media blends.

7. ***Sphagnum* moss (Peat moss):** Its characteristics are good water absorption and aeration. *Phragmipediums* and some other orchids are best grown in a substratum composed exclusively of *Sphagnum* but, for most orchids, peat moss is mixed with tree bark which does not retain water very well. Peat moss can also be used in planting orchids on tree bark where the roots are wrapped in moss and then attached to large or small bark pieces. Other mosses which grow in the woods are definitely not adequate for orchid planting. Generally, *Sphagnum* becomes sour and in need of replacement in the 6 to 9 month range, although that can be extended by blending it with charcoal.

8. **Hardwood charcoal:** It is a very long-lasting medium ingredient having a moderate-to-low moisture holding capacity. Charcoal is added due to its absorption capabilities. It can extract a number of waste substances from the substrate and, to a certain extent, also the nutrients contained in it, which is not desirable. Charcoal does not absorb much water and, similarly to shredded bark, it helps to aerate the substrate. *Vanda* orchid may be grown by using the coarse grade as the

sole medium. It has been observed that the charcoal sweetens the medium by absorbing some of the foul chemicals produced during the decomposition of other media components or excess salts from fertilizers.

9. **Perlite:** It is often referred to as sponge rock for the coarser grades, is expanded volcanic glass, and is a great aerator of blends, but still holds enough moisture to be a fairly good substrate for Semi-Hydroponic culture.
10. **Pumice:** It is similar in nature to perlite, but with a higher density and lower moisture-holding capacity.
11. **Lightweight Expanded Clay Aggregate:** It is a general term for any number of more-or-less spherical terra cotta pellets that were originally designed to be the aggregate in lightweight concrete formulations, but later diverted to horticultural use. There are many brands available, including Aliflor, Hydroton, Prime Agra and others, and they vary greatly in their properties and applicability to orchid culture.
12. **Diatomite:** It is calcium diatomaceous earth mud. It is highly porous, and holds a tremendous amount of water.
13. **Styro foam**: These are the substrates contain fine particles of styro foam which improve aeration. These particles are completely inert (they do not decompose), are long-lasting and do not give out any nutrients as other ingredients in substrate do. Large pieces of styro foam can be placed at the bottom of the pot, which will provide excellent non-decomposing drainage material (small stones can also be used for this purpose).

Mixture of substrate: In general a mixture of fine or middle size bark particles (depending on the orchid size) and *Sphagnum* in a 4:1 ratio is recommended for most orchids. For *Cymbidiums, Paphiopedilums, Zygopetalums* and *Ludisia,* it is recommend to add an additional part of rough peat or coconut fibre. Charcoal may also be used by replacing one part of the bark. Large pieces of styro foam can be used for the purpose of drainage. Drainage material can also be provided by using large pieces of bark or small stones. *Masdevalias* and *Miltonias* require a 2:1 ratio of shredded bark and *Sphagnum*. *Vandas, Ascocentrums, Arachnis* and similar natural and artificially bred types of orchids require only roughly shredded bark. Terrestrial orchids, like *Spathoglottis, Phaius* and *Calanthe,* should be grown in 20-25 cm size pots with 1:1:1 mixture of leaf mould, FYM and sand. For *Paphiopedilum*. A mixture of 2 parts leaf mould, 2 parts loam soil and 1 part each of brick pieces and charcoal is recommended. It is observed that orchid *Rhychostylis gigantea*,perform better on chunks of hard-wood charcoal alone as potting substance. Tree fern fibre also performed better than the other media while coconut husk and over-burnt brick as planting substance had adverse effects on growth and flowering of *R. gigen*tea plants. *Cymbidiums* like a slightly acid potting mix with good drainage facilities. For *Cymbidium* planting mixes, fine bark mix with 20 - 25 per cent perlite, sand, coarse peat and redwood bark are often added to improve the water retention. Larger grades of bark will require more frequent watering, but there is less risk of overwatering. Coir, coconut husk chips, coarse peat or leaf mold and sand may also be used. They can also be grown in mixes containing pumice or lava rock.

Fig. 3.4 : (A-D) Various Substrates for potting

A pot experiment was conducted by Mohapatra and Saravanan (2006) to determine the suitable growing media for *Dendrobium* (*Dendrobium* sp.) cv. Flamme. The treatments were coconut coir, bhusa, rice husk, tile bits, charcoal, cow dung, brick pieces and groundnut shell. Gravel + groundnut shell + cow dung recorded the highest number of new shoots per plant and shoot girth, followed by charcoal + coconut coir + cow dung. Plant height and number of leaves were greatest with coconut coir + cow dung + bhusa. The flowering parameters, such as early flower bud emergence, number of spikes per plant, spike length and number of florets per spike, were also greatest with gravel + groundnut shell + cow dung, followed by charcoal + coconut coir + cow dung. Coconut coir, gravel and cow dung were superior among the growing media.

The production costs force the orchid's growers to look for media that could become an alternative for the basic ones. *Phalaenopsis* plants (ex *in vitro*) were planted into containers filled with different media: New Zealand *Sphagnum* moss, mixture of expanded clay pellets and New Zealand *Sphagnum* moss (v:v=1:1) and expanded clay pellets. Originally, orchids were grown in translucent plastic pots of 7 cm diameter and then, they were transferred to translucent plastic pots of 11 cm diameter and 0.5 dm^3 capacity. Effect of medium on the size of plant, hydratation of leaves tissue and flowering was evaluated. Orchids grown in New Zealand *Sphagnum* moss create a significantly greater mass of the aboveground plant part and the roots. Parameters characterizing the inflorescence and flowers indicate that the most favorable for the orchids is to grow them in *Sphagnum* moss. The least favorable growth conditions were provided by the growing in expanded clay pellets without any additions. In this medium, only 25 to 35% of plants flowered (Trelka *et al.,* 2010).

In order to assess the growth and flowering of ornamental orchids in peat alternative substrates, two cultivars of an ornamental orchid species, *Doritaenopsis* Labois and White Castle, were cultivated in 3 different substrates: pure perlite, perlite+coir dust (3:1), and perlite+rockwool (3:1). Growth responses in terms of fresh and dry weight of shoots and roots were measured after 24 months of culture. Inflorescence size, number of flowers, flower diameter, and days to flowering were evaluated after 8/9 and 20 months (1st and 2nd flowering season, respectively). Plantlet growth response (fresh and dry weight of roots and shoots) was greatest in growing media containing perlite+rockwool (3:1) in both cultivars. Root development (root dry weight, number of roots, and root length) was drastically improved in perlite+rockwool (3:1), while perlite+coir dust (3:1) was found to be unfavorable for the growth and root development of these two cultivars. Growing media had no effect on inflorescence emergence during the 1st and 2nd flowering season in the two cultivars, but significant differences were observed in inflorescence length, number of stalks per plant, number of flowers per stalk, and flower diameter. The inflorescence length, number of flowers per florescence, and flower size significantly increased in plants grown in perlite+rockwool (3:1) during the 1st and 2nd flowering season, in both cultivars(Wu *et al.*, 2005).

A trial was conducted to study the influence of the media and frequency of nutrient application on growth and flowering of *Cymbidium* hybrid 'H.C. Aurora' by Barman *et al.* (2012). Treatments involved growing of *Cymbidium* in different combination of media, i.e., M1-perlite + coco peat + brick pieces; M2-coco peat + brick pieces + leaf mould; M3-perlite + brick pieces; M4- thermocol + coco peat + brick pieces; M5-rotten log + sand + coconut husk + brick pieces; M6-*Castanopsis hysterix* bark + maize cob + leaf mould; M7-leaf mould + sand + coconut husk + brick pieces + charcoal (Farmer's practice) and spraying of water soluble fertilizer (N19P19K19) at 1 g/l concentration at weekly and biweekly interval. The result indicated that among the vegetative characters, maximum number of shoots, number of pseudobulb, pseudobulb girth, length leaf, and floral characters like spike length and number of spike/m^2 were recorded at growing media *Castanopsis hysterix* bark + maize cob + leaf mould. Further frequency of water soluble fertilizer significantly influenced growth and flowering of *Cymbidium*. Maximum number of shoots , spike length and number of flowers/spike were recorded with biweekly spraying of water soluble fertilizer (1 g/l). Further, the growing media significantly influenced the NPK content of leaves. The physical properties, i.e., pH (6.25), EC (0.23 mmhos/cm^2), bulk density (0.52 g/cc) and water holding (49%) of *Castanopsis hysterix* bark + maize cob + leaf mould medium was found to be better for growth and flowering of *Cymbidium* orchid.

The artificial textile fiber, a waste product of the textile industry, is clean and neutral. It could be a promising alternative medium for orchid cultivation. Vegetatively propagated young plants of *Oncidium* Sweet Ears 'Pacific Gold', each with a newly matured second-generation pseudobulb, were planted in artificial textile fiber or a substrate consisting of 90% fine grade California fir bark and 10% long-fiber *Sphagnum* peat by volume. Plants were irrigated with 0.67 g/L 15N-$5P_2O_5$-$15K_2O$ Ca-Mg soluble fertilizer, alternating with clear water, when the substrate became dry. After one year, plants grown in the artificial textile fiber substrate flowered 2 weeks earlier than those in the bark mix. Plants grown in the artificial textile fiber had more spikes with higher degrees of branching and greater flower count than those grown in the bark mix. Growing substrate had no effect on flower size and pseudobulb width(Dai *et al.*, 2010).

Angraecum sesquipedale is an epiphytic orchid originated to Madagascar. It is famous for its long spur. Experiments were conducted to test on five growing media for the growth and flowering of *Angraecum sesquipedale* by Wu and Lai (2010). The results revealed that the plants were tallest and strongest when grown with *Sphagnum* moss compared with other media. The values on leaves number, leaf area, flowering percentage and flower quality were highest. The plants that grown in both bark mixed tree-fern and bark mixed tree-fern, satsuma grew worst, low flowering percentage and worse flowers quality. Using *Sphagnum* moss as growing medium with 0.5 g/L fertilizer solution every week is recommended for excellent potplant production of Angraecum sesquipedale.

Sphagnum moss is the main growing medium for cultivation of *Phalaenopsis* . In recent years, over-exploitation of *Sphagnum* moss, resulting in increased costs and unstable quality, affected the quality of orchids. Polylactic acid, one of the biodegradable polymers, has been developed for many years and was the most widely used material. Dai *et al.* (2010) studied if the polylactic acid could substitute for *Sphagnum* moss. The coconut fiber and peat moss were also compared. The pH and EC of the different media were measured and the effects on the growth of *Phalaenopsis* were studied. The results showed that polylactic acid was a neutral medium; there were significant differences in the leaf growth between polyactic acid and *Sphagnum* moss treatments. All the plants produced one spike successfully in natural flowering season with one spike, no significant difference among treatments. Spike lengths of polylactic acid granules (P1) and pieces (P2) treatments were 89.7 and 91.7 cm, respectively, significantly shorter than that of the *Sphagnum* moss treatment (117.2 cm). The average number of flowers per spike of P1 and P2 were 13.4 and 12.6, respectively, 2-4 flowers less then that of the *Sphagnum* moss treatment.

To clarify suitable media for the mass production of *Phalaenopsis* orchids, five potting media, New Zealand *Sphagnum* moss, coir mixes (CM), CM partially mixed with SM, granular rockwool mixes and perlite mixes were examined using clonal plants of two cultivars, 'Stripe' and 'White Red Lip', at two-month-old and eight-month-old stages. In pots filled with 100% SM, water retention and total porosity were the greatest, while pH and EC were the lowest. In perlite mixes, water retention and total porosity were the lowest. The CM partially mixed with SM yielded significantly greater number of leaves, and fresh and dry weights of both cultivars at the two growth stages. Scanning electron micrographs showed no differences in the shape of stomata or transversal cells of roots among plants in pots with five different media, while leaves of granular rockwool mixes produced abundant wrinkles on their epidermis. The CM partially mixed with SM was a practical potting medium replacing SM, because of its higher potential to support the superior growth of *Phalaenopsis* (Hwang and Jeong, 2007).

In order to evaluate the possibility of using the commercial bark for cultivation of *Phalaenopsis* seedlings, the physical and chemical properties of bark were investigated. Orchid seedling growth under different combinations of culture media, fertilizer application frequency and nutrient solution concentrations also were tested in the screen house. Results showed that the pH value of the bark ranging from 5.0 to 6.12 was not significantly different from that of *Sphagnum* moss. However, the electrical conductivity (EC) and water holding capacity of the bark was lower than those of *Sphagnum* moss. The growth performance of *Phalaenopsis* seedlings in bark and various mixed media was inferior to those in *Sphagnum* moss when commercial fertilizer application frequency and concentration of nutrient solution (667 mg/litre once a week) were applied. When bark and various mixed media for growth of *Phalaenopsis* were tested with various fertilizer application frequency and concentrations of nutrient solution: 667 mg/litre once a week, 333 mg/litre twice a week and 222 mg/litre thrice a week. It was found that the growth performance of plants with 667 mg/litre once a week fertilizer application was superior to others. Under this fertilizer application condition, the growth performance of plants grown in bark with various media was not significantly different from that of *Sphagnum* moss with 667 mg/litre (Chang *et al.*, 2007).

The effects of container opacity and growing media on the rooting and vegetative growth of several clones of *Phalaenopsis* and one clone of *Doritaenopsis* hybrid orchids were studied by Blanchard and Runkle (2008). *Doritaenopsis* cv. White Moon and *Phalaenopsis* cv. Sharon Bay were grown in 12-cm translucent or opaque pots containing bark-based media. After 30 weeks at 29°C, plants in opaque pots formed >7 roots outside of each container, whereas <2 roots per pot developed outside the translucent containers. *Phalaenopsis* cultivars Brother Showpiece, Brother Wild Thing, and Pink Twilight were grown in opaque 12-cm plastic pots containing different ratios of medium-grade Douglas fir bark, medium-grade coconut coir, long-fibred *Sphagnum* moss, coarse perlite, chunky peat, and fine-grade charcoal. Plants were grown for 33 weeks at 29°C with a maximum light intensity of 300 micro mol m-2 s-1. In general, plants grown in a medium consisting of 1:1 (by volume) chunky peat:perlite or 2:1:1 perlite:chunky peat:coconut coir had fewer areial roots than plants grown in a bark-based or 1:1 perlite:charcoal media. Plants grown in the chunky peat:perlite mix or chunky peat:charcoal mixture showed the greatest increase in leaf span. The results confirmed the advantage of using translucent pots and the suitability of chunky peat as an alternative to a bark-based medium.

The efficacy of coir as a supportive base and as an alternative to standard, costly agar-gelled media, was explored as a means for propagating *Cymbidium pendulum* (Roxb.) Sw. *in vitro* through seedling culture. Germination time and morphological differences were compared for embryos cultured on coir fibers soaked in distilled water, in a standard Mitra orchid growing medium with and without agar. Although seeds failed to germinate on coir fibers soaked in distilled water, germination was seen only in the seeds that were grown on the coir fibers soaked in the Mitra nutrient solution, and on agar gelled solution. Time taken for the seeds to germinate and for further growth and development was virtually the same in both the media. Seedlings bearing leaves and roots were obtained in 15 weeks on the coir-based medium. These seedlings were subsequently transferred to the greenhouse in clay pots containing a mixture of coir, charcoal and brick pieces where 90% seedlings survived. The use of coir fiber instead of agar during the initial phase of seed germination and seedling development offers a less costly alternative for cultivating *C. pendulum* and could be used for other orchid species (Aggarwal and Nirmala, 2012).

Pasteurization of Media

Medium may contain harmful disease causing organisms, nematodes, insects and weed seeds, so it should be decontaminated by heat treatment or by treating with volatile chemicals like methyl bromide, chloropicrin *etc.* as mentioned in table 3.1.

Physical propagation facilities such as the propagation room, containers, flats, knives, working surface, benches *etc.* can be disinfected using one part of formalin in fifty parts of water or one part sodium hypochlorite in nine parts of water. An insecticide such as dichlorvos sprayed regularly will take care of the insects present if any.

Table 3.1 : Chemicals with recommended doses for pasteurization of media

Agent	Method	Recommendation
Heat	Steam	30 min at 85°C
Methyl bromide	10 ml/cu. ft. of medium	Cover with gas proof cover for 24-48 hr. Aerate for 24-28 hr before use.
Chloropicrin	(Tear gas) 3-5 ml/cu. ft. of medium	Cover for 1-3 days with gas proof cover after sprinkling with water. Aerate for 14 days or until no odour is detected before using.
Basamid	8.0 g/cu.ft. of medium	Cover for 7 days with gas proof cover and aerate for atleast a week before use.
Formalin	20 ml/l of water (37%)	Apply 2 l/cu.ft. cover for 14 to 36 hr and aerate for at least 14 days.

Care should be taken to disinfect the planting materials before they are moved into the new pots with a fungicide and insecticide combination. Disinfectant solution such as trisodium phosphate or potassium permanganate placed at the entry of the orchid house would help to get rid off the pathogens from the personnel entering the orchid houses. Whatever potting material or mix choose, it must be wetted before use. Otherwise, it will never hold moisture properly and will always dry out. The following points should be kept in mind while preparing potting mixes:-

1. Intend amount of potting material should be pour into a bucket that has about twice the volume of the mixture.
2. Bucket should be filled with hot water. Hot water penetrates the material better than cold water.
3. Let it soak overnight.
4. In the next day, pour out the mix into a colander or strainer.
5. Mix should rinse thoroughly to wash out the dust in it.

Above mix is ready to use. The above media should have following physical and chemical properties which are distinct from field soils:-

1. An ideal medium should have a good balance between physical properties like water holding capacity and porosity.
2. The medium should be well drained.

3. Medium which is too compact creates problems of drainage and aeration which will lead to poor root growth and may harbour disease causing organisms.
4. Highly porous medium will have low water and nutrient holding capacity, affects the plant growth and development.
5. The media reaction (pH of 5.0 to 7.0) and the soluble salt (EC) level of 0.4 to 1.4 dS/m is optimum for most of the orchids).
6. A low media pH (<5.0) leads to toxicity of micronutrients such as iron, zinc, manganese and copper and deficiency of major and secondary nutrients while a high pH (>7.5) causes deficiency of micronutrients including boron.
7. A low pH of the growth media can be raised to a desired level by using amendments like lime (calcium carbonate) and dolomite (Ca-Mg carbonate) and basic, fertilizers like calcium nitrate, calcium cyanamide, sodium nitrate and potassium nitrate.
8. A high pH of the media can be reduced by amendments like sulphur, gypsum and Epsom salts, acidic fertilizers like urea, ammonium sulphate, ammonium nitrate, mono ammonium phosphate and aqua ammonia and acids like phosphoric and sulphuric acids.

Epiphytic orchids can be mounted on tree. Mounting gives them perfect drainage, simulates their natural habitat and can be an easy way to maintain them. Mounting on wood cuttings, fern slabs is ideally practiced for *Phalaenopsis*. For mounting of epiphytes should follow the following steps:-

1. Plant should be placed on a small handful of moistened, squeeze-dried *Sphagnum* moss.
2. Roots of plants should spread around the *Sphagnum* moss.
3. Orchid should be placed on the mount so its centre points down.
4. Orchid should not be positioned with the growing point up. Because it will collect water in the centre of the plant, which can lead to disease that causes the centre and growing point of the plant to rot (and can lead to death).
5. After the orchid is centred properly, it should be wraped either by stainless-steel wire or clear fishing line (monofilament)

around the top and bottom of the moss to hold it in place. In several months, after the new roots have taken hold, wire or line should be removed.

6. The finished mounted orchid is ready to hang in a bright place in a home greenhouse or near a window. Because these mounts drain so rapidly, they need to be watered frequently, sometimes more than once a day during the hot summer months.

Methods of Potting

The simple steps to follow for potting are as under:-

1. The medium is easier to handle if it has been sprinkled with water at the night before actual potting.
2. Remove the orchid from the pot.
3. Remove the old, loose, rotted potting materials and any soft, damaged or dead roots.
4. If the roots are healthy, firm and filling the pot, put the orchid in a pot just one size larger than the one removed it from, placing the older growth toward the back so the new lead or growth has plenty of room. Old clinging roots are severed from the pot with a sharp knife, and the compost, containing the undisturbed plant, is forced from the pot with a screw driver. The roots are shaken free from decomposed material. Dead or long roots are trimmed slightly, since they would be injured in potting. If the roots are rotted and in poor condition, repot the plant in a container of the same or one size smaller than it was removed from. Growing material will stay too damp if placed in too large container, which will result in more of the roots rotting. It is better to add a coarse material like broken clay pots or Styrofoam in the bottom of the pots to improve drainage.
5. Place the plant in the pot so it's at the same depth as it was originally. The new shoot should be level with the pot rim. With the plant held firmly in one hand about one-half inch below the top of the pot, the back of the plant close to the edge of the pot, and the growing end toward the center, potting begins. The potting process for monopodial orchids is very similar to the *Cattleya* process, except that the orchid should be placed in the center of the container, rather than toward the back.

6. Press the fresh potting material into the pot and around the orchid roots with your thumbs and forefingers. The orchid should be secure in the pot so it doesn't wiggle otherwise, the new roots won't form properly. A piece of *Osmunda* is held at a right angle to the top of the pot with the grain also running at a right angle to the top; with a little flip of the potting tool, so that it never touches roots or plant, the chunk slips neatly in place within the pot, to rest vertically and against the plant. This procedure is repeated, from side to side, until all the gaps are filled, each chunk being made shorter as the edge of the pot is approached. If the plant leans to one side, can be straightened by adding potting mix to prop it up. Plants that are held weakly may wobble and roots may be broken. Firm potting also prevents plants from falling out of dry and shrunken compost and keeps water from settling in holes and rotting the roots. It is wise to spread the roots, packing potting mix among them for aeration.
7. Place a wooden or bamboo stake in the centre of the pot, and tie up the new and old leads with soft string or twist ties.
8. Bottom hole of pot should be enlarged to give the drainage and aeration needed by epiphytes.
9. Some growers use a single piece of crock in the bottom, while others fill as much as one-third of the pot with crock bits of brick, coke, haydite or gravel may be used instead; in any case, the hole should not be tightly covered.
10. The potting material should be worked in without letting the potting tool touch the roots. Success depends largely on the root being disturbed as little as possible. A further and equally necessary precaution is to take care that the rhizome lies along the top of the potting material and particularly that the dormant eye is not buried.
11. When the pot is full, the stake is inserted at the back of the plant and the bulbs are tied up. They should be tied firmly enough to prevent a movement that will break off roots as they form, but not tight enough to choke the bulb or to force it so suddenly into an upright position that it will snap from the strain.
12. The label is placed toward the back of the pot and the pot is then set aside in a place especially designated for newly potted plants.

Fig. 3.5 : Various steps of potting

Repotting

Repotting is needed when the plants have over grown the pots or when potting medium has deteriorated. Repotting is done every 2 to 4 years and is best indicated by the plant bulbs filling the pot or the plant not growing well over the past year. A healthy plant should grow 1 to 2 new bulbs each year from each bulb that grew the previous year. An overgrown orchid (*pot-bound*) will dry off much more quickly than one that has plenty of space in the pot. When pot space is limited, there is less potting material to hold onto the water, so the overgrown plant quickly uses it up. When repotting, the roots should be teased apart to remove old potting mix and then dead roots are to be removed. If required a plant may then be divided into two or more plants by twisting and tearing the bulbs apart. For best results each plant should keep as large as possible otherwise flowering would be retarded for 1 to 2 years. A plant has three types of bulbs:-

1. **Old back bulbs without leaves** : These bulbs are not important to the plant and act as a reserve food supply for emergencies. It is advisable to leave one of these on each divided plant. Back bulbs can make new plants but they may take years to flower.
2. **Old bulbs with leaves** : These bulbs support the new growth and may produce flowers for a number of years depending on

the variety. When dividing, the plant must retain at least two old bulbs or have one back bulb attached to be able to re flower next year.

3. **New leads or bulbs** : These are the youngest bulbs on the plant and it is from these that the flowers and most new growth come. It should be divided with at least 1 old bulb and one back bulb to ensure the plant flower in the following year.

The ideal time to repot most of the orchids is when the plant starts new growth, usually right after it flowers. With certain orchids like the *Cattleyas*, swelling at the base of the plant might be seen, which is the beginning of the new lead or shoot that will form the next stem, leaf, and flowers. It occur when orchids are putting out new roots. Pot-bound *Cymbidiums* will often be underwatered which can reduce the vigor and blooming capability. When a *Cymbidium* fills the pot and starts to grow against or over the edge, it should be repotted even if the mix is not broken down. *Cymbidiums* generally need to be repotted about every three years under normal conditions, whereas, *Vandeceous* orchids and *Paphiopedilum* should not be disturbed unless very necessary. *Cymbidiums* should be repotted just after the old blooms have fallen off (spring seasion). If the old mix is broken down, the plant should be repotted as soon as possible to minimize root rot. Broken down mix tends to hold too much water and will reduce the air movement around the roots which often causes the older roots to die back. If the mix is very soft, the plant probably needs repotting. Finger may be inserted in the drainhole of a plastic pot to test the condition of the mix. If the mix is soft enough so a finger can enter the mix easily, it should repot. Vigorous plants will often outgrow the pot and may even break the pot before the mix starts to break down. How well a plant tolerates repotting, that can generally be judged by the plant's growth rate, slow growers being less tolerant of disturbance than others.

During repotting large plants may either be divided into muliple plants or kept together and potted up into larger pots for multispike specimens. Plant should take out of the pot and remove the old bark for repotting. Dead or compacted roots from the bottom of the root-ball should be removed. Between plants cuttings tools should be sterilized. The dormant bulbs (back bulbs) can be removed without destroying the strength of the division. These can be potted up to resprout and bloom in two to four years. A pot size which will allow the plant to grow unrestrained for three to four years should be selected. During placing the plant in the pot, bulbs should be positioned so they sit just a little into the surface of medium and so that the newest growth is in the center of the pot, with room to grow. Plants should be potted firmly by pouring the medium around the roots while tapping the sides of the pot and pushing down on the top with thumbs or a potting stick. Most large *Cymbidium* plants have leafless bulbs (backbulbs) toward the center of the plant. The green divisions of the plant will be connected to the backbulbs, and the location of the backbulbs usually marks the natural dividing points for the plant. Overgrown or potbound *Cymbidiums* are often difficult to remove from the pot. With clay or redwood pots, the root ball can be separated from the pot by working a long thin knife down to the bottom of the pot and cutting it away from the pot. Plastic pots will sometimes deform from the pressure of the roots inside and the only way to remove the plant is to cut the pot down one side. If the roots are a pale yellow or white color and firm, the root system is in good health. Brown and soft roots indicate the plant has either been in the pot too long or it has been kept too wet and the roots have rotted as a result. Some loss of roots is normal, but the loss of too many will slow down the growth of the plant. Once, the individual sections are separated, clean as much of the old mix and dead or broken roots from the rootball as is practical without damaging the good roots and the new roots. Be careful with new growths or spikes which may be emerging from the base of the psuedobulbs. Some root pruning may be necessary to fit the plant easily into the new pot. This does not hurt the plant, and will encourage new root growth. A mound of moist potting mix should be placed in the centre of the pot. Roots should be spread over the mound of mix and work mix in among the roots. Roots should not be balled in the centre of the pot. Pot should be filled with mix and should be shaked or pat the pot vigorously or tap the pot firmly on the potting bench to help the mix settle into the roots. Tamp the mix in around the edge of the pot firmly with the potting stick. The mix should tight in the pot, not loose. Root growth in the pot will usually fill some, if not all, of this gap in time. The last step in potting the division is to put a tag in the pot to identify the plant.

Temperature Management

The different orchid species are classified according to the temperature in which they grow and can thrive. *Cymbidium* orchids flower grows best if given a spot in bright light. In fact, it can be safe even bringing them outdoors to a shady spot for the summer. This can be cool-climate orchids, intermediate-climate orchids, and warm climate orchids. The more popular individual orchid plants are arranged in these categories as follows:

The warm climate orchids: These orchids occur naturally in the tropical areas. Warm climate orchids require warm, humid conditions and slightly cooler winter temperatures. They include the *Aerangis, Angraecum, Ascocenda, Calanthe, Catasetum, Cattleya, Paphiopedilum* (lady slipper orchid) *Phalaenopsis* (moth orchid) *Vanda* and *Vanilla*. These orchids prefer warmer temperatures, day-time temperature high of 29°C with corresponding night-time lows not to go below 15.5°C.

The intermediate climate orchids: The orchids classified as intermediate-climate orchids grow naturally in climates that are neither hot nor cold. They occur naturally between temperate and tropical zones. The optimum environment for these orchids has a temperature that is subtropical. They prefer a day-time temperature of 27°C and night-time does not belows 13°C. In this group most of the orchids prefer bright, but not direct sunlight and will be most comfortable in a semi-shaded area. *Oncidium* and their intergeneric hybrids, warm-tolerant *Cymbidium*, ever-popular *Miltonia* and *Miltoniopsis* orchid are notable examples of intermediate climate orchids. Amongst the *Paphiopedilum* orchids, *Maudiae, Cochlopetalum, Multiflorals, Brachypetalums*, green-leaf species and *Parsvisepalum* orchids are the main hybrids. *Lycaste* and *Anguloa, Bulbophyllum* and *Cirrhopetalum* are other intermediate climate orchids.

The cool climate orchids: The orchid plant that comes to mind immediately is the *Cymbidium* orchids. These are found to be the largest potted orchid. The following examples are naturally cool-climate orchids: *Odontoglossum Alliance* and the different types of *Odontoglossum, Disa uniflora* and its hybrids, *Pleione, Zygopetalum, Dendrobium, Masdevallia, Brassia* and *Coelogyne* orchids. The examples include *Odontoglossum* and *Masdevallia*.

Terrestrial orchids that resort under the cool-climate are the *Cypripedium*, Oriental *Cymbidium, Diuris Pterostylis, Goodyera, Ludisia* (Jewel Orchid), *Epipactus* and *Spiranthes* orchids. They need 24°C day temperatures with 10°C night temperature.

Table 3.2 : Temperature for orchid houses

Months	Warm Growing Orchids (Temperature°C)		Intermediate Growing Orchids (Temperature°C)		Cool Growing Orchids (Temperature°C)	
	Day	Night	Day	Night	Day	Night
January	18-21	15	15-18	12	10-12	7
February	18-21	15	15-18	12	10-12	7
March	18-21	15	15-18	12	12-15	10
April	18-21	15	15-18	12	12-15	10
May	21-23	18	18-21	15	15-18	12
June	23-26	21	21-23	18	15-18	12
July	23-26	21	21-26	18	15-21	12
August	23-26	21	21-26	18	15-21	12
September	23-26	21	21-23	18	15-18	12
October	21-23	18	18-21	15	15-18	12
November	18-21	15	15-18	12	12-15	10
December	18-21	15	15-18	12	10-12	7

Temperate orchids may be considered as temperatures tolerant having wide range of tolerance from -10C to 39°C, require a certain periods of low temperature for blooming. Summer day temperatures of 24°C to 29°C and night temperatures of 10 to 17°C are ideal. In winter, day temperatures of 17 to 24°C and night temperatures of 7 to 12°C will encourage blooming. However, heat-tolerant miniatures are a few exception having flower-spikes in the autumn and differences between day and night temperatures plays a key role. Indoor conditions or climates that stay persistently warm without cooling off at night (below 11°C in winter) are not suited for blooming standard temperate Orchids.

During heat spells, it is essential to boost the humidity by watering and misting more frequently. During cold spells, temperate orchids can take 0°C but should be given some protection in case the temperature drops further. Damage to spikes is occur at -3°C and to the plant at -4°C. To guard against effects of the cold, plants should be moved up against the house or under a tree. Warmer night temperatures for any prolonged period can cause developing buds to turn yellow and drop off; however, once flowers have opened, no longer require the cool and may be brought indoors for show. The specific temperature requirement of *Cymbidium* are as under:

Effect of temperature in Vegetative phase: During the vegetative phase, plantlets grow best at temperature of 18°C at night and temperature of 20-25°C during the day. During the winter period (late October to late February): 16-18°C at night and 18-20°C during the day. Once the plants have reached the size indicating that they can start flowering within the coming season, they should be subjected to different temperatures according to their flowering period.

Johnson and Kane (2012) investigated the role of simulated south Florida temperatures and illumination (dark and 12 h photoperiod) in regulating germination and seedling development using asymbiotic seed germination assays of *Bletia purpurea*. The slowest germination was observed at simulated winter temperatures (22/11°C). Illumination initially inhibited germination and development, but resulted in equal or greater development by week six. Germination under 22/11°Cwas strongly inhibited by illumination, indicating an interaction between temperature and light sensing systems. Yao *et al.* (2011) evaluated the conditions of best temperature and light intensity of greenhouse/ structure growing *Phalaenopsis* by the measurement of daily carbon balance with CO_2 flux using eddy covariance technique for the purpose of improving orchid management in greenhouse cultivation. The results indicated that the maximum carbon balance can be reached at the day/ night temperature of 25/25°C on leaf basis, and in range of 24-27°C on the base of greenhouse as a whole. The relationship between night temperature and CO_2 fixation was stronger than day temperature and respiration. The light and dark reactions of carbon fixation for crassulacean acid metabolism (CAM) plant were separated. The suitable light intensity range for *Phalaenopsis* leaves was 300-500 micro mol m^{-2} S^{-1} from the measurements of daily carbon balance and fluorescence parameters. In addition, higher daily carbon balance occurred in 1 day/ night cycle per day (148.3 micro mol CO_2 m^{-2} day^{-1}) compared with 2 cycles (127.1 micro mol CO_2 m^{-2} day^{-1}) and 3 cycles (92.4 micro mol CO_2 m^{-2} day^{-1}). Latent heat was not significantly correlated with temperature and light intensity because the pad-fan interfere the vaporing of water in greenhouse.

Table 3.3 : Temperature requirement for early flowering

Phase	Period	Day Tem. (^{0}C)	Night Tem.(^{0}C)	Ave. daily Tem.(^{0}C)
Bud initiation	Dec. - Jan.	13 - 15	11 - 13	13
Spike elongation and growth	Feb.-August	20 - 24	20 - 16	20
Flowering	Sept-Nov.	18 - 22	15 - 18	18

Table 3.4 : Temperature requirement for mid flowering

Phase	Period	Day Tem. (^{0}C)	Night Tem.(^{0}C)	Ave. daily Tem.(^{0}C)
Bud initiation	Jan. - April	13 - 15	10 - 12	13
Spike elongation and growth	July - Dec.	20 - 24	20 - 16	20
Flowering	May - June	18 - 20	13 - 15	16

Temperature Management

The temperature of an orchid house can easily be regulated by using appropriate heating system. The amount of heat an orchid house will require during coolest day should be known in order to provide adequate heating equipment.

Table 3.5 : Temperature requirement for late flowering

Phase	Period	Day Tem. (^{0}C)	Night Tem.(^{0}C)	Ave. daily Tem.(^{0}C)
Bud initiation	Nov. - Feb.	12 - 15	11 - 13	13
Spike elongation and growth	April - Oct	20 - 24	20 - 16	20
Flowering	Feb. - Mar	14 - 16	12 - 14	14

The actual size of the system decided what form of heating to use. There are special considerations in the heat applied, air directed to air to influence the plant canopy and heat applied to floor to influence the root system. Heating equipment in addition to providing the optimum temperatures for crop growth and yield also can be employed in combination with controlled venting to dehumidify the orchid house. In cold climates of Himalayan Region at high altitudes the heating of orchid houses is advisable for getting better produce. Heating systems can be of the following types:-

Odontioda is a cool-growing, sympodial epiphytic genus of orchids originating from the Andes Mountains of South America. Several hybrids are commercially grown as potted flowering plants for their brightly colored flowers and compact growth habit. Blanchard and Runkle(2008) quantified how constant and fluctuating day/night temperatures influenced inflorescence initiation, time from visible inflorescence (VI) to flower, and pseudobulb development. *Odontioda* George McMahon 'Fortuna' and Lovely Penguin 'Emperor' were grown at constant temperature set points of 14, 17, 20, 23, 26, or 29 ^{0}C and day/night (12 h/12 h) temperatures of 20/14, 23/17, 26/14, 26/20, 29/23, or 29/17 ^{0}C. Plants were grown in glass greenhouses under a 12-h photoperiod and a maximum irradiance of 500 micro mol.m^{-2}.s^{-1}. Within 6 weeks, heat stress symptoms such as leaf necrosis were observed on plants grown at a day temperature of 26 ^{0}C or greater regardless of the night temperature. After 20 weeks, 90% or greater of both clones had a VI when grown at a constant temperature of 14 or 17 ^{0}C. Plants grown at a constant temperature of 17 ^{0}C had the greatest pseudobulb diameter with a mean increase of 3.5 to 4.0 cm. In all treatments, a minimum pseudobulb diameter was required for uniform inflorescence initiation; pseudobulbs with a diameter of 5.5 cm or greater developed a VI in 93% of plants. Data for time from VI to open flower were converted to a rate, and a thermal-time model relating temperature with inflorescence development was developed. The base temperature and thermal time for VI to flower in George McMahon 'Fortun' and Lovely Penguin 'Emperor' were estimated at -0.1°C and 1429°C.d-1 and 0.8°C and 1250°C. d^{-1}, respectively. This information could be used by commercial orchid growers to assist in producing potted flowering *Odontioda* orchids for specific market dates.

Commercial production of the pansy orchid, *Miltoniopsis*, requires an understanding of how to regulate the flowering process. Lopez and Runkle(2008) quantified constant and fluctuating day/night temperatures and pseudobulb maturity influencing time to visible inflorescence (VI) and time from VI to open flower. Average time from the end of vernalization to appearance of the first VI was 89 and 8 days for plants with pseudobulb diameters <=1.5 or >=3.1 cm, respectively, regardless of forcing temperature. Average time from VI to flowering was 122, 97, 78 or 62 days at constant temperatures of 14, 17, 20, or 23^{0}C, and 93 or 76 days at day/night temperatures of 20/14 or 23/17°C. Flower diameter increased from 7.4 to 8.6 cm as forcing temperature decreased from 23 to 14°C, but temperature had no effect on flower number and inflorescence height. We conclude that pseudobulbs of

Miltoniopsis Augres 'Trinity' must be sufficiently mature (diameter >=3.1 cm) before cooling for rapid flowering and a high flowering percentage. Once plants are induced to flower, subsequent development is a function of the average daily temperature.

Constant air temperatures above 26°C inhibit flowering of most *Phalaenopsis* hybrids, and a 25/20°C day/night temperature regime is used commercially to induce flowering. However, the relative promotion of flowering by constant versus fluctuating day/night cool temperatures (<25°C) has not been well described. *Phalaenopsis* Miva Smartissimo x Canberra '450' and Brother Goldsmith '720' were grown at constant temperatures of 14, 17, 20, 23, 26, and 29°C, and day/night temperatures of 20/14, 23/17, 26/14, 26/20, 29/23, and 29/17°C. Plants were grown in glass greenhouses with a constant photoperiod of 12 hours, and maximum instantaneous irradiance was <=150 micro mol m-2s-1 (8100 lux). After 20 weeks at the various temperature set points, 80% of plants of both cultivars had visible flower spikes when grown at 14, 17, 20, 23, 20/14 and 23/17°C. None of the plants were reproductive within 20 weeks when grown at temperature setpoints of 26, 29, 29/17 and 29/23°C. These results suggest that temperature during the day and night both influence flowering of these two *Phalaenopsis* orchid hybrids (Blanchard and Runkle, 2005). Low temperature decreased the growth rate and the endoreduplication transition rate. A longer growth period compensated for the decreased growth rate so that the distribution of ploidy levels similar at both growth temperatures. Low temperature decreased growth rates and endoreduplication transition rates for both orchid species. These results are consistent with previous studies that endoreduplication cycle may use the same molecular machinery for their DNA replication as that used during the S-phase of the mitotic cell cycle. In both species, the average C value is positively correlated with the average cell fresh weight. However, 15°C increased the ratio of average C value to average cell fresh weight in P. aphrodite but decreased the ratio in O. varicosum. Although ploidy level plays a role in final cell size, growth temperature modulates the effect ploidy level on final cell weight (Ching *et al.*, 2007)

Phalaenopsis orchids require a day temperature of 26°C or less to initiate inflorescences, whereas the night temperature has little or no effect on inflorescence initiation. Newton and Runkle (2009) determined the duration of high temperature required each day to prevent inflorescence initiation of four *Phalaenopsis* and *Doritaenopsis* clones. In Years 1 and 2, mature potted plants were grown in separate greenhouse sections with five daily durations at 29°C: 0, 4, 8, 12, or 24 h. The high temperature was centered in the 16-h photoperiod (0600 HR to 2200 HR) and the remainder of the day was at 20°C. Exposure to 29°C for 8 h or longer inhibited inflorescence initiation of *Phalaenopsis* Miva Smartissimo x Canberra 'Mosella' and *Phalaenopsis* Brother Pink Mask x Brother Success 'Explosion', but *Phalaenopsis* Baldan's Kaleidoscope 'Golden Treasure' and *Doritaenopsis* 'Newberry Parfait' required exposure to 29°C for 12 h or longer to inhibit inflorescence initiation. Flowering was completely suppressed only when high-temperature exposure time was continual for *Doritaenopsis* 'Newberry Parfait' and *Phalaenopsis* Baldan's Kaleidoscope 'Golden Treasure' and 12 h for *Phalaenopsis* 'Mosella'. Plant leaf span generally increased as duration of exposure to 29°C increased, but high-temperature exposure had few or no significant effects on flowering characteristics of flowering plants. These studies indicate that as few as 8 h of high temperature can prevent flowering of some *Phalaenopsis* hybrids, whereas others require greater than 12 h of high-temperature exposure.

Daytime air temperatures above 26°C inhibit flower initiation of *Phalaenopsis* regardless of the night time air temperature. This opens possibilities for energy saving by lowering the night time air temperature during the vegetative growth phase. As chilling injury can occur after a few hours exposure to cool temperature conditions, growers prefer to cultivate *Phalaenopsis* at daily temperatures between 25 and 30°C. A better understanding of the low night temperature effects on the photosynthetic physiology of *Phalaenopsis* would improve production of these orchids in terms of greenhouse temperature control and energy use. Therefore, the *Phalaenopsis* hybrid 'Hercules' was subjected to nighttime air temperatures of 27.0, 24.0, 21.1, 18.0, 15.4 and 12.8 ^{0}C, while during the day temperatures of about 27.5°C were maintained. Photoperiod was set to 9 h with non-saturating white fluorescent light (217.3+or-2.8 micro mol PAR m^{-2} s^{-1}). The chlorophyll fluorescence temperature response curves revealed a significant reduction of the photochemical quenching (qP) and quantum efficiency of PSII electron transport (Phi PSII) when plants were exposed to night temperatures below 21.1°C, but maximum quantum yield of PSII photochemistry (Fv/Fm) suggested no PSII damage for these night temperatures. Early morning determined leaf acidity, which can be used as a measure of nocturnal malic acid production, and leaf water potential were lowest at night temperatures of 21.1 and 24.0°C. However, maximum net CO_2 uptake in the late night (2 h before daytime) was observed between 15.4 and 21.1°C. Furthermore, PSII photochemistry required only 5 h for full recovery of all night temperatures imposed. Our results suggest that

Phalaenopsis can tolerate relatively cool night temperatures during the vegetative growth stage (Pollet *et al.*, 2011).

Light Management

Light is essential for orchids in many ways - it provides the energy necessary for the manufacture of carbohydrates and it helps the mature plant to flower. Light requirements of orchids vary from genus to genus The most popular orchid species are the *Phalaenopsis* (Moth Orchid), *Cymbidium* and *Dendrobium* while *Vanilla* Orchid is the most highly produced orchid species. Some tropical varieties would like longer of sunlight while other orchids prefer shorter periods of time exposed to direct sunlight. Other orchids prefer shorter periods of exposure to sunlight in the winter and in summer require about 12 to 14 hours of sunlight. Very few orchids will tolerate direct sunlight, except may be for an hour or two after sunrise and an hour or two before sunset. If orchids grown at a south or southeast or southwest location, in a sunroom or in a orchid house, there should be some shade, at least for the brightest part of the day. A balanced approach should be to put the orchids in the morning sunlight more and avoiding the mid day sun since it can wither the orchid and might caused its death if exposed to the midday sun too long. Without proper light there will be weak growth and no flowers in orchids. Good light also brings out the best in orchid colours; shady conditions result in greener, muddier-colour blooms. The foliage should be yellowish-green in colour. However, too much light will result in a pale yellow colour and if burned, a black spot at the arch of the leaf can be seen and if severely burned, the leaf may bleach white. If too shady, plants will be dark, lush green but will bloom less or not at all.

Light requirements of orchids are variable depending on the species of the plant. Natural light comprises ultra-violet, violet, blue, green, yellow, orange, red and far-red light and there also is invisible infra-red light. The green and yellow parts of light are important for our vision, but these colours are not important for plants. Blue, red, far red and infra-red light are the important parts of light for plants. The blue and red rays of light are necessary for plants to photosynthesize. Red light affects maturation, ripening and dormancy of plants. Far red is necessary for plants to grow. Far red and infra-red affect cell activity stimulate plant growth and development of new leaves and roots.

Heat drives transpiration, which results in the upward movement of water and minerals within the plant. As water evaporates from the leaves it is replaced from the roots. Without light and transpiration plants cannot grow. Thus light is one of the most important climatic factors in growing orchids. Light is measured differently depending on what part of the light spectrum is being measured. The spectrum is measured in lumens. The lumen is the metric unit of light intensity and the term lux refers to the number of lumens per square metre of surface of area and Klux stands for kilo lux or 1,000 lux. The total light spectrum coming from the sun is 400 to 1100 nanometer wavelengths and is also measured in units of watts/m^2. Generally on a bright sunny day the light level might be 100 Klux in comparison to 8 Klux in a cold winter day. Since the total amount of light is same at any given area, the places away from the light source have larger the illuminated area and less amount of light will fall on any given square meter of area. Light meter can measure the light level needed by orchids.

Daily Light Integrals is a direct measure of the number of particles of light delivered to the plant. The orchid light requirement and growth is closely correlated with daily light Integrals. Under low light conditions, plants try to increase light interception by increasing the area of individual leaves and thus leaves tend to be thin and pliable. In contrast, individual leaf size decreases and leaf thickness increases under high light conditions. These leaves are thick and have accumulated a lot of starch. However, under high light conditions, plants will have more leaves. So, even though they are individually smaller leaves, the leaf area of the entire plant increases as daily light integrals increases.

Very low light conditions (5.5 to 11 Klux) results in poor plant growth and flowering during this condition plant lacks sufficient energy to produce a high quality plant and there may be insufficient light to support flowers, so flowering can be delayed, flowers may be very small, few flowers may be produced, or the plants may stay entirely vegetative. Supplemental lighting is very beneficial under very low light conditions. Under cool growing conditions plant quality can be enhanced but in contrast, high temperatures during low light conditions result in poor quality growth thus supplemental lighting is beneficial under such conditions. Commercially acceptable plant growth of orchids is under moderate light conditions, 22 to 44 Klux. Potted orchids perform very well under moderate light condition. It is relatively easy to manage watering under moderate light conditions compared to higher light. Once the plant has sufficient light to support flowers, increasing the light level further has little effect on time to flower. The potential benefit of supplemental lighting is limited under moderate light conditions.

The quality orchids are usually produced under high light conditions, 44-65 Klux. High light conditions provide extremely well-branched bushy plants and high flower numbers. Root growth is proportional to shoot growth, so high light conditions also produce excellent root systems. In very higher light levels (65 to 110 Klux) plant temperatures can be cooled outdoors due to increased air movement, lower relative humidity and thermal cooling due to the exposure to the open sky. Excessively high light may result in a change in leaf orientation and shape. Plant grown under excessively high light produces vertical and curled leaf blade in order to avoid light interception. Excessive light can also result in heat stress or sunburn of some species. Sunburn is most likely to occur when a plant has not been acclimated to high light conditions. Plant water use and evaporation increases as sunlight increases, so water management can be more difficult under high light conditions. Root death can also occur on the south side of dark-colour containers due to excessively high soil temperatures resulting from direct outdoor sunlight.

Temperate orchids require filtered sunlight during the spring, summer and fall. In warm summer climates, a high level of light may cause reduced blooming, thus additional shade may be necessary. In cooler climates, *Cymbidiums* prefers full sun. The determination of amount of light is done by size and shape of the orchid leaves, plants that have long, slender leaves have the greatest need for light and should be placed in an area of strong light for most of the day. If the leaves are softer and wider, the orchid will tolerate less light and may even like a shady spot. The growth of orchid would also indicate its need for more light or less. If growth of the new leaves is same, size and shape as the older one, then they are getting the right amount of light. If they are elongated, soft and dark, they need more light.

In general exposure to direct sunlight is detrimental to the orchids. Orchids need bright filtered light (55% shade) all day or full morning sun. *Cypripedium* and *Phalaenopsis* needs from 2.15 to 3.35 Klux of light and hence are to be kept in the comparatively more shaded regions of the orchid house. However, genera *Vanda, Arachnanthe etc.* thrive well under 85 Klux of light. In such cases as much of sunlight as the plant can withstand without causing injury to the leaves, can be beneficial to plant. In between these two extremes, the optimum orchids generally require is between 32.5- 65 Klux. During the growth season *Cymbidiums* treasure ample sunlight or almost 50 percent shade. It can be moved in open air in the summer time, except they are not to be set up into direct sunlight. If they are grown in an orchid house then 50 per cent to 60 percent shading is needed during the summer months to prevent leaf burn and overheating. For outside growing in summer, early morning sun is ideal but has to provide some shade against the strongest sunlight. Some hours of slight morning sunshine and shaded afternoons encompasses the complete environment. On the basis of light requirements orchid plants may be grouped into three groups:

1. **Low Light** = 11 - 22 Klux, two hours of filtered sunlight per day. These plants can be grown indoors under indirect light or outdoors under shade such as *Phalaenopsis* & *Masdevallia.*
2. **Medium Light** = 22 - 32.5 Klux, four hours of sunlight per day. Outside, moderately sunny windows or under lights would be suitable such as *Miltonia & Odontoglossum* required medium light whereas, *Cattleya, Vanda, Oncidium* & *Cymbidium,* needs medium to high light.
3. **High Light** = 32.5 - 43.3 Klux, six hours of sunlight per day. In a sunny window or outdoors would be ideal for these plants such as *Dendrobium.*

The light requirement for various growth stages of *Cymbidium* is as under:-

Vegetative phase

During the vegetative phase *Cymbidium* require exposure of 20 to 30 Klux of light for vigorous growth. During the winter months (and also for purposes of extending day length), providing them with lighting (35-40 Klux) during this stage of growth is recommended. If light intensity exceeds 30 Klux, it will have to be reduced by a light whitewashing or using a shading system.

Spike initiation phase

As soon as the plants become larger, the economic benefits of providing lighting for growth decrease. The maximum light intensity for *Cymbidium* depends on temperature and humidity and is should be from 35 to 50 Klux. If light intensity exceeds 50 Klux, it will have to reduce by a light whitewashing or using a shading system.

Bud initiation phase

Light is important for the initiation of flower buds. At least 30 Klux should be provided for bud initiation. For spikes initiation light intensity should be below 50 Klux (or more than 750 Watts/m^2).

Flowering

Exposing the crop to a minimum of 30 Klux during bud initiation is important for the initiation of flower spikes. During flowering, expose to a maximum of 35-50 Klux (the exact amount depending on temperature and humidity) is found beneficial. Reducing light intensity by means of whitewashing, *etc.* is particularly important for spike elongation.

Symptoms of Over Lighted And Under Lighted Orchid Plants

Symptoms of low light

- The leaves of the orchid plant are coloured dark green and they have no brightness or added glow in them.
- The orchid plant has never bloomed or it has rarely bloomed.
- The growth of the plant is incremental or very little growth is shown.
- The orchid plant lacks sturdiness, unhealthy, and lacks any strength.
- The root system is thinly spread and very little.

Symptoms of strong light

- The leaves may exhibit burn spots.
- The spots turn yellow or exhibits brown dry patches on the leaves. They could also be rough and slightly raised.
- The leave should feel cool to the touch and not be hot.
- A plant exposed to strong light will eventually wither and die.
- The plant growth is too little and the leaves lack lustre and turn yellow.
- On some orchids, the colour is bleached out of the leaves. In other varieties the leaves turn red.
- The leaves starts falling off and possibly will turn black

Artificial Light Management

Artificial light should be provided if there is insufficient natural light for orchids. In most cases where winter becomes a limiting factor to sunlight, the exposure to artificial lights becomes a necessity to continuous growth in the orchid plant. There may be an adjustment period of two years to developing under florescent lamps/artificial lights for some orchids after that they blooms more regularly and frequently depending upon varieties of orchids and lighting requirements. High artificial light intensity can intensify the development of seedlings. In their natural habitat some orchids are exposed to a bright tropical sun while others live within shady forests. It is better to simulate this environment with the same lighting levels they would normally receive in nature. Fluorescent as well as incandescent lights have their own spectral emission characteristics, may be used for proper growth and flowering of orchids. Florescent lights produce the spectral curve of photosynthesis having red and blue rays, purplish in colour and low in yellow and green spectrum with very high illumination intensity. On the other hand incandescent, are rich in far reds, stimulating photoperiodic responses. Only low levels of intensities 0.55 to 1.1 Klux or even as low as 25 - 110 lux are used only for short periods for light management in most of cases. The ratio between the two types of illuminations received by the plants is important in the production of flowers, and this ration varies with species. Usually best growth and flowering have been obtained by using a combination of red (6000- 7000 Angstrom) and blue (4000- 5000 Angstrom) fluorescent bulbs with 5 to 10 Per cent visible radiation from incandescent bulbs. Reflectors may also used to increase the intensity of light when tubes are mounted. Fluorescents should be kept at near the plant but incandescent should not be kept so near because of heating up of the plant and atmosphere. With the help of artificial light management it is possible to grow orchids in cellars where no sunlight penetrates. When electric lights are used in orchid houses, care should be taken and green lights of low intensities may be preferred for this purpose, because they interfere the least with the metabolic processes of the plant. The majority of orchids are day-neutral i.e. they are not influenced by day-length. Because of this, any amount of extra illumination given could only be beneficial to plants at initial stage. By giving 24 hours of fluorescent illumination, plantlets developed from meristem culture, are made to grow and attain maturity faster. Shade cloth is commonly used to reduce natural light &temperature into a orchid house. Shade cloth varies in 30, 40, 50, 60, 70, 80, and 90 percent shading.

High light requiring plants (25 to 30 per cent shade), such as *Cattleya,* its intergenernic hybrids, *Vanda* and *Ascocenda* can also stand higher temperatures. These orchids should be placed on the west side of the orchid house or hang them high where they can receive well light for long time.

Low light or shade requiring plants (70 to 85 per cent shade), such as *Phalaenopsis* ,*Phaius, Paphiopedilum* and *Ludisia* also need cool temperatures. These orchids should be placed near the evaporative cooler or wet wall. Medium light requiring plants (20 to 60 per cent shade), such as *Dendrobium, Doritaenopsis* and *Oncidium* needs good light full time. These plants should be placed on the bench where they can receive full light. Arrangement should be made in such a way that low light requiring orchids are placed on the shady end, mid light orchids in the middle and high light at the sunnier end.

Prevailing wind, orchid house orientation to the sun, trees and buildings influences the sunlight and temperature. The temperatures should not go above 32°C for low and mid light plants. High light requiring plants can withstand higher temperatures as long as humidity is high.

Newer to the artificial-light choices are high-intensity-discharge lights. They emits more light than fluorescent lamp and these are efficient in production of light in places where orchids requiring higher light intensities than fluorescent lamps and a greater working distance between the lights and plants. These lamps may be of 100, 250, 400, 600 and 1,000 watts, emitting 11 to 100 Klux. The amount of light penetrating the orchid house is determined by the glazing material used geographic location and surrounding.

Sophro*LaelioCattleya* and *Cymbidium* plants, grown in a shaded greenhouse kept between 18 and 28°C, were exposed to full sunlight (FS, 0.61 $kW.m^2$) on 21 October, while others were grown under 70% (high radiant flux), 40% (control) and 10% (low radiant flux) sunlight from May to October. Activities of superoxide dismutase (SOD), L-ascorbate peroxidase (APX) and catalase (CAT) were higher in Sophro*LaelioCattleya* than *Cymbidium*, with enzyme activities increasing slightly during the day in the former and at the beginning of the night in the latter. FS reduced SOD activity, but significantly enhanced CAT and slightly promoted APX activities, in Sophro*LaelioCattleya* leaves. The same radiant flux on *Cymbidium* leaves resulted in a large decrease (attributed to light stress) in SOD and CAT activities, while a slight decrease was observed in APX activity. One month after the onset of shading, SOD and CAT activities in Sophro*LaelioCattleya* leaves at the high radiant flux decreased and remained at low levels. CAT activity, however, recovered slightly with further cultivation. In *Cymbidium* leaves under a high radiant flux, SOD, APX and CAT activities decreased during the first three months of the experiment. At the low radiant flux, enzyme activities in leaves of both plants increased until further cultivation, with the exception of SOD activity in Sophro*LaelioCattleya*. Foliar chlorophyll contents in both species decreased with increasing radiant flux, exhibiting high positive correlations with enzyme activities. Ratios of chlorophyll a to chlorophyll b and beta -carotene contents of Sophro*LaelioCattleya* grown at the high radiant flux were low after one month of shading, but became equal to those of the control after three months. In *Cymbidium* leaves, the ratios and beta -carotene contents decreased slightly at the high radiant flux. At the low radiant flux, chlorophyll and beta -carotene contents increased. Thus, Sophro *LaelioCattleya* plants have a greater potential to adapt to high radiant fluxes than *Cymbidium* plants. However, these two orchids readily acclimatize to low radiant fluxes but not to high levels of radiant flux(Li *et al.*, 2001).

The effects of light generated by red and blue light-emitting diodes (LEDs) on callus induction from protocorm-like body (PLB) segments, callus proliferation and PLB formation from callus in *Cymbidium* orchid were investigated. The cultures were placed in 'LED PACK 4' incubators under different ratios of red and blue LED light (100% red LEDs, 75% red LEDs+25% blue LEDs, 50% red LEDs+50% blue LEDs, 25% red LEDs+75% blue LEDs, and 100% blue LEDs) at 45 micro mol m-2 s-1 with a 16-h photoperiod or put under plant growth fluorescent (PGF) light with the same light intensity and photoperiod. Among the treatments, 100% red LEDs was the most effective for callus induction from PLB segments. Callus proliferation was best in the 75% red LEDs+25% blue LEDs treatment but was not significantly different from that in PGF light. The highest PLB formation from callus was obtained in 25% red LEDs+75% blue LEDs. The results suggested that LEDs are the effective light source for callus induction, callus proliferation and PLB formation from callus in *Cymbidium* orchid (Lee *et al.*, 2004).

The response and adaptation of two orchids: *Tangtsinia nanchuanica* and *Cephalanthera falcata*, to variable sunlight was investigated by Guo *et al.* (2003) using an open gas exchange system (LI-COR6400). In any variable light environment from 800 to 200 micro mol photons/ m2second1, the two orchids could keep an almost constant net photosynthetic rate (Pn), transpiration rate (E), stomatal conductivity (Con), water use efficiency (WUE) and internal CO_2 concentration (Ci) and showed good adaptation to the large light variation. In conditions of shade less than 200 micro mol photons/m^2 second1, Pn declined, E and Con decreased, but Ci increased more then 20%. Results indicated that T. nanchuanica and C. falcata could adapt to the variable light environment in the understorey of *Pinus massoniana* forests very well.

Ventilation Management

Adequate air circulation is very important in orchid culture. Orchids must have free circulating air and high humidity around them. Air movement in a growing environment ensures a more uniform air temperature and dramatically reduces disease problems by preventing the leaves from staying wet for long time.

The ventilation system provides the means by which the orchid house air is circulated, mixed and exchanged. It allows uniform climate and helps to distribute heat from the heating system as well as removing heat from the orchid house when cooling is required. In combination with the heating system, ventilation also provides a means for dehumidifying the orchid house environment. Ventilation is required throughout the year; however, the ventilation required varies with the outside environment. During the winter months ventilation is required primarily for dehumidification, as warm, humid air is exhausted, cool dry air is brought in. The important consideration when bringing in cold air is proper mixing with the main mass of greenhouse air in order to minimize the negative effects of the cold air contacting the plants. Heating an Orchid house is fairly simple, but cooling one is much harder. Ventilation is also used to reduce excess humidity in an orchid house and to bring in fresh supplies of carbon dioxide. The simplest system is a full length ventilator 30-60 cm wide along the length of the roof ridge combined with air inlets in the side or end walls. Inlets are hinged to operate by hand as needed. This system needs continuous checking and readjusting during the day because temperature extremes can occur. Thermostatic controls are available to open and close these vents automatically. Another system uses exhaust fans and inlet louvers on opposite sides of a greenhouse to move the stale humid air out and allow fresh air to enter. Exhaust fans large enough to replace all the air in an orchid house every 1 to 1½ minutes will provide summer cooling. These fans keep the inside temperature 6°C more than outside, depending on the relative humidity of the outdoor air. It is much easier to cool an orchid house in a climate with low relative humidity. Adding a wet wall of pads opposite the fans will increase cooling even more. This is valuable in areas with hot summer weather, low relative humidity and for plants requiring cool temperatures. Orchid house shading paint can be applied to the cover at the beginning of the summer and should be washed off during winter. Various types of screens can also be mounted inside or outside an orchid house to provide shade. Orchid house cooling requirements are calculated based on square feet of floor area. Usually, about a 4°C difference between inside and outside air temperature is observed. An air exchange rate of about 8 cubic feet per meter is sufficient to keep this temperature difference. If an orchid house is 10 x 20 feet, it would have 200 square feet of floor area. This Orchid house would require combination of fans that could manage 200 x 8 or 1,600 cubic feet per meter to cool adequately. It also evenly distributes the gas that is produced by the plants in the dark and used by the plants to produce their food during the daylight hours. It is not wise to create gale-force winds in growing area, but do want to produce enough airflow to cause the leaves of the orchids to very lightly sway in the breeze. Two of the most effective methods for providing airflow in orchid house is ceiling fans and oscillating fans. The general recommended fan size is of 0.9 to 1.1 cubic meters per minute per square meter of floor area with a velocity of lesser than 1 meter per second across the plants. Equipments used for ventilation purpose are as follows:

To examine the effectiveness of super-elevated (10,000 micro mol mol-1) CO_2 enrichment under cold cathod fluorescent lamps (CCFL) for the clonal propagation of *Cymbidium*, plantlets were cultured on modified Vacin and Went (VW) medium under 0, 3,000 and 10,000 micro mol mol-1 CO_2 enrichment and two levels of photosynthetic photon flux density (PPFD, 45 and 75 micro mol m-2 s-1) by Norikane *et al.* (2010). Under high PPFD, 10,000 micro mol mol-1 CO_2 increased root dry weight and promoted shoot growth. In addition, a decrease in photosynthetic capacity and chlorosis at leaf tips were observed. Rubisco activity and stomatal conductance of these plantlets were lower than those of plantlets at 3,000 micro mol mol-1 CO_2 under high PPFD, which had a higher photosynthetic capacity. On the other hand, plantlets on Kyoto medium grown in 10,000 micro mol mol-1 CO_2 under high PPFD had a higher photosynthetic rate than those on modified VW medium; no chlorosis was observed. Furthermore, growth of plantlets, in particular the roots, was remarkably enhanced. This result indicates that a negative response to super-elevated CO_2 under high PPFD could be improved by altering medium components. Super-elevated CO_2 enrichment of in vitro-cultured *Cymbidium* could positively affect the efficiency and quality of commercial production of clonal orchid plantlets.

Lopez and Runkle (2005) summarizes world orchid production and reviews how environmental factors regulate growth and flowering of several commercially important orchid genera: *Cattleya*, *Cymbidium*,

Dendrobium, Miltoniopsis, Phalaenopsis , and *Zygopetalum*. These genera primarily flower in response to relatively low temperatures, and, for some species and hybrids, flowering is promoted when the plants are also exposed to short photoperiods. Effects of light and temperature on growth and development are summarized for these genera, and implications for controlled production are discussed.

Pigment composition and CO_2 fixation were studied in the leaves, protocorms and floral parts of *Cymbidium, Paphiopedilum, Dendrobium* and several other orchids, grown at controlled temperatures but under natural (low) light in winter by Kirichenko *et al.* (1989). The pigment system of the flowers had lower chlorophyll-a:-b and chlorophyll-a + -b:carotenoid ratios than that of the leaves, and the chlorophyll and carotenoid levels in the flowers were only about 15% of those of the leaves. The level of photosynthetic activity in the floral parts, however, amounted to 40-90% of the activity level in the leaves. The rate of CO_2 assimilation in the sepals and ovaries of *Paphiopedilum* leeanum was higher than that in the leaves. The flowers and pseudobulbils retained high photosynthetic activity throughout their ontogeny; the unpollinated flowers had a life span of several months.

The antioxidant properties of Sophro*LaelioCattleya* and *Cymbidium* exposed to near ultraviolet (UV-A) radiation were evaluated. The superoxide dismutase activity in both plants and flavonoid content in *Cymbidium* showed strong responses to UV-A, UV-A was effective in promoting growth and formation of carotenoids and chlorophyll in both plants (Zhao *et al.*, 2005). Potted mature *Phalaenopsis* cv. Joseph Hampton (clone Diane) plants were placed in each of 4 growth chambers with 0, 8, 60 or 160 micro mol m^{-2} s^{-1} photosynthetic photon flux (PPF) for 12 h daily and at 20/15°C day/night air temperature by Wang (1995). Plants under 160 and 60 micro mol m-2 s-1 began spiking (producing an elongating reproductive bud protruding through the base of its subtending leaf) in an average of 28 and 34 days, respectively. None of the plants placed under 0 or 8 micro mol m^{-2} s^{-1} started spiking within 6 weeks. These plants, following return to a greenhouse, spiked and flowered 8 weeks later than those receiving 160 micro mol m^{-2} s^{-1}. In a second experiment, plants were placed in each of 3 growth chambers and kept in complete darkness at 20/15 degrees day/night for 2, 4 or 6 weeks before exposure to 160 micro mol m^{-2} s^{-1}. Air temperature was maintained at 20/15 degrees day/night for an additional 6 weeks and then raised to 25/20 degrees to accelerate flowering. Plants exposed to 2, 4 and 6 weeks of darkness required 45, 60 and 77 days, respectively, to reach spiking. However, all plants spiked at similar times (31-35 days) after lighting began. Anthesis occurred at progressively later dates for plants placed in darkness for increasing durations, but plants in all treatments required 123 days to reach anthesis following their exposure to light. Flower count and size were not affected in either experiment.

Humidity Management

Humidity is moisture that is suspended in the air. The temperature of the air determines how much moisture it can hold. Warm air holds more moisture than cool air, so as the temperature rises, the percentage of moisture in the air decreases. Because of this, more moisture needs to be added to maintain the desired level of humidity. When the air lowers, the air gives up its moisture in the form of rain. The outside air might have a comfortable level of humidity, but when that air is heated, the relative humidity drops, causing the air to be very dry inside the house during the dry months. A humidifier can help to maintain the desirable level of humidity. For instance if outdoor temperature is 0°C, the maximum amount of water that a cubic meter of air can hold at this temperature is 5 grams. By heating this air to 25°C provides a relative humidity of 23 per cent. The humidity of orchid house can be measured by an instrument known as hygrometer. Orchids lose water through pores (stomata) found on their leaves. The drier the air, faster the orchid will lose its moisture. Keeping the humidity high reduces the speed of water loss (dehydration). Orchids prefer a location where there is a range of humidity from 40 per cent to 70 per cent. A fluctuation of humidity is desirable and gives the plants a chance to dry out. A 40 per cent drop during the day for short periods can be beneficial to orchids.

Despite the fact that orchids are mainly tropical plants, their photosynthesis is processed differently compared to other plants of tropical ecosystems. This type of photosynthesis is technically called CAM. CAM plants (mostly succulent plants found in desert areas) are more economical in handling water, by applying other mechanisms during their daily rhythm, surviving dry spells far better. This type of plant typically 'work' even during the night, since the synthesis of sugar is in process even when they receive no light. Therefore, it is nothing wrong with watering the orchids in the evening. Humid warm atmosphere is most essential for the growth of most of the tropical orchids, which do not have well established root system. It is a good idea to have a water tank or pool in the centre of the orchidaria to maintain humidity, which should not be less than 30 per cent at night and 80 per cent during day time.

Cymbidium can easily tolerate lower RH levels. If the RH drops below 40 per cent, shading and humidification or roof sprayers can be used. Roof sprayers can lower greenhouse temperatures by a maximum of 3°C and increase RH by a maximum of 8 per cent. As long as the leaf temperature of *Cymbidium* is lower than around 27°C, the plant will continue to absorb water and transpire, even at a lower RH. When orchids get enough humidity, they grow lushly and their leaves have a healthy shine. Insufficient humidity can stunt an orchid's growth and in severe cases, it can cause brown tips on leaves. It also cause buds falling (known as bud blast), leaves wrinkling and drying of the sheaths (the tube like structures that surround the developing flower buds), which can result in twisted or malformed flowers. During the winter, homes, especially those in cold climates with forced-air heating systems, usually have a relative humidity of about 15 percent. For greenhouses, this process is a relatively simple matter. High humidity levels that would be no problem in an orchid house but which can peel the paint, plaster, and wallpaper off the walls of houses. Evaporative-pad humidifier (in which fans blow across a moisture-laden pad that sits in a reservoir of water) is found the best type of humidifier. It is usually better than a mist humidifier because it doesn't leave on orchids with a white film (from the minerals in the water being deposited on the leaves). To further increase the humidity level, try to grow the plants on top of a waterproof tray filled with pebbles. Water should be added to the tray so that the level is just below the surface of the pebbles, then put the plants on top of this bed of damp gravel. With these units, fans blow across a moisture-laden pad that sits in a reservoir of water. Evaporative-pad humidifiers are first choice for orchid house humidification because

1. They are reasonably priced and readily available.
2. They don't spray the room with droplets of water that can carry mineral deposits and bact*Eria.*
3. They circulate air at the same time.
4. They only increase the humidity by about 50 to 60 percent. This is a level that is beneficial to plants, but not sufficient to cause moisture damage to the house.
5. They require no plumbing and very little maintenance – just change the moisture pads one or two times a season.

Misting is another way to increase humidity. Cool-mist humidifiers can be effective for small areas, but with constant use, they can cause deposits of minerals on leaves and be a bact*Eria* carrier. This works very well, but in order for it to be effective, should do it several times a day, because the water usually evaporates very quickly. A problem with misting is that, if water source is mineral-laden, orchid's leaves may become encrusted in white – not only is this unsightly, but it keeps light from penetrating to the leaves. A benefit to misting is that it can clean the dust from the leaves.

One of the best, low-cost ways to raise the humidity level of orchid's environment is to grow plants over a humidity tray. It should be remembered, that orchids never want to sit in standing water. A humidity tray is something that can either make or purchased from an orchid specialty supplier relatively cheaply. One of the main benefits of using a humidity tray is that it can increase the humidity directly around orchid while keeping the rest of our home at normal levels.

Mist Systems

Both high and low-pressure mist systems are used for cooling and adding humidity to the orchid house. Mist systems can be employed in both mechanically and naturally ventilated orchid house. Mist systems work by forcing water through nozzles which break up the water into fine droplets. This process allows the droplets to evaporate fairly quickly into the air. As the evaporation of water requires heat from the environment, the air is cooled. Misting systems must be carefully controlled to provide the required cooling without increasing the relative humidity beyond optimum levels for plant performance or allowing free water to form on the plants. If the quality of the water used for misting is poor, there is the possibility of mineral salts being deposited on the leaves and flower, which could result in reduced flower quality and yield loss.

Oncidium are planted in open protected culture or under shading nets in subtropical regions. Because of the lack of cooling techniques for the inside microclimates, the production period is limited, and the quantity and quality of flowers is unstable. Without ventilating, high humidity and temperature induces severe growth problems. In this study, mechanical ventilation fans with intermittent misting were applied to modify the internal microclimate of the *Oncidium* greenhouse. Nine experimental setting of misting intervals were tested. The misting efficiency is defined and served as the performance index. The misting intervals of 3 min on to 3 min off or 5 min on to 5 min off had the higher efficiency. The cooling technique can keep the interior air temperature lower than that of outside air temperature. The temperature of plants could be maintained close to the greenhouse air temperature. The intermittent misting function with the best evaporative cooling efficiency was controlled with interval operation. This technique can be applied to other protected structures of *Oncidium* and other orchids (Shen and Chen 2012).

Carbon Dioxide Supplementation

Carbon dioxide (CO_2) plays an important role in increasing crop productivity. Actively photosynthesizing orchid will quickly deplete the carbon dioxide from the orchid house environment. Generally in summer, even with maximum ventilation, carbon dioxide levels in orchid house typically fall below the ambient levels of carbon dioxide. Optimum carbon dioxide concentration in orchid house atmosphere should approximately be 700 to 800 parts per million. Carbon dioxide supplementation via combustion may be done to introduce carbon dioxide into the orchid house.

Yoon *et al.* (2009) assessed the effect of growth at either 400 micro mol mol^{-1} (ambient) or 1000 micro mol mol^{-1} (elevated) CO_2 and 0 g L^{-1} (deprivation) or 30 g L^{-1} (supplementation) sugar on morphological traits, photosynthetic attributes and intrinsic elements of the CAM pathway using the CAM orchid *Phalaenopsis* 'Amaglade'. The growth of shoot (retarded) and root (induced) was differently affected by CO_2 enrichment and mixotrophic regime (+sugar). The Fv/Fm ratio was 14% more in CO_2-enriched treatment than at ambient level during *in vitro* growth. At elevated level of CO_2 and sugar treatment, the content of Chl(a+b), Chl a/b and Chl/Car was enhanced while carotenoid content remained unaltered. During *in vitro* growth, gas-exchange analysis indicated that increased uptake of CO_2 accorded with the increased rate of transpiration and unchanged stomatal conductance at elevated level of CO_2 under both photo- and mixotrophic growth condition. At elevated level of CO_2 and sugar deprivation, activities of Rubisco (26.4%) and PEPC (74.5%) was up-regulated. Among metabolites, the content of sucrose and starch was always higher under CO_2 enrichment during both *in vitro* and ex vitro growth. Our results indicate that plantlets grown under CO_2 enrichment developed completely viable photosynthetic apparatus ready to be efficiently transferred to *ex vitro* condition that has far-reaching implications in micropropagation of *Phalaenopsis*.

The CO_2 exchange rates of inflorescences, capsules, pseudobulbs and leaves were studied by Dogane and Ando (1990) during the flowering and capsule development periods to determine the Carbon dioxide balance between photosynthetic and reproductive organs. The inflorescence used large amounts of Carbon dioxide for respiration and as a tissue component; the leaf on the same shoot only supplied 6% of the used Carbon dioxide. Consequently, the process of flower development and maintenance until fading was mainly dependent on Carbon dioxide sources from the pseudobulbs or Carbon dioxide translocated from older leaves. In contrast, during the period of capsule development, the leaf on the same shoot could fix 42% of the Carbon dioxide required for capsule development. In addition, Carbon dioxide flowed from fading flowers after pollination and the photosynthetic activity of developing capsules contributed to the Carbon dioxide balance in the pollinated plants.

Yao *et al.* (2011) observed that the maximum carbon balance can be reached at the day/night temperature of 25/25°C on leaf basis, and in range of 24-27°C on the base of greenhouse as a whole. The correlation coefficient between daily mean temperature and carbon balance was 0.65, calculated by annual eddy flux data. The relationship between night temperature and CO_2 fixation was stronger than day temperature and respiration. The light and dark reactions of carbon fixation for crassulacean acid metabolism (CAM) plant were separated. The suitable light intensity range for *Phalaenopsis* leaves was 300-500 micro mol m^{-2} S^{-1} from the measurements of daily carbon balance and fluorescence parameters. In addition, higher daily carbon balance occurred in 1 day/night cycle per day (148.3 micro mol CO_2 m^{-2} day^{-1}) compared with 2 cycles (127.1 micro mol CO_2 m-2 day-1) and 3 cycles (92.4 micro mol CO_2 m^{-2} day^{-1}). Latent heat was not significantly correlated with temperature and light intensity because the pad-fan interfere the vaporing of water in greenhouse.

Ichihashi *et al.* (2008) investigated CO_2 uptake in *Phalaenopsis*. CO_2 uptake at night (Phase 1) changed with temperature. Maximum CO_2 uptake was observed around 20°C. CO_2 absorption at night increased in proportion to CO_2 concentration. Rate of CO_2 absorption was higher at 20°C than at 25°C. CO_2 uptake in the late afternoon (Phase 4) showed a maximum around 20°C. CO_2 absorption in Phase 4 increased in proportion to CO_2 concentration but stomata conductance decreased under high CO_2 levels. CO_2 absorption response curve to light intensity in Phase 4 was not saturated till 500 micro mol.m^{-2}.s-1 PPFD when CO_2 level was 2000 ppm, and then the stomata conductance showed very low values. Under various water and humidity condition when *Phalaenopsis* was irrigated weekly, CO_2 uptake in Phase 1 was the largest at one day before watering, but that of Phase 4 was the largest at one day after watering. Total CO_2 uptake in all Phases was stimulated the most at 70% relative humidity and suppressed drastically at 30% relative humidity. These results indicated that higher relative humidity was likely the most important factor for high CO_2 absorption in *Phalaenopsis*.

A laboratory study was conducted by Li and Run (2003) to examine the effects of long-term CO_2 enrichment on sucrose phosphate synthase (SPS) and sucrose synthase (SS) and on the subsequent growth and related physiological changes following transfer to ambient air, and to determine to what extent SPS and SS activities are related to sucrose and starch accumulation under both elevated CO_2 and subsequently ambient CO_2 conditions in the tropical epiphytic orchid hybrid *Oncidium goldiana*. Treatments comprised: high CO_2, with a CO_2 concentration approximately twice the ambient (750+or-200 micro l/litre); and ambient CO_2 (~380 micro l/litre) fed with ambient air. In the elevated CO_2 treatments, CO_2 was added day and night. After two months of treatment, plants grown at elevated CO_2 were transferred to a chamber fed with ambient air. An upregulation of both leaf SPS and SS activities was observed in CO_2-enriched plants, along with a higher carbohydrate and lower nitrogen content. Following transfer to ambient CO_2, leaf SPS and SS activities, together with leaf sucrose and starch content, dropped sharply in plants enriched with CO_2 for two months. The activities of SPS and SS in leaf extracts were closely associated with assimilation rates. SPS activity was also closely associated with sucrose accumulation and SS activity closely associated with starch accumulation.

Nutritional Management

The term essential element was proposed by Arnon and Stout (1939). They concluded three criteria must be met for an element to be considered essential. These criteria are:

1. A plant must be unable to complete its life cycle in the absence of the mineral element.
2. The function of the element must not be replaceable by another mineral element.
3. The element must be directly involved in plant metabolism.

The number of elements considered essential for the growth of higher plants now varies from 16 to 20 or more, depending upon the definition of essentiality. Other scientists such as Nicholas (1961) believed that an element should be considered essential if its addition enhances plant growth even though it merely substitutes for one of the 16 elements that Arnon declares to be essential. For example, sodium can substitute in plant nutrition for some potassium and vanadium for some molybdenum, Nicholas would consider both sodium and vanadium as essential, but Arnon would not. On the basis of the criteria used, Arnon specifies 16 elements and Nicholas 20 elements as being essential for the growth of higher plants. Three other debated nutrients are nickel (urea transformations), cobalt (N_2 fixation) and silicon. These criteria are important guidelines for plant nutrition but exclude beneficial mineral elements. Beneficial elements are those that can compensate for toxic effects of other elements or may replace mineral nutrients in some other less specific function such as the maintenance of osmotic pressure. The omission of beneficial nutrients in commercial production could mean that plants are not being grown to their optimum genetic potential but are merely produced at a subsistence level.

Since some are needed in relatively large quantities and others in very small amounts, the former are referred to as "major" elements and the latter as "minor" or "trace" elements, or as micro nutrients. The terms "major" and "minor" do not refer to the relative importance of the functions of the elements in plant growth and for this reason the term "trace" element is preferable for the latter class. Carbon (CO_3^{-2}, HCO_3, CO_2), hydrogen (H^+, OH^-), and oxygen (O_2, OH^-) are supplied by air and water. The six macronutrients, nitrogen (NO_3^- , NH_4^+), phosphorus (HPO_4^{-2}, H_2PO_4), potassium (K^+), calcium (Ca^{+2}), magnesium (Mg^{+2}) and sulfur (SO_4^{-2}) are required by plants in large amounts. The rest of the elements are required in trace amounts (micronutrients). These trace elements include boron (H_3BO_3, $B_4O_7^{-2}$), chlorine (Cl^-), copper (Cu^{+2},Cu^+), iron (Fe^{+2} , Fe^{+3}), manganese (Mn^{+2} , MnO_4), sodium (Na), zinc (Zn^{+2}), molybdenum ($HMoO_4^-$, MoO_4^{-2}), and nickel (Ni). Beneficial mineral elements include silicon (Si) and cobalt (Co). The beneficial elements have not been deemed essential for all plants but may be essential for some.

Nitrogen

Nitrogen is a major constituent of several important substances, which occur in plants. N is biologically combined with C, H, O and S to create amino acids, which are the building blocks of proteins. Amino acids are used in forming protoplasm and nucleus, the site for cell division and thus for plant growth and development. It is of outstanding importance among the essential elements in that nitrogen compounds comprise from 40 to 50 per cent of the dry matter of protoplasm. Since all plant enzymes are made of proteins, N is essential for all of the enzymatic reactions in a plant. It is a major part of the chlorophyll molecule and is therefore necessary for photosynthesis. It is a necessary component of several vitamins. For this reason nitrogen is required in relatively large quantities. Proteins are compounds of nitrogen whilst chlorophyll, the green coloring matter of the leaves, also contains the element. Chlorophyll is the green pigment that plays an important part in photosynthesis. If nitrogen is limiting, chlorophyll molecules cannot be synthesized. The plant loses its green color. Nitrogen is a necessary element for leaf growth and blossom formation. As a result, the N-deficient plant does not produce required carbohydrates. An adequate supply of nitrogen encourages the formation of the growth hormones cytokinin, auxin and gibberellic acid but a reduction in its availability affects the concentration and balance of these hormones and encourages the formation of the senescence hormone, abscisic acid. With a nitrogen deficiency the growth of plants is retarded as fewer structural and enzyme-functioning proteins are produced. Certain compounds of nitrogen are very mobile in plants, and this enables them readily to mobilize supplies of the element at vital growing points and to transfer stored supplies to points where they are most required. Such transference is common from old tissues to young growing points when supplies of the element are short. This mobility and re-utilization of nitrogen explains why deficiency symptoms of the element always appear first in the older parts of plants and why growing points are the last to be affected. The mobility in soil is dependent on the chemical form of the element used.

Deficiency Symptoms

- Stunted growth may occur because of reduction in cell division.
- Pale green to light yellow colour (chlorosis) appearing first on older leaves, usually starting at the tips. Depending on the severity of deficiency, the chlorosis could result in the death and/or dropping of the older leaves. This is caused by the translocation of N from the older to the younger tissues.
- Reduced N lowers the protein content of seeds and vegetative parts. In severe cases, flowering is greatly reduced.
- N deficiency causes early maturity in some crops, which results in a significant reduction in yield and quality.

Nitrogen is applied as fertilizer in organic forms as animal manures, composts, and green manure crops. Inorganic N fertilizers are available in liquid, granular, and slow-release formulations. Ammonium-N fertilizers delay leaching because NH_4+ is held on the soil exchange complex. When NH_4^+ is converted to NO_3^-, N is more subject to leaching loss. Urea fertilizer leaches readily until it is converted in the soil to $NH_4^+$by the naturally occurring enzyme urease, usually within a few days of application. Urea is a cheap form of nitrogen often used in fertiliser. Solutions of urea in water have low conductivity. Urea in its better, low biuret grades makes a good nitrogen source for plants grown at temperatures that are above 16°C. It is not a good source of nitrogen for cool growing crops, particularly during the winter months because bact*Eria* populations, needed for nitrifying urea, are in low population. Urea requires Bacterial action before its nitrogen is available to plants. Organic N is not available to plants until it has been converted to an inorganic form by soil bact*Eria*. Uptake inhibited by high phosphorus levels. N/K ratio is important: high N/low K favours vegetative growth; low N/high K promotes flowering and fruiting. Nitrogen is generally made available to plants in fertiliser solution as ammonium ions, nitrate ions or urea. Most of the nitrate ions that enter the roots from the surrounding solution are carried upwards through the sap-conducting vessels (the xylem) into the cytoplasm of the leaves. Excess nitrate is stored in the vacuoles. For their main function as a nitrogen nutrient, nitrate ions are first converted to nitrite ions in the cytoplasm with the aid of an enzyme (nitrate reductase) containing molybdenum and iron. These nitrite ions are then transported to the chloroplasts where an iron-containing enzyme (nitrite reductase) catalyses their conversion to ammonium ions. Because of their toxicity to plants, even in quite low concentrations, the ammonium ions are rapidly converted to several nitrogen compounds, one of which (glutamine) undergoes further enzyme-controlled reactions to produce many different amino-acids. Finally in ribosomes (mainly in the cytoplasm) twenty of these amino-acids are combined together in the genetically coded sequences defined by DNA and guided by RNA to form the wide variety of proteins required for their structural and enzymatic roles in the plant. If ammonium ions are present in fertilisers, their toxic effect on the plants is avoided because bacterial action on the root surface converts the ions to amino acids that are then translocated from the roots to the leaves for protein synthesis. Many simpler nitrogen compounds are synthesised in plants from nitrate and ammonium ions. These include compounds (guanine, adenine, cytosine, thymine and uracil) required for the formation of DNA.

Excess N may cause K deficiencies. With excess nitrogen the leaves become darker green, wider and longer but reduced in thickness. This sometimes causes the plant to become droopy. The roots are generally smaller and more bunched than normal.

Bare-root seedling plants of a white-flowered *Phalaenopsis* hybrid (P. amabilis x P. Mount Kaala Elegance) were grown in 5 potting media under 3 fertilizer levels (0.25, 0.5 and 1.0 g/litre) from a 20N-8.6P-16.6K soluble fertilizer applied at every irrigation. The 5 media included (by vol.) (i) perlite:Metro Mix 250:charcoal (1:1:1); (ii) perlite:composted pine bark:vermiculite (2:2:1); (iii) composted pine bark; (iv) perlite:Metro Mix 250:charcoal (3:3:1); (v) and perlite:rockwool (1:1). During the first flowering season, plants in medium (i) had slightly fewer but larger flowers and thicker stalks (section of the inflorescence between the base and oldest flower) than those in medium (v). Medium had no effect on stalk length. Two media ((iv) and (v)) resulted in root systems that were inferior to those in the others. Fertilizer level had no effect on flowering date or flower size. Regardless of medium, increasing the fertilizer application from 0.25 to 1.0 g/litre increased flower number, stalk diameter and length, and leaf production following flowering. During the second flowering season, media had limited effect on plant performance, but increased fertilizer promoted earlier inflorescence emergence and flowering. Higher fertilizer rates also caused a linear increase in the number of flowers and inflorescences per plant, and in stalk diameter, total leaf count and leaf size during the second season(Wang and Gregg, 1994).

Mou *et al.* (2012) studied the effects of adding nitrogen (N) (0, 105, 210, and 420 mg.L^{-1}) on the vegetative growth and reproduction of *Paphiopedilum armeniacum*. N enrichment improved leaf area and lengthened the leaf lifespan during the vegetative stage. The effects of N application on flower size were minor. The intermediate N level of 210 mg.L^{-1} (MN) increased the seed capsule weight, seed germination rate, and improved the growth of seedlings that developed from seeds of MN-treated plants. N fertilizer exerted little influence on ramet emergence and ramet number per plant, but a low N concentration of 105 mg.L^{-1} promoted the leaf number and leaf area of ramets. Appropriate N levels for P. armeniacum in production and cultivation should be determined according to different production objectives.

An experiment was conducted to study the effect of varying levels of N and P on yield and quality of cut flowers in orchid *Dendrobium* cv. Sonia 17 by Nair and Sujatha (2010). The results indicated a significant influence of the treatments and months on different vegetative and yield parameters. The maximum number of canes per plant (6.73) was recorded in T4 (1:6:1 NPK level). The number of spikes per plant (1.8) was the highest in T8 (2:6:1 NPK level) during January. However the treatment T4 (1:6:1 NPK level) recorded the maximum total number of spikes produced per plant per year (10.93) and was on par with T8 (2:6:1 NPK level) which produced 10.80 spikes. During October, the treatment T4 (1:6:1 NPK level) produced the maximum number of flowers per spike (15.53). The longest spikes (72.10 cm) were produced during April in T4 (1:6:1 NPK level). Flower diameter was maximum (10.10 cm) in June in T8 (2:6:1 NPK level). Vase life was not significantly affected by the treatments during any of the months. The observations were compared with the benchmark data for *Dendrobium* cv. Sonia 17 and were found to be superior with special reference to production and the quality of spikes. The yield and quality could be sustained without any drastic decline round the year.

Two Disa hybrids ('Unidiorosa' and 'Kewensis') were subjected to different ammonium levels in nutrient solutions under varying environmental conditions by Pienaar and Combrink (2007). Plants were cultivated in a controlled environment and growth was evaluated after 266 days. Plants were fertigated according to guidelines provided by Naaldwijk for *Cymbidium* orchids, modified for different ammonium/ nitrate ratios with the same amount of total nitrogen. Two levels of shading were provided (56% and 69%) and two root medium temperatures tested. Where most greenhouse crops need less than 10% of their total N supply as ammonium, *Cymbidiums* need about 20%. 'Unidiorosa' and 'Kewensis' performed best with 40% and 60% of the applied N in the ammonium form respectively. No significant difference was found between the two shading levels, probably due to the cool greenhouse conditions. Lowering of the average mid-day root medium temperature from 21.4°C to 16.8°C had a negative effect on root growth and plant mass, but produced longer leaves and increased leaf area.

Experiments were conducted to determine how nitrogen (N), phosphorus (P), and potassium (K) rate and fertilizer termination time affect the growth and flowering of a *Dendrobium* nobile Lindl. hybrid, *Dendrobium* cv Red Emperor 'Prince'. Nitrogen, P, and K were tested in separate experiments as a factorial combination of five rates and three termination dates (1 Sept., 1 Oct., and 1 Nov. 2005). Nitrogen and K rates were 0, 50, 100, 200, and 400 mg.L^{-1}. Phosphorus rates were 0, 25, 50, 100, and 200 mg.L^{-1}. For all nutrients, ending fertilization on 1 Sept. resulted in greater or similar pseudobulb thickness compared with ending fertilization on 1 Oct. or 1 Nov. Pseudobulbs grew taller as the

N rate increased, peaking at 100 and 200 mg.L^{-1}. There were interactions between the N rate and fertilizer termination time on all reproductive characteristics. For all fertilizer termination times, flower number increased once N was applied. When ended on 1 Nov., 200 and 400 mg.L^{-1} N caused a delay to reach anthesis. All P rates resulted in taller plants with equally more nodes when compared with 0 mg.L^{-1}. As the K rate increased from 0 to 100 mg.L^{-1}, plant height increased, with no further increase at higher rates. The number of leaves remaining increased as N and K rates increased up to 200 mg.L^{-1}. Total flower number and flowering node number increased as the K rate increased to 100 mg.L^{-1} (terminated on 1 Sept.) or 50 mg.L^{-1} (terminated on 1 Oct. or 1 Nov.). In the fourth experiment, only N was ended at four termination times, whereas all other nutrients continued to be supplied until flowering. Control plants received all fertilizer elements until flowering. The duration of N application did not affect vegetative or flowering characteristics. No aerial shoots were observed as a result of prolonged application of N at all rates. In summary, 100 mg.L^{-1} N, 25 mg.L^{-1} P, and 100 mg.L^{-1} K are recommended for optimal vegetative growth and reproductive development of *Dendrobium* cv Red Emperor 'Prince' (Bichsel *et al.,* 2008).

Naik and Barman (2006) investigated the influence of foliar application of nitrogen on flowering of *Cymbidium* hybrid Show Girl grown in polyhouse and revealed showed that the spike length and rachis length significantly increased over control with increasing concentration of nitrogen and found maximum length of 45.0 and 30.25 cm, respectively in the treatment N20:P5:K5 at 0.3%. However, number of flowers/spike and number of spikes/clump significantly increased over control with increasing concentrations of nitrogen and found maximum of 5.5 and 2.75, respectively in the treatment N15:P5:K5 at 0.3%. The treatment N20:P5:K5 at 0.3% recorded highest nitrogen content in leaf (10.1 g/kg) while potassium content (15.0 g/kg) was highest in treatment N5:P5:K5 at 0.3%. A positive linear relationship was found between nitrogen content of leaves and spike length, rachis length, number of flowers/spike and number of spikes/clump.

Increased nutrient availability is generally believed to be harmful for terrestrial orchids. Hejcman *et al.* (2010) observed that *Dactylorhiza maculata, Platanthera bifolia* and *Listera ovata* survived decades-long application of Ca and CaN fertilizer, but not CaN together with *P. Cover* of *D. maculata* was highest in areas of Ca treatment. The number of flowers per plant of L. ovata was significantly higher in the CaN treatment compared with the control indicating that enhanced N availability stimulated its growth. Long-term N application (NH_4NO_3

at an annual rate of 100 kg N ha^{-1}) was not fatal to the studied orchids if not accompanied by P application. It is more likely that enhanced N availability can cause loss of studied orchids only where the availability of P in soil is high enough to enable the growth of productive species that can better compete for light on eutrophic soils.

Nair and Sujatha (2010) conducted an experiment to study the effect of varying levels of N and P on yield and quality of cut flowers in orchid *Dendrobium* cv. Sonia 17. The results indicated a significant influence of the treatments and months on different vegetative and yield parameters. The maximum number of canes per plant was recorded in T4 (1:6:1 NPK level). The number of spikes per plant was the highest in T8 (2:6:1 NPK level) during January. However the treatment T4 (1:6:1 NPK level) recorded the maximum total number of spikes produced per plant per year and was on par with T8 (2:6:1 NPK level) which produced 10.80 spikes. During October, the treatment T4 (1:6:1 NPK level) produced the maximum number of flowers per spike. The longest spikes were produced during April in T4 (1:6:1 NPK level). Flower diameter was maximum in June in T8 (2:6:1 NPK level). Vase life was not significantly affected by the treatments during any of the months. The observations were compared with the benchmark data for *Dendrobium* cv. Sonia 17 were found to be superior with special reference to production and the quality of spikes. The yield and quality could be sustained without any drastic decline round the year.

Gower Ramsey *Oncidium* orchids for cut flower production, and potted Aloha Iwanaga *Oncidium* were fertigated at the same N rate with four NO3$^-$N:NH_4^{+}-N ratios (1:0, 10:1, 6:1, and 4:1). The greater the proportion of NH_4^{+}-N the higher were the N and P contents in plant tissues. However, K, Ca and Mg contents were decreased. The N content was lowest when the ratio of NO_3^-N:NH4^{+}-N was 1:0. The flowering quality of Gower Ramsey and Aloha Iwanaga was best when the NO_3^-N:NH4^{+}-N ratio was 6:1(Lin *et al.*, 1999).

A trial was conducted by Barman *et al.* (2012) to study the influence of the media and frequency of nutrient application on growth and flowering of *Cymbidium* hybrid 'H.C. Aurora'. Treatments involved growing of *Cymbidium* in different combination of media, i.e., M1-perlite + coco peat + brick pieces; M2-coco peat + brick pieces + leaf mould; M3-perlite + brick pieces; M4- thermocol + coco peat + brick pieces; M5-rotten log + sand + coconut husk + brick pieces; M6-*Castanopsis hysterix* bark + maize cob + leaf mould; M7-leaf mould + sand + coconut husk + brick pieces + charcoal (Farmer's practice) and spraying of water soluble fertilizer (N19P19K19) at 1 g/l concentration at weekly and biweekly interval. The result indicated that among the vegetative characters, maximum number of shoots, number of pseudobulb, pseudobulb girth, length leaf, and floral characters like spike length and number of spike/m^2 were recorded at growing media *Castanopsis hysterix* bark + maize cob + leaf mould. Further frequency of water soluble fertilizer significantly influenced growth and flowering of *Cymbidium*. Maximum number of shoots, spike length and number of flowers/spike were recorded with biweekly spraying of water soluble fertilizer (1 g/l). Further, the growing media significantly influenced the NPK content of leaves. The physical properties, i.e., pH (6.25), EC (0.23 mmhos/cm^2), bulk density (0.52 g/cc) and water holding (49%) of *Castanopsis hysterix* bark + maize cob + leaf mould medium was found to be better for growth and flowering of *Cymbidium* orchid.

Phosphorus

This element is closely concerned with the vital growth processes in plants as it is a constituent of nucleic acid, and nuclei in which this occurs are essential parts of all living cells. Hence a deficiency of this element will also be expected to result in greatly restricted growth. Phosphorus is also of importance in seeds and in connection with the metabolism of fats. Compounds of phosphorus are also concerned with the processes of respiration and with the efficient functioning and utilization of nitrogen. This relationship to nitrogen probably accounts for the fact that several of the symptoms of phosphorus deficiency are identical or similar to those which result from a deficiency of nitrogen. It is of special importance in the processes concerned in root development and the ripening of seeds and fruits. DNA (Deoxyribose Nucleic Acid) and RNA (Ribonucleic Acid), which are responsible for replication of a plant in the same form through generations, are phosphorus compounds composed of long sequences of four different building blocks called nucleotides. Each nucleotide in DNA contains a phosphate group, a sugar (deoxyribose) and one of four nitrogen compounds (adenine, guanine, cytosine and thymine) while those in RNA are similar but sugar is ribose and uracil is present instead of thymine. Nucleotides also serve another equally important function in plants by acting as chemical agents for the storage and transfer of energy between substances during reactions. The most important, adenosine triphosphate (ATP), containing three phosphate groups, participates in a great number of reactions during photosynthesis, respiration and other plant processes. Many other complex phosphorus compounds are involved in the 50 or more reactions which give rise to the final products of both photosynthesis and respiration. For example in utilising the light energy stored during photosynthesis carbon dioxide must first react with a suitable compound to begin the process, function is performed by ribulose bis phosphate. When reserves of protein are being deposited during the formation of seeds (bulbs and tubers) most of the phosphate is converted to phytic acid, a compound with six phosphate groups. Phytic acid becomes associated with calcium, magnesium, potassium, zinc and iron as phytates. These substances regulate the deposition of starch during the development of the above-mentioned storage organs of the plant. When growth begins again, the proteins release amino-acids that are then translocated with other mineral nutrients (phosphate, calcium and magnesium) to the developing roots and shoots. Seeds have the highest concentration of P in a mature plant and are required in large quantities in young cells, such as shoots and root tips, where metabolism is high and cell division is rapid. P aids in root development, flower initiation, and seed and fruit development. It has been shown to reduce disease incidence in some plants and has been found to improve the quality of certain crops.

Because P is needed in large quantities during the early stages of cell division, the initial overall symptom is slow, weak, and stunted growth. It is relatively mobile in plants and can be transferred to sites of new growth, causing symptoms of dark to blue-green coloration to appear on older leaves of some plants. Older leaves turn a darker green followed by a purplish tint starting from the leaf margins and leaf tips dry off. The plants are often darker green than normal and a reddish colouration is occasionally seen due to the pigment, anthocyanin. In ornamental plants flowering and fruiting may be affected. Lack of P can cause delayed maturity and poor seed and fruit development. Phosphorus-deficient plants are retarded in growth because fewer phosphorus-containing compounds essential for photosynthesis and respiration are formed. A deficiency also affects the plant hormone balance thereby delaying flower initiation and decreasing the flower count. Deficiency symptoms are more prevalent during cold, wet conditions.

Excess phosphorus hastens the maturity of a plant. Root growth is more extensive than normal.

Second to N, P is often the limiting element in soils. Phosphorus is supplied to plants as water-soluble compounds such as calcium dihydrogen-phosphate. It enters the roots as phosphate ions. To correct the deficiency apply solid or liquid phosphate fertilisers. Phosphorus is relatively immobile in soil and does not leach. It is stored in seeds and fruit. It is most readily available to plants between a pH of 6 and 7.5 (unavailable in very acid or alkaline soils). Phosphorus absorption is reduced at low soil temperatures.

An experiment was initiated to determine the effect of a low N, high P and K fertilizer, applied during flowering, on growth and flowering in *Phalaenopsis* by Wang (2000). Plants received N:P:K at 100:44:83 mg/litre from a 20N-8.8P-16.6K soluble fertilizer; control plants continued to receive this dose of fertilizer. On September, plants were given high P (N:P:K at 30:398:506 mg/litre) at each or every fourth irrigation. The high P treatments, regardless of the frequency of application, had no effect on the date of emergence of the flowering stem (spiking), anthesis or flower size. All plants treated with the high P fertilizer had fewer flowers than controls. Continuous application of adequate N appears to be more important than low N and increased P for best flowering. In a separate experiment, terminating fertilization completely on 1 or 29 September, or 27 October, or when the flowering stems were emerging (1 October) reduced flower count (<=19 compared with 24). Flower longevity was reduced by 12 days when fertilization was terminated on 1 September. Flower size was unaffected. Discontinuing fertilization prior to late November reduced flower count. Withholding fertilization for extended periods resulted in red leaves, loss of the lower leaves, and limited production of new leaves.

A trial was conducted to study the influence of the media and frequency of nutrient application on growth and flowering of *Cymbidium* hybrid 'H.C. Aurora' by Barman *et al.* (2012). Treatments involved growing of *Cymbidium* in different combination of media, i.e., M1-perlite + coco peat + brick pieces; M2-coco peat + brick pieces + leaf mould; M3-perlite + brick pieces; M4- thermocol + coco peat + brick pieces; M5-rotten log + sand + coconut husk + brick pieces; M6-*Castanopsis hysterix* bark + maize cob + leaf mould; M7-leaf mould + sand + coconut husk + brick pieces + charcoal (Farmer's practice) and spraying of water soluble fertilizer (N19P19K19) at 1 g/l concentration at weekly and biweekly interval. The result indicated that among the vegetative characters, maximum number of shoots, number of pseudobulb, pseudobulb girth, length leaf, and floral characters like spike length and number of spike/m^2were recorded at growing media *Castanopsis hysterix* bark + maize cob + leaf mould. Further frequency of water soluble fertilizer significantly influenced growth and flowering of *Cymbidium*. Maximum number of shoots, spike length and number of flowers/spike were recorded with biweekly spraying of water soluble fertilizer (1 g/l). Further, the growing media significantly influenced the NPK content of leaves. The physical properties, i.e., pH (6.25), EC (0.23 mmhos/cm^2), bulk density (0.52 g/cc) and water holding (49%) of *Castanopsis hysterix* bark + maize cob + leaf mould medium was found to be better for growth and flowering of *Cymbidium* orchid.

The research was conducted to find out the distribution of different nutrients in the leaves of naturally growing plants of various orchid species like *Bulbophyllum, Coelogyne, Dendrobium, Eria* and *Paphiopedilum* during active vegetative stage by Naik and Barman (2007). Most recent fully developed leaves were collected from the plant and analyzed for nutrient composition. The total N content varied from 2.0 to 8.7 with a mean of 3.81 g/kg dry matter, total P content varied from 0.3 to 2.3 with a mean of 1.02 g/kg dry matter, total K varied from 3.2 to 14.3 with a mean of 8.66 g/kg dry matter, total Ca varied from 14.3 to 38.3 with a mean of 23.97 g/kg dry matter, total mg varied from 2.4 to 28.4 with a mean of 13.29 g/kg dry matter and total S content varied from 0.16 to 0.86 with a mean of 0.35 g/kg dry matter in the leaves of orchid species. The average N:P:K ratio among the various orchid species found to be 3.73:1:8.5. The water content of orchids showed a positive correlation with its P, K, Ca and Mg content and a negative correlation with N and S content. The linear regression analysis indicated that water content of orchids were significantly influenced by K, Ca, and Mg content of orchids at the level of 41.74, 65.67 and 21.44%, respectively.

Orchids are similar to the other non-orchidaceous plants in their requirements for minerals, except that they generally take a longer time to show mineral deficiency. Analysis of tissues shows that the mineral content of orchids is in the same range as those reported for non-orchidaceous plants. Nitrogen, potassium and phosphorus are the three major elements that have received greater attention in the studies involving the use of inorganic fertilizer application. The effectiveness of application of one element depends on the presence of other elements. There is evidence to indicate that a combination of organic and inorganic fertilizers gives better orchid growth. The rate of mineral uptake by orchid roots is relatively low when compared to the other crop plants (Naik *et al.*, 2009).

The study was conducted to find out the chemical composition and correlation between nutrient content of epiphytic orchids and associated substratum in the subtropical belt of Sikkim by Naik *et al.*, 2006, Naik *et al.*, 2007. The total nitrogen content of orchid species varied from 0.11% (*Vanda cristata*) to 0.28% (*Dendrobium fimbriatum*), the total phosphorus content varied from 0.11% (*Vanda cristata*) to 0.16% (*Dendrobium fimbriatum*), the total potassium content varied from 0.57%

(*Vanda cristata*), to 1.42% (*Dendrobium* aphyllum), the total calcium content varied from 2.1% (*Dendrobium chrysotoxum*) to 3.66% (*Luisia filiformis*), the total magnesium content varied from 1.1% (*Dendrobium chrysotoxum*) to 3.0% (*Vanda* critata), and the total sulphur content varied from 1% (*Vanda cristata*) to 2.3% (*Dendrobium* densiflorum). Linear regression analysis indicated that N, P, K, Ca, Mg and S content of orchid species were significantly influenced by N, P, K, Ca, Mg and S content of its associated substratum at the level of 83.0, 70.1, 97.5, 86.3, 80.2, and 63.3%, respectively.

Potassium

Unlike all the other major elements, potassium does not enter into the composition of any of the important plant constituents, such as proteins, chlorophyll, fats and carbohydrates, concerned in plant metabolism. The element is present in all parts of plants in large or fairly large proportions. It seems to be of special importance in leaves and at growing points, as these are especially rich in potassium. Probably the whole of the potassium in plants is present in soluble form, and most of it seems to be contained in the cell sap and cytoplasm. Potassium ions play an essential role, not by being part of the structure of plants, but by exerting a regulatory function on reactions and processes. Potassium ions activate over 50 enzymes in plants. This activation has been likened to a "lock-and-key" effect. For example the enzyme, starch synthetase, assists in the rapid conversion of glucose to starch in amyloplasts but only if potassium ions are present to modify the shape of the enzyme (activate). Similarly these ions activate the enzyme, nitrate reductase, as it converts the nitrate ion to nitrite ion in a step towards synthesis of proteins. It is important in many enzymes that are essential for photosynthesis Potassium ions play an important role in the steps involved in the conversion of amino-acids to proteins. These ions also protect plants from disease by promoting the growth of thicker outer walls of leaves. In addition they regulate the level and balance of growth and flower initiation (senescence) hormones in plants. The most important function of potassium ions in plants is in the control of the movement of water into, and out of, cells by a process called osmosis. The vacuoles of cells contain a high concentration of dissolved substances of which potassium ions are the most prominent positively charged ion. This results in water moving through the cell-wall membrane into the concentrated solution of the vacuole thereby building up a pressure within the cell. This pressure can be maintained because of the rigidity of the cell wall. The resultant turgidity of the cells plays a major part in producing the rigidity of the plant as a whole. This same mechanism is observed in two other important plant processes:

- The outer surface of leaves contains holes (stomatal pores) whose size can be increased by the swelling of adjacent sausage-shaped guard cells. When light falls on a leaf, potassium ions move from cells surrounding the stomata into the two guard cells. Water then enters the more concentrated solution and the resultant pressure build-up causes guard cells to bow outwards. This leaves an opening through which carbon dioxide from the air can enter for photosynthesis. The oxygen which is formed moves out through this opening. At night the potassium ions move away from the guard cells and the stomate closes. The non-growth hormone, abssicisic acid assists in bringing about this closure.
- Cell extension occurs because potassium ions accumulate in the central vacuole of the cell with other substances and water moves in by osmosis. As the vacuole swells to occupy 80-90 per cent of the whole cell volume, the cell wall extends uni-directionally under the influence of the auxin.

It promotes vigor and disease resistance, helps development of root system, improves plant quality, and increases winter hardiness due to carbohydrate storage in roots. Increases protein production, and is essential to starch, sugar and oil formation and transfer and in water relations. Potassium is also required for flower and fruit formation and thickening of cell walls.

The great mobility of potassium in plants, its special importance for and its reutilization by young tissues, and its apparent functions as a regulator of plant processes on a large scale are in harmony with the observations that, when potassium is moderately deficient, the effects are seen first in the older tissues and progress from these towards the growing points, but, when the deficiency is acute, growing points are severely affected, and die-back and general collapse of the plants commonly occur. Plants lacking potassium are more susceptible to fungal diseases, probably due to an accumulation in plant tissue of unused sugars and amino-acids which provide the food for the attacking organism. Under these conditions the roots are readily affected by rotting. Some potassium-deficient plants are more susceptible to frost damage. Lack of potassium causes leaf margin chlorosis, followed by necrosis from the outside to the midvein. Mature leaves show a browning and drying of the upper surface and puckering on the margins. Darkening appears between the leaf veins. Bronzing and dying of leaf margin. Chlorotic spotting with brown spots between veins of leaf occurs. Stunted internodes and roots. The stalks are thin and shortened. The fruits may fail to develop full colour.

Potassium (potash - K_2O) is generally immobile in soil. To correct the deficiency apply sulphate of potash. Inorganic fertilizer K is sold in various formulations including liquid, granular, and slow-release. Relatively small amounts of K become available from decomposition of manures and crop residues. It has been shown in many instances that the potassium content of plants is frequently much higher than is necessary for healthy growth, and it is generally considered that luxury (i.e., unnecessary) absorption of potassium often takes place. An excess of potassium can affect the uptake and availability of magnesium and calcium to a plant.

Vaz and Kerbauy (2000) studied the effect of nutrients on floral spike induction of Psygm*Orchis* pusilla by changing the total mineral salt concentration of standard Vacin and Went medium. Inorganic N ($NH4^+$ and NO^{3-}), P, K and Ca concentrations were investigated separately. A negative correlation between the total mineral salt concentration and the number of floral spikes was observed. A low concentration of N stimulated flower formation and this effect was more pronounced than that observed for the lowest total salt level used. Presence of $NH4^+$ in the medium was essential for plant growth and development, while absence of $NO3^-$ depressed flowering. Flower formation was also enhanced by P and K reduction to a half strength level, whereas the highest Ca concentration showed some floral stimulation. Plants grown under nutrient deficiency exhibited more yellowing than those cultivated on complete medium. Anthesis was favoured by removal of ethylene using a $KMnO^{4-}$trap.

The experiments were conducted by Chiang and Chen(2012) to investigate the effect of six fertilizers on growth and flowering of two *Doritaenopsis* cultivars, Dtps. Sogo Vivien 'Sogo F858' (F858) and Dtps. Happy Valentine x P. Otoglade (PA). Eighty milliliters of following fertilizers, Peters 20-20-20 (PT), Poly-feed 19-19-19 (PF), Taiwan Fertilizer 15-15-15 (TF), OMEX 20-20-20 (OX), Micro-Bao 19.7-21-20.7 (MB), and Floriculture Research Center fertilizer 8-9-14.5 (FRC) were fertigated to each pot respectively at concentration of 0.5 or 0.35 $g.L^{-1}$, when the moss volumetric water content was between 28% and 32%. Control plants were watered only. After four months, fertigation with FRC resulted in the lowest moss pH, and the moss pH with two cultivars decreased with time. Moss fertigated with FRC fertilizer always had the highest EC value, followed by TF. The EC of moss was higher with 0.5 g. L^{-1}fertilizer than that with 0.35 g. L^{-1}. Plants fertigated with FRC had more shoot growth, but less root growth. Higher fertilizer concentration (0.5 g. L^{-1}) resulted in better plant growth than 0.35 g. L^{-1}fertilizer. Plants with 0.5 g. L^{-1}fertilizers produced wider and longer leaves, with heavier fresh and dry weights. PT was the most expensive among the fertilizers tested, but plants with PT did not grow better than other fertilizer treatments. All F858 plants supplied with PF produced two spikes, regardless of fertilizer concentration used. Plant with FRC had the best vegetative growth and flowering, possibly due to increased potassium and iron in leaves, and also increased potassium, calcium, magnesium, zinc, and iron in roots. The results indicated that plants flowered earlier, had more flowers and heavier flower fresh and dry weights when PF, TF or FRC were applied.

Potential effects of $KClO_3$ on *Phalaenopsis* orchid flowering were investigated by Li *et al.* (2006). Increasing concentrations of 2, 4, 8 and 16 mmol/L KClO3 delayed spike emergence by 5, 6, 18 and 26 days, respectively. Moreover, they reduced final spike length by 2.1%, 4.0%, 16.2% and 46.1%, respectively. Nonetheless, application of $KClO_3$ at 4 and 8 mmol/L advanced the time to appearance of the first open flower by 13 and 24 days, respectively. Use of 8 mmol/L $KClO_3$ also increased the number of floral buds by 16%. Treatments with $KClO_3$ tended to reduce flower size. Overall, the data suggest that application of $KClO_3$ at an appropriate concentration (e.g. 8 mmol/L) can increase the number of floral buds and advance the time to *Phalaenopsis* orchid flowering, but may reduce flower size.

Calcium

Calcium occurs in plants chiefly in the leaves and the amounts present in seeds and fruits are relatively low. One of its main functions is as a constituent of the cell wall, the middle lamella of which consists largely of calcium pectate. This function appears to be of fundamental importance since, if calcium is replaced by any other of the essential elements, such as magnesium or potassium, the organic materials and mineral salts in the cells are readily leached through the walls. It provides a base for the neutralization of organic acids; it is concerned with activities of growing points (meristems), especially with root tips; it may be of importance in nitrogen absorption. Although a large proportion of the calcium contained in the plant may be soluble in water it does not appear to move freely from the older to the younger parts of plants, and hence young tissues contain lower proportions of calcium than older ones. This may explain why calcium deficiency effects begin at the tips of shoots. It has a major role in the formation of the cell wall membrane and its plasticity, affecting normal cell division by maintaining cell integrity and membrane permeability. It is an activator of several enzyme systems in protein synthesis and carbohydrate transfer. It combines with anions including organic acids, sulphates, and phosphates. It acts as a detoxifying agent by neutralizing organic acids in plants. Calcium is necessary for cell elongation and division, protein synthesis, root and leaf development, and plant vigor.It influences intake of other nutrients and increases calcium content of plants. Its deficiency is not often observed in plants because secondary effects of high acidity resulting from soil calcium deficiency usually limit growth, precluding expressions of Ca deficiency symptoms.

When calcium is deficient, terminal bud dies, young leaves are hooked, because Ca^{++} is not easily translocated inside the plant. Dying back occurs at tips and margins, foliage may become distorted. Stalk dies off at the terminal bud. Root systems may be damaged by the root tip death. The growing tips of roots and leaves turn brown and die. Without adequate Ca, which in the form of calcium pectate is needed to form rigid cell walls, newly emerging leaves may stick together at the margins, which causes tearing as the leaves expand and unfurl. This may also cause the stem structure to be weakened. Buds and blossoms fall prematurely in some crops. Symptoms are more prevalent on the fruits than on the leaves.

Calcium is immobile in plants and relatively immobile in soil. Calcium is often limited in acidic soils that receive abundant rainfall. Sometimes it is difficult to differentiate between calcium deficiency and magnesium toxicity. Excess Ca may cause boron or magnesium deficiencies.

Magnesium

Magnesium is essential for photosynthesis, proteins and chlorophyll formation. It is the central atom in the green pigment chlorophyll found in the chloroplasts of leaf cells. In the photosynthesis, certain wavelengths of light are absorbed by the chlorophyll to begin the many reactions which ultimately replace carbon dioxide from the air with oxygen while forming the high energy sugar, glucose. Magnesium ions activate some of the enzymes involved in these reactions especially those in which ATP are used as a source of energy. Magnesium ions play a major role in activating those enzyme-controlled reactions using ATP that occur during plant respiration. In this process glucose and oxygen react to release energy:

- Mostly as heat to maintain plant life throughout day and night;
- While the remainder constitutes to be entrapped in still more ATP (Adenosine Tri Phosphate) and other high-energy chemicals which ultimately lead to the formation of many important plant substances including chlorophyll, amino-acids and nucleotides.
- Magnesium also plays a part in the formation of DNA and RNA; in the activation of the enzymes that assist in the conversion of ammonium ions to amino-acids and, together with calcium, as phytate during the storage of starch in seeds and bulbs.
- Magnesium is also regarded as a carrier of phosphorus in the plant.

It moves freely within the plant and is taken from the older leaves to supplement new growth. Lower leaves are first affected, yellowing from the tip downwards. Mottling and yellowing between veins of older leaves while veins remain green. Yellow areas may turn brown and die. Yellowing may also occur on older leaves. Leaves may turn reddish purple due to low P metabolism. In severe cases, symptoms may appear on younger leaves and cause premature leaf drop. Plant stems are stiff, thin, and woody. Symptoms may be similar to N deficiency and are most often found in sandy soils that are low in organic matter and receive moderate to heavy rainfall. Symptoms occur most frequently in acid soils and soils receiving high amounts of K fertilizer or Ca.

Deficiencies occur on sandy acid soils. To correct the deficiency apply magnesium sulphate (Epsom salts) or dolomite (magnesium calcium carbonate $MgCO_3$, $CaCO_3$).

Deficiencies most likely on leached sandy soils and where high levels of N and K have been applied. Magnesium deficiency is rarely a problem in most soils. Excessive magnesium, on the other hand, can induce potassium deficiency due to interference with K uptake and utilization.

Zinc

It is required in the synthesis of tryptophan, which in turn is necessary for the formation of indole acetic acid in plants. It is an essential component of several metallo-enzymes in plants (dehydorgenases) and therefore is necessary for several different functions in plant metabolism. The enzyme carbonic anhydrase is specifically activated by Zn.It has a role in RNA and protein synthesis. Interveinal chlorosis occurs on younger leaves, similar to Fe deficiency. However, Zn deficiency is more defined, appearing as banding at the basal part of the leaf, whereas Fe deficiency results in interveinal chlorosis along the entire length of the leaf.

It is not a very mobile nutrient so a foliar spray of zinc and manganese is recommended for fruit set. Zinc is mobile in plants; mobility in soil decreases as pH increases. Availability is reduced by high pH, low levels of organic matter in mineral soils, soil compaction, excessive rates of P, and low temperature and wet soil. Organic matter can both inhibit and stimulate zinc uptake. Deficiency is more common in sandy soils. Spotted older leaves, usually yellow to white between veins (interveinal chlorosis). Early twigs loss of plant. Distorted or puckered leaf margins. Reduction in size of leaves (little leaf) and internodes of plants occurs.

Iron

Many fertiliser mixtures contain this micronutrient in the form of pale green iron sulphate crystals. However, unless the pH of the nutrient solution is below 5, very little iron is absorbed through the roots because iron ions, react with water producing an almost insoluble substance. Iron chelates are far more effective. They dissolve in, but are unaffected by water and on contact of the nutrient solution with the root surface they are reduced to Iron ions which are immediately bound to certain chemicals in the sap (e.g. citric acid) and carried throughout the plant. In plant growth iron mainly functions as an electron-transfer agent by alternating between the Iron (II) and Iron (III) states - ferrous and ferric states. Several iron-sulphur proteins are present in plant cells, the most important being ferredoxin. In photosynthesis this compound plays a key role in the presence of chlorophyll, in releasing oxygen from water. In the reaction leading to protein synthesis ferredoxin transfers electrons to reactants in enzyme-controlled reactions which convert nitrite ions to the amino-acid, glutamine. It also takes part in the conversion of sulphate ions to sulphide ions in the series of reactions that ultimately form the two sulphur-containing amino-acids, cysteine and methionine. Other iron-proteins also act as enzymes for reactions during plant growth. One such hemeprotein (peroxidase), catalyses the formation of lignin in cell walls. Other hemeproteins (cytochromes) provide electrons for the final steps of respiration which Oxygen forms water in a reduction reaction. Yet another protein of this type (catalase) is involved in the synthesis of chlorophyll and also plays a part in a reaction releasing oxygen during photosynthesis. Iron is closely concerned with chlorophyll formation but is not a constituent of it. Its role appears in this connection to be that of a catalyst. As a result of this function of iron, chlorosis is invariably an outstanding symptom when the element is deficient. Iron may also act as a catalyst, in the role of an oxygen carrier, in respiration. A point of great importance in connection with iron is its relative immobility in plant tissues. Its mobility seems to be affected by several factors, such as the presence of manganese, potassium deficiency and high light intensity. There is evidence that the amount of chlorophyll is related to active (readily soluble) iron in plants. It will thus be seen that so-called iron deficiency in the plant may in fact usually mean iron immobility. Lack of mobility may also account for the fact that iron deficiency is first shown in the younger tissues. Interveinal chlorosis in younger leaves. The youngest leaves may be white, because Fe, like Mg, is involved in chlorophyll production. Young leaves are yellow with contrasting narrow green veins; older basal leaves remain darker green. Exposed leaves are bleached and will eventually exhibit apical or marginal scorch. Leaves may be small, symptoms will be severe in cold, wet springs. Shoot length is usually normal, but diameter will be small; twig dieback and defoliation will occur when the deficiency is severe. Usually it observed in alkaline or over-limed soils. Iron is immobile in plants and mobility decreases in soil with increasing pH. Deficiencies can be corrected by lowering pH. Excessive P may induce deficiency.

Manganese

Manganese is usually supplied to plants as Manganese sulphate. It is absorbed through roots from the nutrient solution as manganese ions, which are then transported in the sap of the xylem to the growing leaves. The functions of manganese are regarded as being closely associated with those of iron and as being concerned with chlorophyll formation. Mn is necessary in Photosystem II, where it participates in photolysis. It fulfils its main function during the first stage of photosynthesis. In the presence of chlorophyll in the chloroplasts of leaf cells a Manganese-chloride protein brings about the light-induced reaction in which water is split with the release of Oxygen. Hence, when manganese is deficient, chlorosis is a common symptom. It may decrease the solubility of iron by oxidation and hence, an abundance of manganese within the plant may lead to iron deficiency and chlorosis. It is a constituent of pyruvate carboxylase. Mn activates indole acetic acid oxidase, which then oxidizes indole acetic acid in plants.

A Manganese deficiency restricts the formation of chlorophyll. Symptoms first appear as chlorosis in young tissues. Manganese deficiency causes yellowing between the veins of new foliage. Greenish-grey specks appear at the lower base of younger leaves. The specks may eventually become yellowish to yellow-orange. From its role in photosynthesis, an inadequate supply of Manganese reduces the amount of sugars and indirectly affects the formation of amino-acids and therefore protein synthesis. An excess of Manganese can reduce the uptake of Magnesium and Calcium.

Manganese is immobile in plants; mobility in soil decreases with increasing pH. Soils very high in organic matter or poorly drained are deficient at pH 5.8 to 6.5. For other soils, deficiency usually occurs between pH 6.5 and 8.0, especially where soil has been heavily limed. It is toxic in very acid soil. Excessive water, poor aeration, and excess heavy metals influence Mn uptake.

Manganese increases availability of Ca, Mg, and P and is necessary for chlorophyll synthesis and photosynthesis. To correct the deficiency apply manganese sulphate.

Sulphur

Sulphur is generally received by plants as sulphate ions. These enter the roots from the surrounding solution and most are conveyed unchanged in the sap-conducting vessels (xylem) to the leaves. Its functions in connection with proteins and chlorophyll are doubtless account for the similarity of its deficiency effects to those due to deficiency of nitrogen. In the chloroplasts sulphate ions are converted to sulphide ions, using energy supplied by ATP with the assistance of an iron-sulphur enzyme, ferredoxin. A similar reaction occurs in the roots. Most of these sulphide ions are rapidly converted into the two sulphide-containing amino-acids, cysteine and methionine that are required with other essential amino-acids for protein synthesis. Traces of cysteine are used to form an important sulphur compound (coenzyme A) that is involved in the step during respiration in which carbon dioxide is formed. The sulphur-containing vitamin B1 (thiamine) coupled with phosphate takes part in this reaction. Small amounts of methionine are converted to a compound (S-adenosylmethio-nine) from which can be formed lignins to strengthen cell walls, brightly coloured compounds (anthocyanins) and chlorophyll. Two iron-sulphur proteins and ferredoxin (in which they are similar) participate in the initial reactions of photosynthesis in which absorption of light by chlorophyll releases oxygen. Deficiencies result in yellow leaves and stunted growth. Sulphur can be used to lower the pH of alkaline soils. A sulphur deficiency in plants, although not very common, affects protein synthesis and, to a lesser extent, photosynthesis. In most plants sulphur is not easily translocated. As a result, the deficiency symptoms are a light green to yellow chlorosis over the whole of new leaves including the veins. The leaves may not grow to their normal size. Growth rate is retarded and maturity is delayed. Plant stems are stiff, thin, and woody. Symptoms may be similar to N deficiency and are most often found in sandy soils that are low in organic matter and receive moderate to heavy rainfall.

Organic sulfur is converted into available sulfate sulfur by soil bact*Eria*. Deficiency, when it occurs, is most likely on sandy, low-organic matter soils. Because enough sulphate is present in most soils, sulfur deficiency is fairly uncommon. Inorganic (mined) gypsum (17per cent sulfur), elemental sulfur (30-99 per cent S), ammonium sulfate (24 per cent S), potassium sulfate (17per cent S), super phosphate (12 per cent S) may be used for contol the sulphur deficiency.

Boron

Boron is an important constituent in cell-walls and plasma membranes of growing tissue including the roots. It is necessary in the synthesis of one of the bases for RNA formation and in cellular activities. It has been shown to promote root growth. It is essential for pollen germination and growth of the pollen tube. It has been associated with lignin synthesis, activities of certain enzymes, seed and cell wall formation, and sugar transport. It can delay the onset of calcium deficiency effects but cannot replace calcium; it tends to keep calcium soluble; it may act as a regulator of potassium/calcium ratios, and of the absorption of nitrogen; it may be concerned with the oxidation-reduction equilibrium in cells. Boron is extremely immobile in plants and is not translocated to new growth, but moves readily in soil. Boron is necessary to increase the yield and quality of Orchids and is associated with calcium utilization. Generally, B deficiency causes stunted growth, first showing symptoms on the growing point and younger leaves. The leaves tend to be thickened and may curl and become brittle. Deficiency is more apparent during drought stress. It causes internal breakdown and external necrosis of stems and roots. It reduces flowering and failure to set seed.

Boron is present in fertiliser mixtures either as sodium borate ($Na_2B_4O_7)^3BO_3$). Plants can tolerate only a narrow range in concentration (about tenfold) between deficiency and toxicity limits - about 0.5 to 6 ppm of boron in the nutrient solution given to the plants. Deficiencies often occur when pH is between 6.0 and 7.5, on deep sandy soils, and when high rates of N, K, and Ca are used. Excess of boron causes blackening or death of tissue between veins.

Molybdenum

Very small amounts of ammonium molybdate are necessary in fertiliser mixtures. The dry tissue analysis of healthy plants shows only about one part of molybdenum in ten million is present, much lower than for any other essential micronutrients. Molybdenum enters the plant as molybdenum ion. It is a structural component of the enzyme that reduces nitrates to ammonia. Without it, the synthesis of proteins is blocked and plant growth ceases. Seeds may not form completely, and nitrogen deficiency may occur if plants are lacking molybdenum. It is a necessary component of two major enzymes in plants, nitrate reductase and nitrogenase, which are required for normal assimilation of N. It is required by some soil microorganisms for nitrogen fixation in soils. Deficiency symptoms resemble those of N because the function of Mo is to assimilate N in the plant. Older and middle leaves become chlorotic, and the leaf margins roll inwards. In contrast to N deficiency, necrotic spots appear at the leaf margins because of nitrate accumulation. Deficient plants are stunted and flower formation may be restricted. Deficiency signs are pale green leaves with rolled or cupped margins. Molybdenum is mobile in plants and in soil. It is less available at a lower pH. Generally no deficiency over pH 6.0 was observed.

Copper

Copper is usually present in fertiliser mixtures as blue crystals of copper sulphate and enters the roots as either copper ions, or in a chelated form (with amino-acids). About fifty per cent of plant copper is present in chloroplasts where it performs most of its functions, generally as an electron-carrier between the copper (II) and copper (I) states. During photosynthesis a copper-protein (plastocyanin) together with another enzyme (plastoquinone) transfer electrons to reactants linking two series of reactions that take place in chloroplast membranes. Although plastoquinone contains no copper it requires a copper enzyme (laccase) for its formation. In most plants a copper-zinc protein (copper-zinc dismutase), rather than the corresponding manganese compound, assists in the removal of the highly reactive oxygen species, O^{2-}, formed during photosynthesis as previously mentioned. In the membrane of chloroplasts another copper enzyme (phenolase) plays a part in synthesising both lignin and the brown substances which exude when plant tissue is damaged thereby preventing fungal growth.

It is relatively immobile in soil and plants. Its deficieny is more in organic than mineral soils, and often in sandy than heavy soils. It is deficient at high pH. A copper-deficiency results in younger leaves being stunted in growth due to reduced photosynthesis. Reduced growth, distortion of the younger leaves, and possible necrosis of the apical meristem. Leaves distorted because of insufficient lignin deposited in the walls of the sap-conducting cells. Multiple sprouts occur at growing points, resulting in a bushy appearance. Young leaves become bleached, and eventually there is defoliation and dieback of twigs. A copper-deficiency results in the non-viability of pollen. Excess copper may affect the uptake of iron by plants.

Chlorine

Chloride ions assist in achieving a charge balance with positive ions in plant solutions and by entry into vacuoles, help to maintain the turgidity of cells. It has been shown to increase the water content of the tissues and affect carbohydrate metabolism, leading to an accumulation of starch in the leaves. The element is present in plants as chloride, and is wholly soluble. Chloride ions activate as enzyme (asparagine synthetases) for one of the series of reactions involved in the conversion of ammonium ions to amino acids. Cl is essential in photosynthesis, where it is involved in the evolution of oxygen. Deficiency symptoms first appear on younger leaves as chlorosis, wilting, stubby roots and bronzing. Odours in some plants may be decreased due to its deficiency. Most plants absorb up to 100 times very small amount of chloride that is essential for optimal plant growth. Deficiency seldom occurs because Cl is found in the atmosphere and rainwater.

Sodium

Sodium is not strictly an essential element it cannot be expected to have a specific role in the metabolic activities of plants. It is regarded as a conserver of potassium and as being able partly to replace that element in its role. In no instance, however, has it been shown that sodium can wholly replace potassium where the latter is acutely deficient. Sodium seems to affect the water relations of plants and often enables crops to withstand drought conditions which would otherwise produce severe adverse effects. It involved in osmotic (water movement) and ionic balance in plants.

Nickel

It is required for the enzyme urease to break down urea to liberate the nitrogen into a usable form for plants. Nickel is required for iron absorption. Seeds need nickel in order to germinate. Plants grown without additional nickel will gradually reach a deficient level at about the time they mature and begin reproductive growth. If nickel is deficient plants may fail to produce viable seeds.

Cobalt

The demand for cobalt is much higher for nitrogen fixation than for ammonium nutrition. Deficient levels could result in nitrogen deficiency symptoms.

Silicon

It is found as a component of cell walls. Plants with supplies of soluble silicon produce stronger, tougher cell walls making them a mechanical barrier to piercing and sucking insects. This significantly enhances plant heat and drought tolerance. Foliar sprays of silicon have also shown benefits reducing populations of aphids on field crops. Improved leaf erectness, stem strength and prevention or depression of iron and manganese toxicity has all been noted as effects from silicon. Silicon has not been determined essential for all plants but may be beneficial for many.

Factors Affecting Availability of Nutrients

Each type of plant is unique and has an optimum nutrient range as well as a minimum requirement level. Excessive nutrient uptake causes poor growth because of toxicity. Therefore, the proper amount of application and the placement of nutrients are important. If only one nutrient is missing from the soil solution the plant will not grow and develop normally. This notion was postulated by Justus von Liebig in his Law of the Minimum. The Law of Minimum maintains that yield is proportional to the amount of the most limiting growth resource. The interrelationship between a nutrient's availability and soil pH, air and soil temperature, available moisture, excesses of other nutrients, soil organic matter and soil mineral content are complex. Soil testing is the best way to accurately determine whether primary nutrient levels are optimum for plant survival and growth. There are several factors which can inhibit proper nutrition of orchid plant:-

pH: One factor which can introduce nutrient deficiencies in orchid plants is the substrate pH. The general pH range for Orchid plants is 5.4 to 6.8, but maintaining the pH from 5.6 to 6.2 is recommended. Poor uptake of nutrients, particularly boron, copper, iron, manganese, and zinc can occur if the pH of a substrate is above 6.5. Certain macronutrients like calcium and magnesium can become less available at pH below 5.4.

Cation Exchange Capacity

The Cation exchange capacity refers to the ability of the growing medium to hold exchangeable mineral elements within its structure. These cations include ammonium nitrogen, potassium, calcium, magnesium, iron, manganese, zinc and copper. Peat moss and mixes containing bark, sawdust and other organic materials all have some level of cation exchange capacity.

Soil Structure. Soil particle size and chemistry affect the way in which water and nutrients are retained and taken up by the plant. Soils high in loam or clay have a higher water and nutrient retention compared to sandy soils where nutrients are easily leached out.

Roots. undeveloped or insect damaged roots are ineffective in taking up nutrients.

Working equipment. Improperly working fertilizer equipment can cause nutrients to be less than optimum in the substrate. The significant problem with equipment failure is the introduction of multiple nutrient deficiencies.

Water Balance. Under or overwatering the plants may lead to nutrient deficiencies. Constant saturation of the substrate can lead to macro and micronutrient deficiencies. As oxygen levels are inhibited by overwatering, root growth can be limited and slow water uptake by plant. Elements such as calcium are transported via water flow and deficiency symptoms can develop rapidly on new growth. Also the inactivity of root systems due to saturated conditions can lead to inefficient uptake of iron or phosphorus.

Soluble salts. Soluble salts refer to the total dissolved salts in the root substrate at any given time and are measured in terms of electrical conductivity (EC). When the EC of the root zone is too low, plant growth is stunted and mineral deficiencies are observed. Low salts are usually due to too many clear water irrigations. Deficiencies among orchid plants like lower leaf yellowing (nitrogen), lower leaf purpling (phosphorus), and lower leaf interveinal chlorosis (magnesium) are common when values are below 0.75 mS/cm.

Mineral antagonisms. When certain elements are provided in excess to plants, uptake of other nutrients may be hindered. One example of a mineral antagonism is the nitrogen-potassium interaction. Other types of antagonisms are the potassium-calcium-magnesium interaction. Any one of these elements in excess can cause a decrease in the uptake of the others. Excess phosphorus can cause a decrease in uptake of zinc, iron, and copper.

Environmental Conditions. Extremes of temperature will inhibit nutrient uptake. Temperature can also play a role in the introduction of nutrient deficiencies. Purpling of the lower foliage is the common symptom.

Disease. Organisms like *Pythium* feed on the nutrients in roots causing an inefficient uptake of minerals. Iron deficiency (upper foliage interveinal chlorosis) can occur if root rot pathogens infect the root system. Foliar diseases, particularly fungal diseases can cause chlorosis of leaf tissue, a direct reflection of harvesting nitrogen from plant cells.

Mycorrhizae and Nutrient Availability

Mycorrhizae are symbiotic associations of roots and fungi that enhance plant nutrition. Mycorrhizae (fungus roots) are modified roots, consisting of mutualistic associations of fungi and roots. The fungus benefits from a hospitable environment and a steady supply of sugar donated by the host plant. Fungus provides several potential benefits to the host plant. First the fungi increase the surface area for water uptake and selectively absorb phosphate and other minerals in the soil and supply them to the plant. The fungi also secrete growth factors that stimulate roots to grow and branch. The fungi produce antibiotics that may help protect the plant from pathogenic bact*Eria* and fungi in the soil. Almost all plant species produce mycorrhizae. This plant-fungus symbiosis may have been one of the evolutionary adaptations that made it possible for plants to colonize land in the first place. Fossilized roots from some of the earliest land plants include mycorrhizae. Mycorrhizal fungi are more efficient at absorbing minerals than roots, which may have helped to nourish pioneering plants, especially in the nutrient-poor soils present when terrestrial ecosystems were young. Mycorrhizae take two major forms: ectomycorrhizae and endomycorrhizae. In ectomycorrhizae, the mycelium forms a dense sheath over the surface of the root. Some hyphae grow into the cortex in extracellular spaces between root cells. Hyphae do not penetrate root cells but form a network in the extracellular spaces to facilitate nutrient exchange. The mycelium of ectomycorrhizae extends from the mantle surrounding the root into the soil, greatly increasing the surface area for water and mineral absorption. Compared with "uninfected" roots, ectomycorrhizae are generally thicker, shorter, more branched, and lack root hairs. Ten percent of plant families have species that form ectomycorrhizae. Endomycorrhizae have fine fungal hyphae that extend from the root into the soil. Hyphae also extend inward by digesting small patches of the root cell walls, forming tubes by invagination of the root cell's membrane. Some fungal hyphae within these invaginations may form dense knotlike structures called arbuscles that are important sites of nutrient transfer. Roots with endomycorrhizae look like "normal" roots with root hairs, but the microscopic symbiotic connections are very important. Endomycorrhizae are found in more than 85 per cent of plant species, including important crop plants such as corn, wheat, and legumes. Roots can be transformed into mycorrhizae only if they are exposed to the appropriate fungal species. In most natural systems, these fungi are present in the soil, and seedlings develop mycorrhizae. However, seeds planted in foreign soil may develop into plants that show signs of malnutrition because of the absence of the plant's mycorrhizal partners. An epiphyte is an autotrophic plant that nourishes itself but grows on the surface of another plant, usually on the branches or trunks of trees. Epiphytes absorb water and minerals from rain, mostly through their leaves. Examples of epiphytes are staghorn ferns, some mosses, Spanish moss, and many species of bromeliads and orchids. There is growing evidence that the molecular biology of root nodule formation is closely related to mechanisms that first evolved in mycorrhizae. The nodulin genes activated in the plant during the early stages of root nodule formation are the very same genes activated during the early development of endomycorrhizae. Mutations in these genes block development of both root nodules and mycorrhizae. The signal transduction pathways of both systems share some components. Experimental application of plant hormones called cytokinins to root cells of legumes activates expression of early nodulin genes even in the absence of Bacterial or fungal symbionts. "Infection" by either symbiont causes cytokinin concentrations to increase naturally. Cytokinins may be one of the links between the "I'm here" announcement of the microbes and the changes in gene expression in the plant that lead to structural modifications of the roots. Further evidence comes from the similarities in the chemical cues used by both microbes. The Nod factors secreted by *Rhizobium* bact*Eria* are related to chitins, the same compounds that make up the cell walls of fungi. A reasonable hypothesis is that root cells have a family of closely related receptors that detect their particular Bacterial and fungal symbionts. Mycorrhizae evolved very early, probably over 400 million years ago, in the earliest vascular plants. In contrast, the root nodules in legumes originated only 65-150 million years ago, during the early evolution of angiosperms. The common molecular mechanism in the root's two major symbiotic relationships suggests that root nodule development was at least partly adapted from a signaling pathway that was already in place in mycorrhizae. On sandy soils small amounts of fertiliser should be used frequently, to ensure a steady supply of nutrients to the plant. On heavier soils apply fertiliser in greater quantities, but less often. Plants can also suffer from nutrient toxicity if too much of a particular nutrient is applied. Make sure that fertiliser is watered in after application, to avoid burning of the plants.

In nature, orchids obtain their supply of inorganic nutrients like calcium, magnesium, iron, potassium, nitrogen and traces of manganese, boron, copper, zinc *etc.* from the tree on which they are growing and also from atmosphere and decaying vegetables and dropping of birds. However, under controlled conditions they have to be supplied with all major and minor nutrients. Taking into consideration the special need of different orchids, a large number of fertilizer mixtures, both solid and liquid, are available in market. Liquid fertilizers are much more quickly absorbed and can be applied more frequently. As the orchids are slow growing, slow release fertilizers like osmocote can be used to get very good result. Usage of fertilizers should also depend on stage of growth. During vegetative growth, large quantities of nitrogen are required while during flowering, nitrogen should be reduced and amount of phosphate increased. It have been getting excellent results by using slow release fertilizer mixtures (NPK 20:20:20) with trace elements and coconut water (20-25%) applied every week for three successive weeks followed by a 10:20:30 (NPK). Orchids are relatively light feeders compared to other plants.

Some orchids, like members of the genus *Disa*, are exceptionally sensitive to feed and are injured at modest feed strengths. *Odontoglossums* are particularly sensitive to moderate feed strengths. The important factor is not to exceed a certain total value of dissolved salts. The concentration of salts in a nutrient solution, held by the mix, affects a parameter known as the "salt index" or osmolarity. As the osmolarity of the substrate solution increases less water transport occurs from the root. Typically, orchids are watered and allowed to partially dry, between each watering, thus a further increase in osmolarity occurs as water evaporates from the substrate. With excessive substrate osmolarity a plants water loss from transpiration will exceed the water replaced by the roots. This results in several problems. A common problems is leaf tip die-back, a condition where the leaf tips dies because of lack of water. Warm days, a poor root system and excessive drying out of the substrate in the presence of high feed rates will exasperate leaf tip die-back.

Bio-fertilizer and Growth or Orchid Plants

Fertilizer for crops may be classified as chemical fertilizer, bio-fertilizer, green manure and compost. Bio-fertilizer contains microbes or enzymes, and is available in the forms of powder, granule or liquid. When bio-fertilizers are applied on the seeds, seedlings or soils, not only the effectiveness of nutrient uptake can be improved, but also the amount of beneficial microbes can be replenished to maintain soil in good eco-environment. In general, the advantages may be grouped into 3 categories: (1) nitrogen fixation, to fix atmospheric nitrogen into ammonia, and convert into nitrate which can be used by crops, (2) solubilization of soil minerals such as phosphorus, calcium and iron to be used by plants, (3) promotion of the nutrient uptake by roots and plant growth. The bio-fertilizer composition comprises of a symbiotic organism, which is selected from the group consisting of *Rhizoctonia* sp. (BCRC930076 and BCRC930077), a growth substance and a medium, which is used to mix and evenly disperse the symbiotic organisms and the growth substance.

The bio-fertilizer can be applied to promote orchid seed germination in test tubes or be applied in seedling growth after deflasking to promote growth, preventing disease, and increasing the beneficial components toward human health of medicinal orchids. In addition, the composition also increases the flowering rate and flowing quality.

An experiment was conducted in order to study the effect of foliar application of organic nutrients on the growth, flowering and flower quality of *Dendrobium* orchid cv. Sakura pink comprising eighteen treatments with four replications. Charcoal, brick pieces and coconut husk have been used as potting media for nourishing the plants. Foliar organic nutrients viz., vermiwash (1% and 2%), panchakavya (1% and 2%), humic acid (0.2% and 0.3%) and tender coconut water (1:1 and 1 : 2 dilution) were sprayed at four days and eight days interval. A treatment comprising recommended dose of chemical fertilizers was used as check and no fertilizer application as control. The results revealed that vermiwash 1% at 4 days interval found to be optimum for flower quality of *Dendrobium* orchid cv. Sakura pink. Even though recommended dose of chemical fertilizers recorded higher values in vegetative growth and flowering characters, it was on par with the plants treated with vermiwash 1% at 4 days interval (Karuppaiah and Sendhilnathan, 2011).

Irrigation System

Orchids originate from the tropical and lush areas of earth and are accustomed to high rain and more humidity levels. During growing season plenty of water is required by Orchid plants. Steady flow of water supply need to appropriately humidify orchids, most orchids can live very well with a humidity level of 80 per cent. The appearance of an orchid can also determine the quantity of water needed. Plants that have big pseudobulbs (thick base of stem that stores water and nutrients), such as *Cattleyas*, usually needs dry condition do not like moist all the time. On the other hand, orchids with no pseudobulbs, such as *Paphiopedilum* and *Phalaenopsis*, require a moist condition. As the orchid need a great deal of irrigation water, it would be wise to ensure that the rainwater that falls on the orchid house does not get wasted. Collecting the rainwater water from the orchid house roof via gutters and stored in containers and storage devices would be most valuable.

Watering of orchids: Water requirement of orchid plants is varying with stage. Juvenile plants need a mild moist condition that can be provided by moist moss or peat. The excess amount of water otherwise causes damping off of orchid seeding. Watering should always be done in the day so that plants dry out by night time. The frequency of watering is depending on the climatic conditions. In general, once watering during the winter and twice in a week during summer season are essential for good growth of orchid. The size of orchid container also determines the frequency to water, regardless of climate conditions. Typically, a 15 cm pot needs water every 7 days and a 10 cm pot needs water every 5 to 6 days.

Water requirements of plants can also affected by the type of potting medium. Bark has a tendency to dry out rapidly however, pot filled with *Sphagnum* moss remain moist for longer time. Moisture content inside the pot can be checked by dipping of finger, bamboo stick or pot a wooden stick for 2-3 cm deep into the pot. The potting medium should always be damp but not soggy and it should not be extremely dry. It should be water only when the potting material becomes reasonably dry. *Cymbidiums can* sustain dry condition longer, though the regular application of a balanced fertilizer will give an increase in growth. *Cymbidiums* should be watered two to three times per week in the spring and summer. It keeps soil moist evenly. *Cymbidiums* should be irrigated with tap water as it promotes leaching of harmful salts, which rather causes die-back of the leaf tips. In the winter, soil should be kept barely moist by watering once a week. *Cymbidium,* though survive in dry conditions but flowered. As a guide use the following.

Summer: During summer season orchid plants watered daily or two to three times/week. During this period plants should be encouraged by a liberal supply of moisture and heated, in order to make their growths. In dry, dusty, hot weather a due supply of moisture should be attended constantly by wetting the floor and walls regularly. In such weather the plants will be greatly benefited by washing the leaves once a-week with a wet sponge. Washing allows cleanses of surface and clears the pores of the leaves also keeps down the population of various insects such as red spider, scale and thrips and also gives a freshness and tidy appearance to the plants.

Autumn and spring: Irrigation or watering should be given once or twice per week and slightly more often if warmer.

Winter: Once per week or two if under cover. During this season most of the orchids especially such species have pseudo-bulbs have a rest hence require irrigation once or twice a week. In the winter plants may be wash by squeezed in a way that water do not adhere on the plant parts.

Cymbidiums have stout roots of 6 mm thickness. These roots act like blotting paper when watered and grow exclusively on trees and in leaf litter, hence it requires special potting mixture and must have specially prepared potting mix. Bring flowering plants under cover to maximize the flower life and more water should be applied as flowers use more water than the plant would normally use. Overhead misting may be used to keep humid air around orchids. For *Cymbidium* it is favorable condition during hot spells. During the winter a cut should be in watering, but plant should not be run dry by keeping the potting culture medium somewhat moist to the feel. Normally, watering once each week is sufficient but, during the hot summer months it may be necessary to water twice weekly. Try to water in the morning, before mid day which will avoid the possible loss of new growths. Knowledge of the type of orchid is very essential for determining the water requirement of plants. Here are some general guidelines:

The influence of gibberellic acid (GA_3) and water regime was evaluated in the flowering induction and quality of two orchid hybrids belonging to the genera *Cattleya* (C.) and *BrassoCattleya* (Bc.). Five GA_3 concentrations (0, 125, 250, 500 and 1,000 mg L^{-1}) were tested through four consecutive leaf applications in adult plants that had already flowered at least once, besides two water conditions (one and four irrigations per week). Applications were performed in October and November for Bc. Marcella Koss and in January and February for C. *Irene* Holguin. Flowering could not be induced in the latter by gibberellic acid. In Bc. Marcella Koss, the application of 250 mg L^{-1} GA_3 combined with decreased irrigation frequency induced flowering in around 83% plants. By using the same GA_3 concentration but frequent irrigation, only 17% plants were induced to flower. The number and size of flowers increased after application of higher GA_3 concentrations. (Cardoso *et al.*, 2010).

Colombo *et al.* (2005) studied to evaluate the efficiency of different vegetal substrates under two irrigation systems, during the acclimatization phase of the *Cattleya* Chocolate Drop x (*Cattleya* guttata x *Laelia* tenebrosa). Orchid seedlings that were cultivated *in vitro* were transplanted to the coconut powder (S1); coconut fibre (S2); defibered xaxim (tree fern fibre S3); and *Sphagnum* (S4). Next, they were acclimatized in two irrigation systems: hand-operated (IM) and intermittent (II). The following variables were assessed 90 days after transplanting: number of roots, areial part height, fresh weight, number of plants with shoots, pH substrate and seedling survival rate. The S2 substrate and II system were the most optimum for the acclimatization of *Cattleya* Chocolate Drop x (*C. guttata* x *L. tenebrosa*) orchid. Droll (2005) describes a mathematical method determining some quantitative recommendations for watering orchids based on measured parameters related to the orchids, the pots, the potting materials and the aeration conditions. It highlights the importance of understanding "moisture diffusion coefficient", a new and useful moisture parameter, in the management of moisture requirements in orchids.

Table 3.6 : Water requirement of commercial genera of orchid

Orchid type	Water requirement
Phalaenopsis (Moth orchids)	Evenly moist year around
Paphiopedilum (Slipper orchids)	Evenly moist and allowed to dry before re-watering
Brassia (Spider orchids)	Moist during active growth period (summer and spring); less moisture requires at dormancy
Cattleya (Corsage orchids)	Sufficient irrigation during active growth, allowed to dry before re-watering, less moisture requires at dormancy
Cymbidium (Boat orchids)	Evenly moist; less water required after growth is complete
Dendrobium, Phalaenopsis type	Allowed to dry before re-watering
Dendrobium, (Nobile type)	Allowed to dry before re-watering; withhold in the winter (2, 3 months)
Masdevallia	Evenly moist year around
Miltoniopsis	Evenly moist year around

Every plant of the family orchidacea needs regulated watering. A simple observation would be to ensure that the plant's root tips do not dry up. *Vanda* specimens require watering on a regular basis, especially during summer. Proper watering, relative humidity, air circulation among the roots, potting compost and light factor plays crucial roles to determine the need of watering. The following guidelines are in consideration during watering.

1. When it is not sure whether the plant is dry enough, just wait for another day before watering.
2. Pots should be enough watered so that the water comes out in the drainage holes. It encourages plant to grow deep roots as well as flush out excess fertilizer that could burn the roots.
3. Direct touch of water to plants should be avoided.
4. Orchids such as nobile type *Dendrobiums* need a dry resting period. But if the pseudobulbs get so dry that they begin to shrivel, just give the plant a little mist with a spray bottle.
5. Orchids should be watered at morning. That allows the leaves and crown of the orchid to dry before the night falls. Otherwise orchid could turn into mushy rot.

6. Potting mixture should be adjusted according to the moisture preference of plant and the desired frequency of watering.

More watering should be provided often when:

- The plant is actively growing
- There is strong wind/air movement
- The temperature is warmer
- The humidity is lower

Less watering should be provided often when:

- The plant is not actively growing
- The temperature is lower
- The humidity is higher

Water Quality: Orchids are susceptible to hard and contaminated water. Hard water contains iron, lime and various salts. The lime containing water produces white marks on the foliage. However, sometime it is beneficial to members of *Gyjpripediums.* The iron containing water may also injurious to orchid plants. Plants if treated by saline water may develop die back of leaf tip. Some plant can tolerate the toxic effect of hard water for example *Phragmipedium* and *Cattlya.* The concentration of salts is expressed by EC (electrical conductivity) and can be measured with a simple EC meter. A high EC indicates a high salt content. The quality of water is top priority because crops react to even a tiny concentration of salts, which causes a reduction in productivity. The pH is other factor determines the quality of water. Generally plants prefer water having 6 to 7 pH value.

Rain water and reverse osmosis water is the purist, while daily supply water and well water can be vary quite a bit in the concentration of dissolved minerals. The quality of water used, whether for spraying or watering, is of great importance. Since tap water has often been chemically treated, generally with chlorine, it should be used with caution. The best water for orchids is rainwater, as it passes through the air, dissolves and absorbs many substances such as dust, pollen and other organic matter, this enriched the nourishment of the plant. Collected salts in tap water could harm the plants, so care should be taken to make flushing every time when plant is watered.

The temperature of the water is also important. If the water temperature and the surrounding air temperature are equal, no harm will result, and slight differences either way can be tolerated by healthy plants. Using too cold water especially in winter time results in fatal or long-term damage, not easily discernible. Orchid root systems are not used for soaking or sitting in extremely wet soil and are very sensitive to over watering. The chlorine in municipal water is not good for plants. Watering of orchids is key to keep them healthy and prosperous for years. This most essential aspect should be immediately addressed in its right perspective. The healthier the roots, the more abundantly do the plants bloom.

Symptom of excessive watering on plants: The thick leaves become yellow, pleated and soft and buds may die prematurely. Excess water makes the plants roots, mushy, soggy and black.

Symptom of Water Stress Conditions

- Leaves become pleated in specimens such as *Miltonias*.
- Shriveled pseudobulbs in specimens such as *Cattleya* and the stems tend to bloat and appear neglected.
- Puckered, soft and droopy leaves in specimens such as *Cattleya*.
- Limp and yellowish leaves in *Phalaenopsis* .
- Premature falling of buds.

Assessment for Under-watering or Over-watering Symptoms

- Plant should be reversed.
- Gently pat the pot in order to lose its compost and allow it to fall on some clean cloth on a table.
- Potting media should not be drop down when invert the pot. Knife should be use for depotting at the edges to loosen the medium.
- The removed potting material, should be analyse critically for mushy and black roots or white and grayish roots.
- In case the roots are healthy and appear normal it would be best to repot the plant orchid in fresh compost.

Water supply and fertilizing: The orchid plant is required to take up large amounts of water and nutrients, with a relatively small root system, and manufacture photosynthesis for a large amount of flower per unit area with a foliar system relatively small in relation to required production. Watering-cans are the simplest way to irrigate the orchids as they provides specific, targeted watering for individual plants, particularly if they are being grown in pots or containers. However, for mass watering, they are too labour-intensive. To overcome the limitations of the traditional watering, various methods like micro-irrigation systems, misters and sprinkler systems are used.

Fig. 3.6 : Watering with organic nutrient extract

Basically the watering system should ensure that water does not fall on the leaves or flowers as it leads to disease and scorching problems.

Micro irrigation system: Micro irrigation system is the best for watering plants in an orchid house. Micro-irrigation refers to low-pressure irrigation systems that spray, mist, sprinkle or drip. The water discharge patterns differ because emission devices are designed for specific applications due horticultural requirements. Micro-irrigation components include pipes, tubes, water emitting devices, flow control equipment, installation tools, fittings and accessories. Micro sprinklers or drip irrigation equipments can be used for orchid crop irrigation.

Sprinkling irrigation system: Sprinkling system pipes is widely used in orchid houses. In this system water in form of fine droplets or mist conveyed to plant body by means of nozzles mounted on a riser by means of pressure created by pump. The pipes could be placed underneath the pots or over hanging condition, which in climates with high air humidity. To make sure that the plants do not remain wet at night, irrigation should be done in the morning. The sprinkling system can also be used to moisten the crop when the weather is sunny and very dry, thus having a cooling effect, increasing the relative humidity and reducing the transpiration. In overhead sprinkler irrigation water is pumped under pressure and sprayed down onto the plants from flat spray nozzles. These may be portable, semi-portable and solid set systems and mounted on an overhead network of pipes or even simply mounted on the top of a stake. Overhead sprinkler irrigation systems need a plentiful supply of water at a relatively high pressure. With this system, the foliage of a crop does get wet and leaves remain wet for an extended period of time. When irrigating inside orchid house or within a specified width, the sprinkler risers may be used as upright mounted or inverted mounted. In micro sprinkler system, water under high pressure is forced through nozzles arranged on a supporting stand at about 30 to 60 cm height. This facilitates watering at the base level of the plants.

Drip irrigation system: The orchids need food and extra water at certain stages of development loves to be fed a food dissolved in water, rather than a granular food that may or may not dissolve on time. Drip irrigation is rapidly on the increase in orchid house cultivation. Here, the plants get water individually. Drip irrigation is such a slow process compared to overhead sprayers or hand watering that the soil in the containers is never washed away or splashed out. Due to prolonged wetness plants can develop molds or fungus on the foliage and wet leaves can sunburn in direct sunlight. Drip irrigation provides proper management from of water stress from over or under watering and heat stress from insufficient water on hot sunny days. The water should be clean because the small droplet openings must not get blocked. It is very desirable that the system has a good filter at the place where the water enters it. Drip irrigation system works on a low water pressure of 0.5 or 1 kg/cm^2. This can be achieved by a pump or overhead tank supply. Usually, fertilizer in solution in the correct dosage (fertigation) is administered when drip irrigation is done.

Fertigation system: In fertigation system an automatic mixing and dispensing unit is installed which consists of three systems pump and a supplying device. The fertilizers are dissolved separately in tanks and are mixed in a given ratio and supplied to the plants through drippers or spray nozzles.

Fertilizers: Fertilizer dosage has to be dependent on growing media. Soilless mixes have lower nutrient holding capacity and therefore, require more frequent fertilizer application. Essential elements are easily available with 5.5 to 6.5 pH. In general Micro elements are more readily available at lower pH ranges, while macro elements are more readily available at pH 6 and higher.

Forms fertilizers: Dry fertilizers, slow release fertilizer and liquid fertilizer are commonly used in green houses. Slow release fertilizer release the nutrient into the medium over a period of several months. These fertilizer granules are coated with porous plastic. When the granules become moistened the fertilizer inside is released slowly into the root medium they should. Slow release fertilizer never be added into the soil media before steaming or heating of media. Heating melts the plastic coating and releases all the fertilizer into the root medium at once. The high acidity would burn the root zone.

Liquid fertilizers are available in powder form singly or in. They dissolve in warm water completely.

Fertilizer Application Methods: In constant feed continuous supply of nutrient to plant growth occurs and results in steady growth of the plant. Low concentration at irrigation is much better. This provides fertilization with each watering is referred as fertigation.

In case of intermittent application liquid fertilizer is applied in regular intervals of weekly, biweekly or even monthly. The problem with this is wide variability in the availability of fertilizer in the root zone. At the time of application, high concentration of fertilizer will be available in the root zone and the plant immediately starts absorbing it. By the time next application is made there will be low or non existent. This fluctuation results in uneven plant growth rates, even stress and poor quality crop. Fertilizer injectors injects small amount of concentrated liquid fertilizer directly into the water lines so that green house crops are fertilized with every watering.

General problems of fertigation: Nitrogen tends to accumulate at the peripherous of wetted soil volume. Hence, only roots at the periphery of the wetted zone alone will have enough access to Nitrogen. Nitrogen is lost by leaching and denitrification. Since downward movement results in permanent loss of NO_3^- N, increased discharge rate results in lateral movement of N and reduces loss by leaching.

Phosphorus: It accumulates near emitter and P fixing capacity decides its efficiency. Low pH near the emitter results in high fixation.

Potassium: It moves both laterally and downward and does not accumulate near emitter. Its distribution is more uniform than N&P.

Micronutrients: Excepting boron, all micronutrients accumulates near the emitter if supplied by fertigation. Boron is lost by leaching in a sandy soil low in organic matter. But chelated micronutrients of Fe, Zn can move away from the emitter but not far away from the rooting zone.

Plant Growth Regulator

Any substance or mixture of substances intended through physiological action, for accelerating or retarding the rate of growth or maturation, or otherwise altering the behaviour of plants or the produce thereof are known as plant growth regulators. It does not include plant nutrients, trace elements, nutritional chemicals, plant inoculants and soil amendments.

Plant hormones are organic compounds other than nutrients, synthesize by plants and regulate plant growth and development. They are active in very low concentrations, produced in certain parts of the plants and are usually transported to other parts where they elicit specific biochemical, physiological and morphological responses. They are also active in tissues where they are produced. The term plant growth regulator is usually used to denote a synthetic plant hormone, but most of the synthetic compounds with structures similar to those of the natural hormones have also been called hormones. For instance, the synthetic cytokinin, kinetin is considered a hormone.

Classification of PGRs

Each plant growth regulators have many responses. The effects of different growth regulators overlap and may be stimulatory or inhibitory. Plant growth regulators may be classified into five groups:Auxin, Cytokinins (CKs), Gibberellins (GAs), Ethylene, and Abscisic Acid (ABA). The new plant hormones, Brassinosteroids (BAs) and Salicylates or Salicylic Acids (SAs) are further included into this category.

Growth hormones, Auxins and Cytokinins, stimulate metabolism and regulate growth. Stress hormones (GA's, Ethylene and BA's) are the complementary of growth hormones. The latter group is triggered by a deficit of nutrients, when there are not enough nutrients to allow growth. They allow a plant to survive during stress period and redistribute resources in order to resume growth. Shock hormones are the ABAs and SAs, allow a plant respond to rapidly developing stress situations (ABA) or return to normal conditions after the stress has passed .Flower initiation is controlled by hypothetical hormones called florigens, but these substances are still to be identified.

All plant hormones have a very broad spectrum with complex action. Each hormone performs its specific functions. However, nearly all measurable responses to heredity and environment are controlled by interaction between two or more hormones. Such interactions may occur at various levels, including the synthesis of hormones, hormone receptors and second messengers, as well as at the level of ultimate hormone action. Furthermore, hormonal interactions may be cooperative, antagonistic or in balance.

It have been indicated that phytohormones enter into cells and regulate intracellular processes. In several cases plant hormones mediate between an extern signal and a physiological activity (a cell's response). Plant hormones act partially synergistic, partially antagonistic. Growth and development of plants is regulated by a combination of genetic factors and environment influences. Plants have receptors that sense and respond to a number of environmental cues including photoperiod, temperature, pressure and moisture changes.

Plant growth regulators mediate the effects of environmental cues. Genes code for the enzymes that catalyze the chemical reactions in plant growth and development. The transport velocity within the system of vascular bundles is dependent on the transpiration that is itself dependent on water supply, temperature, and species-specific differences. Signal transduction pathways are important for chemical messaging in plants and both environmental cues and hormones serve as signal stimuli.

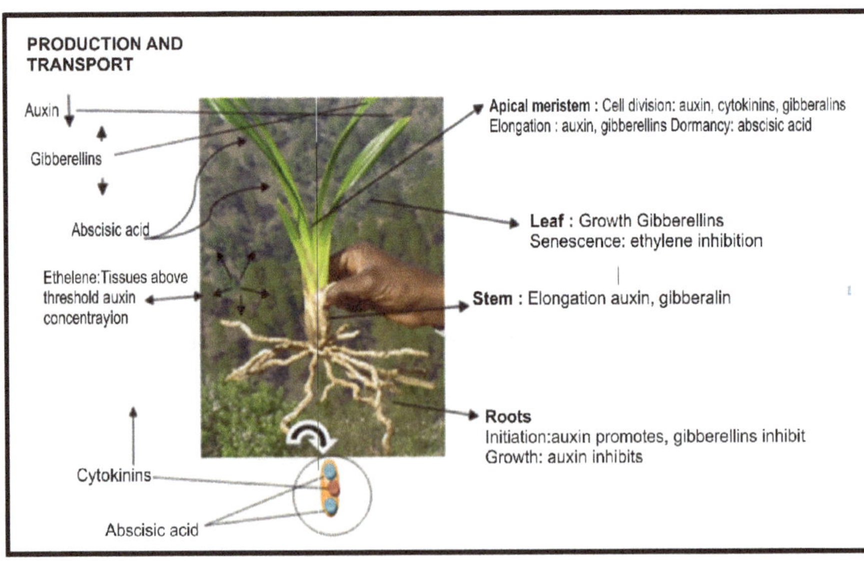

Fig. 3.7 : Production of Plant hormones and action centre

Plants growth regulators can function as signal molecules (ligands) that trigger signal transduction pathways in cells. Such pathways often result in the synthesis of transcription factors that in turn promote synthesis of enzymes that facilitate chemical reactions within the cell.

Auxins

This hormone was named auxin by Frits Went in 1926. Indole acetic acid (IAA) is the natural auxin, synthesized from indole or tryptophan. There is several synthetic auxins. It promotes growth in molar concentrations of 10^{-3}to10^{-8}. They are synthesized in apical meristems of shoots, young leaves, seeds and fruits. Auxins always move down the stem parenchyma cells towards roots by polar (charged) transport (auxin becomes negatively charged) using proton pumps, an ATP requiring process. Auxin enters cells as IAAH passively, or as IAA- via active cotransport. IAAH dissociates within the cytosol and special auxin transport proteins in the basal end of the cell are needed to carry auxin through the plasma membrane to the top of the adjacent cell. Auxin destined for root tissue, however, moves through phloem sieve tubes.

Functions of Auxins

- Auxins are the common rooting hormones. The treatment of keiki root with a minute concentration of the auxin is found effective to initiate orchid roots.
- Auxin produced in apical buds tends to inhibit the activation of buds lower on the stems. This is known as apical dominance. This effect lessens with distance from the shoot tip.
- It promotes elongation and cell enlargement. For cell elongation; proton pumps increase the H^+ concentration in cell walls, which stimulates expansions of proteins that disrupt hydrogen bonds and break cross linkages in cellulose. This facilitates wall expansion when cells take more water.
- Auxins are involved in tropic responses. They migrate away from light, which accounts for the uneven elongation of cells on the shaded side of a plant unevenly exposed to light.
- It stimulates the production of secondary growth by simulating cambium cells to divide and secondary xylem to differentiate. Wound tissues repair is initiated by auxin when portions of vascular bundles are damaged.
- Promotes other hormone production, especially ethylene when auxin concentration increases.

- Promote flower initiation.
- Loss of auxin initiates leaf abscission.
- Auxin also plays a role in fruit development after released from developing seeds.
- Synthetic auxin used as herbicides include 2-4-D and 2-4-5-T.

Cytokinins

Cytokinins are a group of phenyl urea derivatives of adenine, one of the molecules in DNA. The first cytokinin was chemically isolated in 1913. Cytokinins are found in actively dividing tissues of seeds, fruits, leaves and root tips and sites of wound tissue. Studies indicated that root tips are sources of cytokinin production and cytokinins are transported through xylem to rest of the plant.

Functions of Cytokinin

- It promotes cytokinesis. They are found primarily in root meristems embryos and fruits. They migrate from roots to the shoot systems of plants by xylem tissues.
- In tissue culture normally, cytokinins work in conjunction with auxins, which promote elongation and cell expansion. When auxin is added to a tissue culture medium, cells in the absence of cytokinins enlarge but division does not take place. When kinetin and IAA are both added to a tissue culture medium, cells divide rapidly producing many small cells. The ratio of cytokinins and auxin controls the tissue differentiation in the tissue culture.
- Cytokinins can promote the growth of axillary bud that overcome the apparent inhibiting effect of auxin. This is one of the ways in which plants balance root and shoot growth.
- It also retards leaf senescence probably by stimulating RNA and protein synthesis and delaying degradation of chlorophyll. It inhibits dormancy.

Gibberellins

Ewiti Kurosawa is credited with the discovery of gibberellins when he determined that a fungus was responsible for abnormal rice seedling growth, called the "foolish seedling" disease. The fungus secreted a chemical that caused the rice plants to grow abnormally long, and then collapse from weakness. The fungus was *Gibberella fujikuroi*, hence the hormone named. Many seeds contain a variety of different gibberellins. Gibberellins are produced in roots and younger leaves, but have their highest concentration in seeds. Most effects of gibberellins are seen only in agreement with auxins.

Functions of Gibberellin

- Gibberellins work in combination with auxins to promote rapid elongation and division of stem tissue.
- It increases the growth of many plants and seedlings.
- It enhances bolting of biennials to produce flowers during the growing season.
- It reversed the genetic dwarfism.
- After imbibitions of the seed coat, gibberellins break the dormancy of seed. It signals to germination activities of seeds.
- Application of gibberellins to seeds will counter the normal environmental cues, such as exposure to low temperatures. They stimulate RNA to promote synthesis of enzymes that convert stored nutrients (starches) to sugars needed for rapid cell respiration during germination.
- It minimizes a plant's requirements of long day exposure or periods of low temperature in order to flower.
- Abscisic acid can encounter the effect of gibberllin to keep seeds dormant.

Abscisic Acid

Abscisic acid (ABA) is a hormone that functions by inhibiting growth activities in environmental stress condition. It often serves as an antagonist to the other growth promoting hormones.

Functions of Abscisic Acid

- ABA promotes activities of seed dormancy. ABA levels are high when seeds mature, promoting lowered metabolism and synthesis of proteins needed to withstand the dehydration associated with dormancy. Seeds germinate when ABA is degraded by some environmental action. Desert seeds must have the ABA washed out of the seed coat, as do plants of marshy areas; temperate area plants have ABA degraded by light-stimulated enzymes. In other cases breaking dormancy is relative to the ratio of ABA (which keeps seeds in dormancy) and gibberellins (which promote germination).

- Low levels of ABA in maturing seeds promote premature germination.
- It promotes winter bud scale formation on woody plants in preparation for winter dormancy. ABA derivatives, called dormins, are used in commercial nurseries to keep materials to be shipped in dormant conditions. The dormancy can be reversed with gibberellins.
- ABA promotes stomata closure during leaf water deficit conditions by activating K^+ ion transport out guard cells. This involves signal transduction pathways with secondary messengers of calcium. ABA in this case originates in roots and detects low water level in root tissues. ABA moves upward into leaves and activates stomata closure.

Ethylene

Ethylene is the single gaseous plant growth regulator. It is produced in many plant organs, though its effects are most studied in fruits. Ethylene is the signal molecule for signal transduction pathways that activate tissue growth and development activities throughout the plant life (fruit maturation, leaf abscission and senescence). It is synthesized from the methionine amino acid. Toxic substances such as air pollutants also trigger the step for ethylene synthesis. It is also produced in wounded or bruised tissues.

Functions of Ethylene

- Ethylene activates ripening in fruits by signaling chemical reactions that degrade the pectins of the middle lamella, softening fruit and promoting the conversion of stored starches and/or oils into sugars.
- Ripening of some fruit involves a significant increase in the rate of cell respiration with concurrent high O_2 uptake. This function of ethylene (fruit ripening) has extensive agricultural impact.
- Ethylene promotes female flower production, whereas, high gibberellins may promote formation of male flowers.

- It help to germinate seeds by overcoming mechanical stress. When the shoot tip cannot penetrate the mechanical obstacle, ethylene is synthesized, which slows cell wall expansion. The walls thicken (more resistant to the pressure) and the stem curves to bypass the obstacle via horizontal growth. The stem tip pushes upward periodically to test for the obstacle. If the obstacle is encountered, ethylene initiates another cycle of stem thickening and horizontal growth. If no obstacle is encountered, the shoot resumes negative geotropic growth. This growth response is known as the triple response (decrease in elongation, stem thickening and change in orientation of shoot growth to horizontal).
- Ethylene is responsible for initiating the programmed death (apoptosis) of sclerenchyma and xylem vessels and tracheids.
- Ethylene is the direct cause of leaf, flowers and fruit abscission. It promotes the degradation of the cell walls in the abscission zone. The decreased concentration of auxin in leaves signals the production of ethylene in the abscission zone during the end of summer. High concentrations of CO_2 inhibit ethylene production so CO_2 can be used to prevent fruit ripening in grocer warehouses.

Brassinolides (Brassinosteroids)

Brassinolides is a plant steroid discovered in pollen of members of the mustard family, and studied in Arabadopsis. Chemically the brassinosteroids are similar to animal steroid hormones and have similar genes that code for the steroid synthesis pathways.

The functions of brassinosteorids overlap those of other plant hormones and may be additive. Brassinosteoroids activate signal transduction pathways that promote cell elongation and cell division. Brassinosteroids also promote differentiation of xylem tissues. Brassinosteroids can also retard leaf abscission. Absence of brassinolides results in dwarf plants. Its effects overlap with those of auxin and gibberellins.

Salicylic Acid

Salicylic acid is known to activate defence genes for resistance against pathogens (hypersensitive response).

Salicylic acid is a phenolic extract obtained from willow bark and have been used as an analgesic. It is now prepared commercially and is the active ingredient of aspirin.

Oligosaccharins (Oligosaccharides)

Oligosaccharins are short chain of sugars present in cell walls that may have a role in defence against pathogens. When a pathogen attracted on the plant, the oligosaccharide activates a signal transduction pathway leading to plant defence responses. They may also help regulate growth, differentiation and flower development all by activating signal transduction pathways.

Systemin

Systemin is a small peptide found in wound tissues. It may stimulate defence activities as the signal molecule (ligand). Systemin activates the signal transduction pathways that include the jasmonates.

Jasmonates

Jasmonates are a group of fatty acid derivatives. They play a role in seed germination, root growth and the storage of protein especially in seeds. Jasmonates synthesized in response to signal molecules produced in wound areas are involved in the signal transduction pathways that result in secondary metabolites (protease inhibitors) that poison the predator, or synthesis of volatile molecules that attract the predator's predators.

Role of Plants Growth Regulators in Orchid Production

1. Keiki production of *Phalaenopsis*

The propagation of *Phalaenopsis* keikis is the easiest application for the use of plant hormones in orchids. Keikis production may be promoted by enhancing the hormonal content of the meristem tissue. The application of a mixture of the cytokinin (BAP) and vitamins in a lanoline paste can assist the plant by prompting dormant reserve buds into growth which is restrained by the phenomena of apical dominance. A large mature *Phalaenopsis* might support many keikis, but a first bloom seedling should be treated with care. Fifth bud counting from the base of the inflorescence (generally the top bud) should not be treated with hormones because it generally produces a branch to the flower spike. Bract of spikes should carefully removed prior to application of regulators. Regulator should not apply until the last flower on the *Phalaenopsis* spike has opened fully. The purpose of application is to quite boosting of the natural hormones of the plant as they will not occurring in plants until the plant enters into growth cycle after blooming. Small amount of lanoline paste having plant growth regulator should apply to the bud and the surrounding tissues. Small amounts of Vitamin B-1 and Vitamin B-2 should also include in the lanoline paste. The importance of the lanolin is that it sheds moisture, may be massaged into the plant tissues and will not wash off during normal watering or the generally high humid atmosphere of an orchid-growing greenhouse or light-growing area.

In one or two weeks the tiny bud should be emerged out. The time of development of buds greatly varies due to the variation in the growing conditions of the plant, genetics and a host of other variables. The speed of development depends on many characteristics of both the plant and the cultural conditions in which it is being grown.

The following conditions have been found conducive to speedy and healthy keiki growth and development:

- Daytime temperature should be maintained by at least 22°C.
- Nodes number 2, 3, 4 from the base of inflorescence should be treated with regulators.
- Relative humidity of at least 60% should be maintained during keiki production.
- High nitrogen fertilizer should be used during the period of keikis development.

When the keiki has about four roots each of about two centimeters long, should be removed from the flower spike carefully. Root development follows leaf differentiation. A brisk twist will cause the keiki to separate naturally. Any small scar should be protected by a fungicide.

2. Extra spikes production

This can be achieved by applying a mixture of cytokinin and vitamins in lanolin. The regulator can be applied before flowering to second or third nodes on the upper portion of the spike, immediately below the lowest flower bud to promote spike branching. However, hormone paste is applied to one of the upper-most buds on the spent spike to encourage secondary flowering. Some *Phalaenopsis* do not branch at all and the regulator can be used to promote branching. This technique may also be used with the spikes of some *Oncidiums, Angraecoids, Vandas* and *Ascocendas.*

3. Rescue of damaged plant

Adventitious buds formation may be encouraged by growth regulators in the attempt to save damaged plants. This technique has been used to rescue crown damage of *Phalaenopsis* and other monopodial orchids. Adventitious buds on the stem of a damaged crown of *Phalaenopsis* frequently respond to the application of plant regulators because the root development is active and supportive for a plant attempting to survive.

4. Plants from cut flower or broken-off spikes

Phalaenopsis cut flower spikes can be treated with auxin and cytokinins after the flowers have faded to produce keikis on the bare spikes. The spikes must be kept in clean water which is regularly changed to maintain it clean. The basal end of the spikes also recut every few days to keep smooth flowing of water into the spike. Only one or two buds should be treated because the isolated stems have a very limited capacity to supply nutrients to the developing buds.

5. Extra plants from old canes or bulbs

Usually discarded dormant bulbs (back bulbs) can be encouraged to produce new growths after the application of plant regulators. New growths can be produced from old shriveled back bulbs. This is useful for *Cymbidiums* spp and *Dendrobium speciosum* where some good clones are very reluctant to produce new growths. Buds at both the top and base of *Dendrobium* canes can be rubbed with paste of plant growth regulators. The top buds generally responded with numerous keikis or sometimes extra flower spikes while buds at the base of the canes form shoots which then develop into new growths. The same technique also works with dormant buds of *Cattleyas, Oncidiums* and many other genera of orchid. *Catasetum* keikis from application to dormant buds on pseudobulb rings help increase these valuable plants for trading or maintaining own insurance plants in case of damage to mother plant.

6. Weed killers

Plant growth regulators may also be used to control weeds thereby reducing competition of weeds in orchid house (2, 4-D and 2, 4, 5-T). Other compounds such as Dalapon are more effective against monocots than dicots, so they could be used in the opposite situation. In fact these weedkilllers cause plants sensitive to grow too fast, damaging their cells so that their conducting tissue does not develop properly, or they use up their food reserves and die.

The influence of plant growth regulators on axillary shoot regeneration in *in vitro* flower stalk node culture of *Phalaenopsis* and *Doritaenopsis* orchids was studied by Wu and Chen (2008). The results indicated significant influences of different kinds and concentrations of cytokinins on shoot regeneration. Low regeneration rate and occasional adventitious root formation were observed on media without or with low level (1 mg.L^{-1}) of cytokinins. TDZ stimulated nodal explants to proliferate many abnormal adventitious shoots. Higher concentration of TDZ up to 4 mg.L-1 inhibited further development of newly induced shoots. BA at 2 mg.L-1 level was better than other cytokinin treatments for shoot regeneration and enlargement growth. Kinetin was less effective than other cytokinins. Cytokinin combined with ethylene inhibitors in the culture media increased explant survival rate, fresh weight as well as the number of larger shoots, but high concentration of AVG (100 micro M) decreased the fresh weight and inhibited shoot enlargement. Addition of 100 micro M $AgNO_3$ was most helpful for axillary shoot regeneration, but addition of ACC or ethephon was inhibitory. The ethylene content in the culture vessels with ethylene inhibitors in the media was significantly lower than that with ethylene precursors. During the period of shoot regeneration, ethylene accumulation in the vessels decreased and the diurnal changes of accumulation diminished along with culture time.

Nambiar *et al.* (2012) studied the potential effect of benzylaminopurine (BAP) on inflorescence production of a *Dendrobium* hybrid (*Dendrobium* Angel White). D. Angel White plantlets were subjected to spray containing different BAP concentrations. The results indicated that the application of BAP increased the percentage of inflorescence production, induce earlier flowering, and contributed to the differences in inflorescence length and the number of leaves and flowers produced. However, the application of BAP did not significantly influence the size of the flowers. This study showed that BAP is a potential plant growth regulator that can speed up the flowering process of D. Angel White.

Flowers can be classified as sensitive, low or ethylene-insensitive based on the responses to the growth regulator. Most of the climacteric flowers are ethylene-sensitive and in carnation, orchid and Petunia sp., pollination increases the production of ethylene causing petal willing and senescence. The climacteric rise of respiration during the flower senescence is associated with autocatalytic production of ethylene in the petals due to the increase of mRNAs from ACC synthase and ACC oxidase. The inhibitors of ethylene production, AVG and AOA, affect the longevity by extending the vase life of some ethylene-sensitive flowers, such as *Dendrobium* sp. However, by treating these flowers with STS or 1-MCP, blocking the ethylene action, are usually more effective in prolonging the longevity than the inhibitors of ethylene synthesis. Cytokinins and gibberellic acid may have positive affect on the vase life of some flowers, specially in those with low or ethylene-insensitive, like Heliconia sp. and rose. Pre-harvest spray or addition of calcium to the vase solution retards the senescence of roses, but the effect seems to be highly dependent of the cultivar. Flowers have intense respiratory activity, which may deplete the limited reserves of carbohydrates stored in the cut stems. The reduction of temperature reduces tremendously the CO_2 and ethylene production, as well as its action. However, chilling sensitive flowers such as bird-of-paradise, Heliconia, orchids and ginger cannot be stored bellow 10 to 13°C due to intense tissue discolouration (Finger and Barbosa, 2006).

Flowering of *Miltoniopsis* orchids is influenced by a combination of cool temperatures and short photoperiod. To determine if application of plant growth regulators could promote flowering without the need for costly structural modification to control photoperiod or temperature, Matsumoto (2006) used drenches of gibberellic acid (GA_3) (2.5 or 50 mM), N6-benzyladenine (BA) (25 or 50 mM) alone or in combination. BA (25 or 50 mM) treatments promoted new vegetative shoots and decreased the number of plants with inflorescences compared to the untreated control plants. This reduction of flowering and increased vegetative shoot production was alleviated by the addition GA_3 in combination with BA. However, the number of plants with inflorescences remained less than the control. GA_3 hastened Miltoniopis inflorescence emergence during the first flowering season by 10.9 to 14.9 days for Bert Field 'Eileen' and by 48.7 days for Rouge 'Akatsuka'. The number of 'Eileen' inflorescences produced per plant increased from 2.2 to 3.0 with 2.5 mM GA_3 treatment. Flower deformities were not observed in the GA_3 treated plants, and flower size and inflorescence length were unaffected by the GA_3 treatment.

A study was conducted by Vejsadova *et al.* (2009) to determine the effect of plant growth regulators (auxins and cytokinins) on *in vitro* growth and development of endangered orchid species (*Orchis* morio, *Dactylorhiza maculata* subsp. maculata, *D. incarnata* and *Platanthera bifolia*).

NAA depending on the type of saccharide applied in the nutrient medium affected the growth and development of *O. morio*. In the initial development phase, NAA had a positive influence on growth parameters in the presence of glucose, while in the later phase, its effect was higher in the presence of sucrose. In both Dactylorhiza spp., significant effect of auxin on shoot growth was not found. However, NAA stimulated root growth in D. maculata subsp. maculata seedlings. In *P. bifolia* plants, significant stimulation effect of 2,4-D on their growth and development was found after 22 weeks of culture. In all orchid species, significant positive influence of cytokinins (kinetin, benzylaminopurine [benzyladenine], isopentenyladenosine, thidiazuron, zeatin and adenine hemisulfate) on seedling growth was not observed. The results proved specific taxon dependence on the auxin type used.

Leaf explants of two *Phalaenopsis* , P. amabilis and P. Nebula, were used to test the effects of auxins (2,4-D, IAA, IBA, NAA), cytokinins (2iP, BA, kinetin, TDZ, zeatin), GA3, ancymidol, polyamines (putrescine, spermine, spermidine), ACC, $AgNO_3$ and $CoCl_2$ on the amount of direct embryo formation on different leaf locations (the cut end, the adaxial side, the abaxial side and the leaf tip). The results showed that there was a genotypic effect on direct embryo formation induced by cytokinins that 13.32 micro M BA and 4.92 micro M 2iP was the most effective in P. amabilis and P. Nebula, respectively. Besides, explant position highly affected embryogenic competence of leaf cells in both species that the cut end showed highest embryogenic response, the adaxial side was the second, and then the abaxial side and the leaf tip. Altogether, cytokinins tested were all effective in both species, and ACC at 20 micro M had 35% of embryogenic response in P. amabilis. However, auxins, GA3, ancymidol and polyamines were inhibitory in both species (Gow *et al.,* 2008).

Physiological Disorders due to Environmental Factors

An imbalance of water, light and temperature creates symptoms of disorders. For example, overwatering can cause roots to rot, shrivelling of pseudo-bulb. A change in the care program can solve physiological ailments.

1. Yellowing leaves

This is a normal aging process if only old leaves on backbulbs are involved. If newer leaves yellow and soften, that might be due to: too much light, low temperature, lack of nitrogen, loss of roots and excess watering.

2. Mesophyll cell collapse

Mesophyll cell collapse is a condition that affects many different orchid genera. This physiological disorder can be caused to low water temperatures or low air temperatures, which leads to damage to the mesophyll cells within the leaf. For many growers, this is a difficult condition to identify because the symptoms frequently show up 6-8 weeks after exposure to damaging temperatures.

Symptoms

Initial symptoms include localized or wide-spread sunken / yellow areas (Fig. 3.8, 3.9 and 3.10) which later may turn dry and necrotic (Fig. 3.8). Once a leaf is damaged, it is permanent. Injured leaves may be colonized by saprophytic fungi, which can lead some growers to believe the disorder is caused by a fungus, prompting them to apply chemicals unnecessarily.

Fig. 3.8 : Symptoms of mesophyll cell collapse

Fig. 3.9 : Close-up view of mesophyll cell collapse

Fig. 3.10 : *Phalaenopsis* orchid with symptoms of mesophyll cell collapse

Diagnosis, cause and control

The data available suggest that the severity of symptoms is related to several factors: temperature, length of exposure to low temperature and age of leaves .Young leaves appear to be more susceptible to chilling injury, while mature leaves appear resistant. Mesophyll cell collapse occurs more frequently in orchids that have thick, fleshy leaves (such as *Phalaenopsis*); *P. insigne* has relatively thin leaves, and therefore, may not be as prone to this condition. Day/night temperature may play a role in the development of mesophyll cell collapse. When one suspects mesophyll cell collapse, it is important to:

1. Examine temperature records for the previous 6 weeks and look for cool temperature or drastic fluctuation in day/night temperatures.
2. Determine if there are areas in the greenhouse or growing area that are prone to drafts.
3. Make sure that when watering, the water is not too cold, which can also cause mesophyll cell collapse .

3. Blackened areas on leaves

The sudden appearance of brown and then black areas on exposed surface of leaves on a bright hot day may be due to sunburn. Shading should be checked and if blackened areas increase in size, it might be Bacterial or fungal disease.

4. Blackened tips or ends of leaves

This could be caused by: overfeeding, excessive soluble minerals in water and leaf dieback, a fungal infection. For management application of fertilizing should be stopped and flush plain water to plant thoroughly.

5. Shriveled pseudobulbs

A sign that the plant is losing water content, it can be caused by: low humidity, underwatering, dry medium and loss of roots.

6. Deformed flowers

If deformed flowers are produced each year by a plant, it is probably inherent. Destroy the plant. An occasional deformed flower can be caused by:high temperature and low humidity when buds are developing, mechanical or chemical injury to bud.

7. Bud blast

Orchids are sensitive to any sudden changes in their temperature. A newly purchased orchid will suddenly lose its flowers when introduced into a new environment. Healthy orchid plants and buds that shrivel and fall off the stem before they have a chance to open suffer called "bud blast."Bud blast is the withering and dropping of buds from an healthy plant. These buds shrivel up right before they open because of different types of stress. Specific causes of this exasperating event are giving below

- Low humidity
- Hot air from furnaces or cool, dry air form air-conditioners directly blowing on the orchid plant
- Over- or underwatering
- Poor root development
- Temperatures that are too high or too low
- Water standing in the buds or bud sheaths
- Dramatic change in the orchids' environment, like bringing the plants from outside to inside
- Natural-gas leaks in the house
- Ethylene gas from ripened fruit
- Light that's too bright on the developing flower buds
- Pollution, such as smog

Fig. 3.11 : Phalaenopsis orchid suffering bud blast

8. Rapid wilting of flower

If dorsal sepal wilts or dries early, or entire flower goes to sleep shortly after opening, it could be caused by: air polluted by ethylene or other gas, too sudden a change in climate, pollination by an unknown pollinator.

9. Spotting of flowers

Light brown or pinkish dots and spots on flower after opening usually indicate fungus,Botrytis, or sooty mold. Reduce humidity, increase ventilation or air movement, remove spotted and old flowers from growing area to reduce source of infection. Keep water off flowers.

10. Punctures in flowers

Denotes presence of: aphids, soft-bodied sucking insects; or thrips, small chewing insects.

11. Colour mottling of flowers

This could be a colour-breaking virus. Isolate plant and get an expert opinion;destroy the plant if virus infection is confirmed.

12. Loss of roots

Numerous causes, difficult to diagnose: overwatering, black rot, an infection of Pythium, slugs or snails, excessive salt content of water, potting medium old and broken down solution.

Chapter - 4

Post Harvest Management

India ranks 23rd among world exporters of floriculture products and its share in the world exports are at around 0.38 per cent. It is estimated that fresh cut flowers of worth Rs. 100 crores are being exported annually from India. On the other hand the post harvest losses are estimated to be more than 30-35 per cent. The loss of quality in ornamentals may result from wilting or abscission of leaves and petals, yellowing of leaves and geotropic or phototropic bending of scapes and stems. Several factors prevailing during pre harvest, harvest and post-harvest stage affect the post harvest quality and longevity of cut flowers. Each step in the life span of post-harvest flower is associated with a number of coordinated biochemical, physiological, hormonal and structural changes that are strongly modulated by the fluctuations of environmental factors and stresses of biotic and abiotic origin.

Comparison of Cut and Intact Flowers

At the time of flower opening, an increase in the levels of glucose, fructose, sucrose and starch are observed in petals of flowers that remained on the plant and these levels are higher than in the cut flowers. Cell death in intact flowers is therefore, initiated in the presence of high levels of starch and sugars. Such intact flowers shows senescence related mobilization. A drastic reduction is observed in the levels of starch, protein, RNA and DNA, concomitant with a sharp rise in activities of RNase and DNase. At the same time, sucrose, amino acids, and amides are formed. These compounds are exported through the phloem, which remained active after most other petal cells has died. Two-thirds of the potassium ions and more than one-half of the magnesium ions are also exported from the petals. Cell walls are apparently hydrolyzed, as β-glucosidase activity drastically increased. Mobilization in uncut flowers proceeds just as in cut ones. It has been indicated that total (overall) soluble sugar levels in petals of most uncut flowers growing under natural light levels remain high until visible senescence.

The developmental events taking place during senescence also involve physiological changes such as loss of water from the senescing tissues, leakage of ions, transport of metabolites to different tissues and biochemical changes(generation of reactive oxygen species (ROS), increase in membrane fluidity and peroxidation, hydrolysis of proteins, nucleic acids, lipids and carbohydrates).

Degradation of Cell Membrane

Membrane deterioration is an early and characteristic feature of plant senescence, which leads to several structural and functional changes. At the biochemical level, senescence is associated with changes in membrane fluidity and leakage of ions in several different flowers. It has been observed that the changes at the membrane, which include the decrease in all classes of membrane phospholipids and increase in neutral lipids, mainly due to increased action of phospholipases and acyl hydrolases. Another important event that leads to loss of membrane permeability is oxidation of membrane lipids (lipid peroxidation) due to lipoxygenases enzymes. However, in ethylene sensitive plants such as *Phalaenopsis*, the lipoxygenase seems not to play apparent role. On the other hand, in the ethylene insensitive category lipoxygenase promotion of senescence has been proposed.

Due to various oxidative reactions, reactive oxygen species (ROS) are known to be involved in normal plant and petals cell death. The ROS is produced from hydrogen peroxide, thus the hydrogen peroxide level regulating enzymes showed differential expression during senescence. The activity of natural antioxidants decreases in several flowers. During the flower senescence membrane proteins mediating transport and redox reactions are also reduced.

Protein Degradation

Proteases degrade proteins by hydrolyzing internal peptide bonds in plants. Among all the proteases, the cysteine proteases are the most frequent and well characterized. The expression pattern of genes coding for proteases suggests their involvement in the protein hydrolysis in tepals at the late senescence stage. During the senescence, serine protease activity increases up to 2/3 of total proteases activity. Several senescence associated cysteine protease genes have been reported to increase in abundance following ethylene treatment. During senescence of floral parts, the degradation of DNA and RNA is the most common feature.

Pollination and Flower Senescence

Pollination triggers a large number of changes in flowers such as petal and anther senescence/abscission and ovary maturation to contain developing seeds. It is well established that compatible pollination activates a series of post pollination developmental events contributing to reproduction and ovary growth, pigmentation changes and petal senescence. The post pollination increase the activity of hydrolytic enzymes and degradation of macromolecules. These observations suggest that there are some inter organ signals that amplify and transmit the pollination signal to floral organs. It is reported that inter organ signalling and signal amplification involves the regulation of ethylene biosynthetic gene expression and inter organ transport of hormones and their precursors.

Senescence, however, occurs after a second burst of ethylene that is reported only when compatible pollination takes place. An ACC synthase gene is reported to be regulated by pollination and auxin in the stigma. There is also an ethylene independent increase in sensitivity to the hormone after pollination, apart from an increase in ethylene production. Once the process is initiated the evolution of ethylene becomes autocatalytic and accelerates the senescence process. It has been suggests that the promoters of ethylene biosynthetic genes contain some ethylene responsive elements. Based on ethylene sensitivity, flower senescence can be grouped into. Ethylene dependent pathway and ethylene independent pathway. The ethylene sensitive and ethylene insensitive flower senescence processes may have some common cellular components with complex network of post transcriptional modifications. Since each part of flower play different role in their life span, therefore, it requires specific co-ordination of senescence signal among them. Pollen tube showed clear evidence to support an important role for the mitochondrion and involvement of caspases. Sepal senescence shows similarity with leaf senescence because they are green and contains chlorophyll. Degradation of sepal chlorophyll is the first visual sign of senescence.

Factors Affecting Flower Senescence

Senescence might be trigger by biotic and abiotic stresses. It is being the most colourful and susceptible part of plant are easily affected by several environmental factors such as insect mediated pollination, seasonal changes, lack of water and various stresses such as invasion by a pathogen or attack by a predator.

A. Abiotic stresses

Flower senescence is often hastened by several abiotic stresses, like drought, light quality, heat shock *etc.* In these stresses the senescence process is mediated by the evolution of ethylene in ethylene sensitive flowers, while in ethylene insensitive flowers elevated level of ABA might be an important hormonal intermediate signal for responses. A change in light source significantly affects the ABA content of the petal. There is an inverse relationship between ABA level and flower longevity as the higher ABA level the shorter the vase life observed.

1. **Growing conditions:** High light intensities causes scorching and dropping of leaves and abscission of petals. Orchid flowers are very specific in their temperature requirements. It also requires adequate nutrients for good longevity. High nitrogen doses should be avoided as they increase susceptibility to diseases. Flowers damaged by pathogens, insects and pests show high ethylene production resulting in poor vase-life. Basically, those forces which improve flower quality before and after harvest usually improve vase life. Optimum light intensity during growth of the flowers is very important to vase life. Temperature also influences photosynthesis and respiration, which in turn influence carbohydrate accumulation. During hot periods of the year, orchids sensitive to high temperatures, have shorter vase lives because flowers contain low carbohydrate levels. In nutshell high temperature enforces earlier flowering.
2. **Methods of harvesting:** The most important factors for harvest are when, how and where—"when" the plant material is reached to the optimum stage of development and "when" during the day to harvest. Right stage, method and time of harvesting of flowers are of considerable importance to ensure their long vase-life. Each orchid has its own best harvest stage that depends on the use of plant material and available market. Some general rules of thumb for when to harvest are: spike

type flowers harvest when one-fourth to one-half of the individual florets are opened; daisy type flowers harvest when flowers are fully opened. The other "when" is, the best time of day for harvesting flowers. The best time is the coolest part of the day and when there is no surface water from dew or rain on the plants. This usually is in the cool of the morning after the dew has dried. Late afternoon or evening has also possibilities because the plants have stored carbohydrates from the day which will provide a food reserve for the plant material. "How" and "where" go together. Besides knowing at what stage of development to harvest, where and how to cut the flower on the plant also is important. The flower having longest stem should be harvested without harming future production. Longer length of spike usually is better. Sharp cutting utensils should be used to avoid the crush to xylem vessels and block the flow of water to the stem. Clean utensils should be used to avoid harmful microbes to the cut stems. Cutting equipment should be cleaned daily with chlorine bleach disinfectant (1:10 with water). The stems should be cut with sharp knives or secateurs. Hardwood stems should always be given slanting cut to expose maximum surface or to ensure rapid water absorption. Flowers for local markets are generally harvested much more open than those intended for storage and/or long-distance transport. An important quality aspect of cut flowers is the behaviour of the flower in a vase at the consumer. This vase life behaviour varies largely. The variability is not only present in different crops. Also within a crop large differences can be present between cultivars, seasons, and growers and even between lots of the same grower at the same time of the year.

3. **Relative humidity:** It affects the transpiration rate. Higher the humidity in the air less is the transpiration rate and vice-versa. Loss of water from flowers is often associated with a loss of quality, as visual changes such as wilting or shrivelling and textural changes can take place. For fresh market produce, any method of increasing the relative humidity of the storage environment (or decreasing the vapour pressure deficit (VPD) between the commodity and its environment) will slow the rate of water loss. The best method of increasing relative humidity is to reduce temperature. If using refrigeration for cooling, the larger the area of the refrigerator coils, the higher

the relative humidity in the cold room. Another method is to add moisture to the air around the commodity as mists, sprays, or at last resort, by wetting the store room floor. Another way is to use vapour barriers such as waxes, polyethylene liners in boxes, coated boxes or a variety of inexpensive and recyclable packaging materials. Any added packaging materials will increase the difficulty of efficient cooling, so vented liners (about 5 per cent of the total area of the liner) are recommended.

4. **Handling :** Handling is everything else involved in preparing the flowers for market. Exactly how these steps are done depends on the crop, the market and the operation size. These are collectively called handling. These handling steps include:-
    1. Grading
    2. Leaf Removal
    3. Bunching
    4. Recutting
    5. Hydration
    6. Special Treatments
    7. Packing
    8. Precooling
    9. Cold Storage
    10. Delivery to Market

All of the above steps are not needed to all flowers. Whether they are used or not depends on the market, flowers are going to sold. Where and how the steps are done depends on the market and the facilities of the operation. Field handling is limited to leaf removal, grading, bunching, hydrating and packing with immediate transport to market or cold storage for brief holding. Flowers for local retail markets often are packed this way since they are marketed immediately after harvest. All the handling steps can be done in a packing shed. It often makes for a better flow of activities if they are all done in the same place. Some of the steps can only be feasibly done in the packing shed, such as special treatments, precooling, cold storage and recutting. These extra steps usually are done for flowers going to wholesale markets. The packing shed may be an ultra modern air conditioned building or an open air covered porch.

The handling space should:

1) be shaded or covered to keep temperatures lower and prevent direct sunlight on the flowers
2) be well lit so can see well when grading the flowers
3) have a clean water source for preparation of harvest and holding solutions
4) have ample space so all handling activities can be performed smoothly
5) have a cold storage or at least a cooler shade place to store the flowers until they are ready for market

Whether going to handle flowers in the field or in the packing shed, the first step after cutting the stem, should be to place them in water or a harvest solution. This solution may be acidified (pH 3.5), tepid water, citric acid or a floral preservative. The harvest containers should be clean and disinfected after each use. Flowers should never be laid on the bare ground. Harvest container should be placed in cool place until they can be handled or taken to market. The cool place can be a shady area in the field or a refrigerator. Leaves should be stripped from the stem. If the flowers are field handled this can be done before they are placed in the harvest containers or before they are bunched into marketable bouquets. Usually, leaves are stripped from the base to one third height of the stem, or at least the ones that would be in any holding solution.

Grading

The designation of grade standards for cut flowers is one of the most controversial areas in their care, handling and marketing. Objective standards such as stem length, which is still the major quality standard for many flowers, may bear little relationship to flower quality, vase life or usefulness. Weight of the bunch for a given length is a method that has been shown to strongly reflect flower quality. Straightness of stems, stem strength, flower size, vase life and freedom from defects, maturity, uniformity and foliage quality are the factors which should also be used in cut flower grading. Mechanical grading systems should be carefully designed to ensure efficiency and to avoid damaging the flowers. It starts with deciding which flowers to be harvest. Only marketable flowers should be harvested. Marketable flowers are free of blemishes, including both leaves and petals. They should be grouped or graded by stem length if there are differences and also by developmental stage. More mature ones should be sold as soon as possible, while others can be held in cold storage for later sales.

Bunching of flowers

The number of flowers in the bunch varies according to growing area, market and flower species. Groups of 10, 12, and 25 are common for single-stemmed flowers. Bunches are held together by string, paper-covered wire or elastic bands and are frequently sleeved soon after harvest to separate them, protect the flower heads, prevent tangling and identify the grower or shipper. Materials used for sleeving include paper (waxed or unwaxed), corrugated card (smooth side towards the flowers) and polyethylene (perforated, unperforated and blister). Sleeves can be preformed (although variable bunch size can be a problem), or they can be formed around each bunch using tape, heat sealing (polyethylene) or staples.

Damage through multiple handlings can be reduced if grading, sizing and even bunching are carried out in the field or orchid house. Flowers should be graded and bunched before being treated with chemicals or being placed in storage. When the flowers are badly wilted or when labour is not available for grading and bunching, flowers should be rehydrated and cooled until these operations can be carried out. Mixed bunches and single type bunches are both popular. Sleeving or wrapping bunches help to prevent the different bunches and flowers from becoming tangled. Wholesale markets have a set of guidelines for the methods of bunching and packaging flowers. The flowers should be well hydrated but not wet when packed. Most spike flowers need to be packed upright to prevent the tips from curving. Special boxes or hampers are made for these types of flowers. Once bunched, flowers should be hydrated, placed in water for minutes before they are packed dry. The hydrating step should include a step where, after the flowers are bunched, the stems are re-cut under water to eliminate any air bubbles in the xylem that can block the uptake of water. These air bubbles can occur when the flowers were harvested. Once re-cut, the flower can be placed in a general holding solution used to hydrate the flowers or receive a special treatment such as silver thiosulfate. Flowers usually are not packed dry into boxes in the field but are in the packing shed for distant wholesale markets. When flowers are packed into boxes, bunches are sleeved or wrapped and then packed tightly so the bunches do not move or vibrate in transit (causes bruising).

Standard for Orchids cut flowers

This standard applies for cut flower of orchid of all genera and varieties of *Dendrobium* spp. *Oncidium* spp. *Aranda* spp. *Mokara* spp. *Vanda* spp. for commercial purpose:

Quality requirements

Orchids of all standard grades shall contain the following minimum quality standards, unless otherwise specifically provided for each grade with certain acceptable deviations:

- Flowers should be fresh and clean.
- It should be free from visible deformed marks.
- It should have strong and rigid stem.
- It should be free of visible pests.

Orchids should be harvested at an appropriate time, under careful harvest and post-harvest handlings. Packaging shall be done in an acceptable condition.

Classifications

Orchids are classified into three quality grades:

Extra Class. Orchids that fall under this class should be of superior quality, i.e. no fallen flower, free from damage caused by pests of animal or plant origin, free from visible marks affecting the appearance.

Class I .Orchids of this class should be of good quality. They shall be practically free from damage caused by pests of animal or plant origin, practically with little mark that do not affect the quality and storage, as well as orchid presentation.

Class II Orchids that do not fit in any higher classes but having minimum quality standards as mentioned above fall into this class. This excludes those with visible mark of affecting appearance, the quality and storage.

Sizes

Sizes of each species of orchids are based on the length of flower stem, total number of flowers and number of blooming flowers per stem. Size specifications are outlined in Tables 1- 4 below:

Table 4.1 : Sizes of *Dendrobium* spp.

Characters	Extra Long	Long	Short	Extra Short
Length of stem	Not less than 55 cm	Not less than 45 cm	Not less than 35 cm	Not less than 30 cm
Number of flowers / stem	Not less than 12 flowers	Not less than 10 flowers	Not less than 8 flowers	Not less than 6 flowers
Number of blooming flowers / stem	Not less than 7 flowers	Not less than 6 flowers	Not less than 5 flowers	Not less than 4 flowers

Table 4.2 : Sizes of *Oncidium* spp.

Characters	Extra Long	Long	Short
Length of flower stem	Not less than 70 cm	Not less than 50 cm	Not less than 40 cm
Number of blooming flowers / stem	Not less than two fifth of the total number of flowers per stem		

Table 4.3 : Sizes of *Aranda* spp.and *Mokara* spp.

Characters	Extra Long	Long	Short
Length of flower stem	Not less than 50 cm	Not less than 40 cm	Not less than 30 cm
Number of blooming flowers / stem	Not less than four fifth of the total number of flowers per stem		

Table 4.4 : Sizes of *Vanda* spp.

Characters	Extra Long	Long	Short
Length of flower stem	Not less than 50 cm	Not less than 40 cm	Not less than 25 cm
Number of flowers ` / stem	Not less than 12 flowers	Not less than 9 flowers	Not less than 7 flowers
Number of blooming flowers / stem	Not less than 9 flowers	Not less than 7 flowers	Not less than 5 flowers

Tolerances

Acceptable produce quality deviation for each package is as follow:

Quality tolerances

Extra Class: Not more than 5 per cent of orchids that does not meet the specifications of Extra class, but should meet Class I quality.

Class I: Not more than 10 per cent of orchids that does not meet the specifications of Class I, but shall meet Class II quality.

Class II: Not more than 10 per cent of orchids that does not meet the minimum standards, but should not have decayed or inappropriate conditioned flowers.

Sizes tolerances

In all classes, not more than 10 percent of orchids shall vary from the size specifications on stem length of flower each class.

Packaging and orchid presentation

Uniformity

Orchids packed in a package should always show evenness in the form of variety, quality, colour and size. Packaged flowers must represent the whole products.

Packaging

Orchids should be packed in such a way to be properly protected. The packing should not be neither dense nor loose otherwise flowers get harm by during transportation. Packaging materials should be of good quality, clean and have quality to avoid all damages of the flowers.

Labelling

All possible details of the product should be mentioned. The use of printed labelling materials especially paper or detailed stamp is acceptable provided that the ink is safe. Orchid packaging can be categorized into three groups:

Packaging for customers shall contain the following information:

- Type of product, i.e. 'Orchid' or 'Orchid cultivar'
- Quality class and size
- Quantity and/or unit

Packaging for wholesalers shall contain the following information:

- Information of product, i.e. 'Orchid' or 'Orchid cultivar'

- Quality class and size
- Quantity and/or unit
- Wholesaler information, e.g. name and address

Packaging for exporters shall contain the following information:

- Type of product, i.e. 'Orchid' or 'Orchid cultivar'
- Quality class and size
- Quantity and/or unit
- Exporter information, e.g. name, address and country of origin

Language

Label of produce shall be in Hindi/, except those for export purpose which needs both English and/or the official language of the trading partner country.

Certification Mark

The official certification is based on the conditions of the inspection body or certification body authorized by the APEDA.

Sampling

Sampling of the flowers is based on the procedures set by APEDA.

Precooling

Precooling is essential for removing field heat from flowers. It is a step that rapidly brings the temperature of the flowers down from the field temperature (20°-30°C) to a proper storage temperature (1°C). A low temperature slows the respiration rate of the flowers which in turn helps them last longer. Forced-air cooling is the best method for flowers. In forced air cooling, cool air is actively forced with fans through the bunched flower. This can be done when the flowers are in a bucket or when they are packed dry into boxes. The precooling of flowers is a very important step for individuals selling to a large wholesale market, distant markets and if their crop is to be stored for a long time.

Packing and transporting

Packaging ensures the fresh supply of flowers to the consumers. Lower rate of transpiration, respiration and cell division during transportation are essential for long storage life and good keeping quality. Before packing, flowers should be dried. They should be treated with systemic pesticides. Packing must ensure protection of flowers against physical damage, water loss and external conditions detrimental to transported flowers. Boxes made of corrugated fibre boards are good. Flowers sensitive to geotropic bending must be transported in an upright position. The flowers should be transported at an optimal low temperature. The relative humidity of the air during precooling and shipment of cut flowers should be maintained at the level of 95-98 per cent. Lack of light during prolonged transportation particularly at high temperature causes yellowing of leaves in many flowers. Transportation of flowers is usually done by truck, air and sea. For short distance and time period shorter than 20 hours, cut flowers may be transported in insulated trucks without refrigeration after precooling and proper packing.

Air transport is the quickest method of flower transport but temperature is not controlled during the freight. Respiration of cut flowers an integral part of growth and aging, generates heat as a by-product. Furthermore, as the ambient temperature rises, the respiration rate increases. For example, a flower held at 30°C is likely to respire up to 45 times as fast as a flower held at 2°C. The rate of aging can be reduced drastically by cooling the flowers. Rapid cooling and maintenance of the cool chain are thus essential for maintaining quality and satisfactory vase life of cut flowers. Air transport is rapid in compared to surface modes (truck, sea container *etc.*), the response of cut flowers and foliage to temperature leads to their rapid deterioration even during the relatively few hours that they are in transit. It has frequently been demonstrated that transport of flowers in surface modes that permit maintenance of the cold chain results in better turn out than air freight under uncontrolled temperatures. For this reason, air transport is seldom the method of choice when alternative transportation modes that offer good control of temperature are available. The bending of the scapes in these flowers is the result of negative geotropism (bending away from gravity) which is accelerated by storage at warmer temperatures. Flowers held at temperatures warmer than the optimal (0°C) show marked loss of quality with increasing storage temperature. Bent stem tips, loss of lower florets and reduced opening of young buds are seen in flowers that are stored at warm temperatures. The optimum temperature for storage of most of the common cut flowers is near the freezing point 0°C. Some tropical crops such as orchids are injured, however, at temperatures below 10°C.

Packing

There are many shapes of packing containers for cut flowers, but most are long and flat and a full telescoping design (top completely overlaps the bottom). This design restricts the depth of the flowers in the box, which may in turn reduce physical damage of the flowers. In addition, flower heads can be placed at both ends of the container for better use of space. With this kind of flower placement, whole layers of newspaper have often been used to prevent the layers of flowers from injuring each other. The use of small pieces of newspaper to protect only the flower heads, however, is a better practice, since it allows for more efficient cooling of flowers after packing. It is critically important that containers be packed in such a way that transport damage is minimized. Some growers anchor the product by using enough flowers and foliage in the box so that the package, after banding, holds itself firmly. To avoid longitudinal slip, packers in many flower-producing countries use one or more "cleats". These are normally foamor newspaper-covered wood pieces that are placed over the product, pushed down and stapled into each side of the box. Padded metal straps, high density polyethylene blocks and cardboard tubes can also be used as cleats. The heads of the flowers should be placed 5 - 10 cm from the end of the box to allow effective pre-cooling and to eliminate the danger of petal bruising should the contents of the box shift. Another system commonly employed to secure flowers is an elastic strap, anchored in the base of the box that is stretched over the stems of the flowers after packing. Some flowers lend themselves to packing in a manner where the stem bases of some bunch are placed against the ends of the box. The rest of the flowers are packed normally (5-8 cm from the end of the box), and the conical shape of typical bunches means that all the flowers are well secured. Flowers such as orchids are packed in various ways to minimize friction damage during transportation. Frequently the flower heads are individually protected by paper or polyethylene sleeves. Cushioning materials such as shredded paper, paper wool and wood wool may be distributed between the packed flowers to further reduce damage.

Box design and construction

There are hundreds of box sizes used in the flower trade and the quality of card used in the boxes is quite variable. The strength of cardboard falls rapidly at high humilities and this is why boxes constructed with poor quality card may collapse, particularly where there are major changes in temperature and resultant condensation in the postharvest chain. The industry should move to a small number of box sizes designed to fit the standard international pallet, and should use high quality card. White card speaks of high quality, and allows the grower to highlight the product with printed designs. A major flaw in the design of some boxes is that the pre-cooling vents are pre-formed and not sealed. This means that they become 'fast warming' vents when the box is handled at high temperatures.

The flaps on the box provide important extra stiffness to the corners, so if staples are to be used, there should be sufficient number (five) to ensure a rigid double sheet of cardboard. Three up the outside corner, one on the inside bottom corner, and one more or less in the middle of the flap. Many marketing chains no longer permit the use of staples for securing the side flaps of the boxes. Where staples are no longer allowed, replace them with fast-setting cement (hot glue) with similar placement, rather than a piece of tape around the outside edges. Ingenious folded box designs are now widely used in the produce industry and an adaptation for flowers would be worth exploring. For strengthening the box an extra triangular strut should be formed in the corners of the box bottom. Since there is an air-gap between the flowers and the end of the box, the triangle would not reduce packing volume, and would greatly increase corner strength.

Packing systems

Because flower freight is typically based on volume rather than weight, producers commonly over-pack their boxes. Quality flowers result from packages where the box protects the flowers but over packed boxes cannot do that.

Room cooling

Room cooling is a relatively low cost but very slow method of cooling when electricity for mechanical refrigeration is available. When using room cooling, produce is simply loaded into a cold room and cold air is allowed to circulate among the cartons, sacks, bins and bulk load. It is important to leave adequate space between stacks of boxes inside the refrigerated room in order for produce to cool more quickly. About 1 inch (2.5 cm) is sufficient to allow cold air to circulate around individual boxes. Produce in vented boxes will cool much faster than produce packed in un-vented containers. In many small-scale cold rooms, produce has been loaded into the room so tightly that cooling cannot take place at all and despite the high cost of running the refrigeration system, the produce temperature never decreases to recommended levels. Stacks of produce inside the cold room should be narrow, about one pallet width in depth (two or three cartons). Fans should be installed to move the cold air throughout the room. Air circulating through the room passes over surfaces and through any open space, so cooling from the outside to the centre of the stacks is mostly by conduction. It is needed to monitor the temperature of the produce within the packages at various locations in the room to determine that the produce is being cooled as desired.

Hydro-cooling

Hydro-cooling provides fast, uniform cooling for some commodities. The commodity as well its packaging materials must be tolerant of wetting, chlorine (used to sanitize the hydro-cooling water) and water beating damage. The simplest version of a hydro-cooler is a tank of cold water in which produce is immersed. Conveyors can be added to help control the time produce stays in contact with the cold water.

Evaporative cooling

Evaporative coolers can be constructed to cool the air in an entire storage structure or a few containers. These coolers are best suited to regions having lower humidity. A cooling pad of wood fibre or straw is moistened and air is pulled through the pad using a small fan. An evaporative cooler can be combined with a forced air cooler for small lots of produce. Air is cooled by passing through the wet pad before it passes through the packages and around the produce.

Drip cooler

Produce stored in these coolers has a longer shelf life than produce kept at ambient conditions. Decay is a major problem, which can be controlled by washing in chlorinated water prior to cooling. An evaporative cooler located in the peak of a storage structure can cool an entire room of stored produce or other chilling sensitive crops. During the hot summer months in India, this chamber is reported to maintain an inside temperature between 15 and 18 °C and a 95 per cent relative humidity of about.

Cooling time calculations

The time necessary to reach a desired temperature is expressed in terms of a typical cooling curve. The rate of cooling becomes very slow as the temperature of the flowers nears the temperature of the refrigerated air. Consequently, flowers are seldom completely cooled to the temperature of the cooling air. The half-cooling time for forced-air cooling ranges from 10 to 40 minutes, depending on product and packaging. Flowers should be cooled for three half-cooling times (by which time they are about 90 per cent cool). Flowers that are to be transported at ambient temperatures can be packed in polyethylene caskets, foam-sprayed boxes or boxes with the vents resealed. Ice that is used after pre-cooling is only effective if it is placed so that it intercepts heat entering the carton (it must surround the product) and care must be taken to ensure that the ice does not melt onto the flowers or which otherwise causes freezing damage.

Vacuum cooling

Vacuum coolers have been installed in flower marketing terminals (mostly near airports) as a means of removing heat from packed flowers. Vacuum cooling is based on the fact that water boils at lower temperatures at lower pressures. At about 1/10th atmosphere pressure, water boils at 0°C. When flowers are placed at these low pressures, the water in their cells 'boils' and thus removes the heat. Although these systems cool leaves, stems and growing media promptly and efficiently, flower petals often cool somewhat more slowly, because water cannot escape readily from them. Essentially the same seven eighths precooling time (or possibly 10-15 minutes less) is required to vacuum cool flowers as for the forced-air cooling systems. When used at or near airports, these systems serve as a useful 'backstop' for flowers that have not been properly cooled by growers and shippers. However, in some cases it can be many hours before packed flowers arrive at the airport or trucking dock, which suggests that they may have already deteriorated significantly before being placed in a vacuum cooler. Some advantages of vacuum cooling methods include that paper and plastic packaging materials in the boxes do not affect the efficiency of cooling, free water is removed, boxes can be stacked in any manner in the precooler and it can work equally well with dry packs and wet packs for cut flowers and greens and for potted plants. The main disadvantage is the high capital cost of the equipment. In a centralized facility this high cost can be offset by the high throughput of boxes that these systems can achieve.

Irradiation

Flowers that are exported are often subject to quarantine inspection for the presence of harmful diseases and pests. It has been suggested that ionizing irradiation might be used to sterilize or kill insect pests without damaging the flowers. The nature of ionizing radiation is such that it leaves no residue after treatment and if insects could be controlled without any effect on flowers or foliage this would provide an excellent quarantine treatment. Insects are relatively more susceptible to ionizing radiation than plants. Doses are as low as 0.2 Gy sufficient to sterilize most harmful insects. Although flower does not suffer from obvious damage below 0.7 - 1 Gy doses of radiation.

Cold storage

Cold storage is recommended for all flowers that will not be marketed. Low temperatures slow the respiration rate of the flowers and prolong the vase life of the flowers. In general, temperatures should be 0^0 to 4^0C and have a relative humidity of 85 to 90 per cent for most flowers. Flowers should never be stored with fruits and vegetables. Some fruits and vegetables produce ethylene that can dramatically shorten the life of the flowers. Once flowers are bunched into marketable units they should be placed in cold storage. Long-term cool storage of flowers may be useful in extending the market options for some flower crops. Cool storage enables quality stems to be held for longer periods before sale and ensures that the flowers still have a good vase life when they reach the market-place. It had been observed that cool storage of *Cymbidium* orchids at 5-8°C for 1 month resulted in an extension of the marketing window for these flowers by 2 weeks compared to flowers that were cool stored for 1 week. *Cymbidiums* can withstand temperatures down to 1°C, although after 1 month of cool storage, 2-4 days extra saleability is obtained if storage temperatures are between 5 and 8°C. Only first-class blooms should be used for long-term cool storage because any minor blemishes become more obvious in storage.

Refrigerated storage

The most common system for handling harvested flowers is refrigerated storage, which involves the following steps:

1. Flower stems should be cut with a sharp knife or shears to prevent crushing of stem and water-conduction cells.

2. The cut flowers should be placed in a preservative solution as soon as possible to prevent wilting. The flowers should not be allowed to be out of water while they are waiting to be transferred to the storage or grading rooms. If cut in the field, buckets containing solution can be brought out on trailers to hold the harvested flowers. Flowers cut in the greenhouse should not be left in the sun or out of water for more than a few minutes. One person should be assigned to carry these flowers to the grading room or storage cooler immediately.
3. As soon as flowers arrive at the storage room they should be placed in preservative solution inside the refrigerated storage room. If wilted, they should be placed in a warm preservative solution at room temperature until turgid. They should then be placed in the cooler.
4. The temperature of the refrigerated room should be 0.5°C-4°C. The lower the temperature, the better, because the respiration rate falls off with diminishing temperature. Low respiration rates have an effect similar to that resulting from adding sucrose to the preservative solution in that they conserve carbohydrates within the flower. A temperature range of 0.5°C - 4°C is usually encountered in flower coolers.
5. Air should be gently circulated inside the cooler only to the extent necessary to insure uniform temperatures in all areas. Unprotected flowers placed in a direct air stream will be desiccated. Flowers immediately adjacent to a cooling coil may freeze even though the air temperature is above freezing. Since the coil itself is below the freezing point, radiant heat is lost from the flower to the coil and the flower can be colder than the surrounding air.
6. Potential sources of ethylene gas should be avoided by keeping fruit and vegetables out of the cooler. Discard old flowers. Wash the inside of the cooler periodically.
7. Replace the preservative solution at two-to seven-day intervals. The preservative should be checked periodically for bacterial growth, which is apparent when the solution becomes cloudy. In spite of the bactericides in preservatives, microorganisms will develop and need to be eliminated periodically.

Dry storage

Dry storage is used when flowers must be held for periods longer than one to five days. Only the best-quality flowers should be dry stored. Flowers should be cut and packaged for storage immediately without being placed in water. Standard cardboard flower boxes are suitable, but a lining of polyethylene film should be placed in them to cover the flowers and seal in moisture. Desiccation can be a problem in long-term storage, especially when an absorbent container such as cardboard is used.

A common problem of dry storage is the presence of free water on the flowers, which encourages the development of disease. While flowers freeze only at temperatures below -1°C, the free water will freeze at 0°C. Resulting ice crystals on the petals can be injurious. Boxes and flowers packed at warm temperatures develop condensation (free water) as the plants and air inside are cooled. Because of the polyethylene barrier, the water cannot escape. Disease, enhanced by this moisture, is a common cause of failure in dry storage. Boxes of Most flowers freeze at -2.7 to1.6°C, so it is essential that the temperature stay above this point. Flower life expectancy is lessened at 0°C and drops rapidly at temperatures above that point.

Flowers removed from dry storage need to be hardened. Cut one-half inch from the bottom of each stem. Place the flower in a preservative solution inside a 3°C - 4°C cooler. Allow the flowers to become fully turgid before marketing them; this will take 12 to 24 hours. When properly handled, dry stored flowers should have reasonable quality and the same longevity as fresh flowers. Poor temperature control or disease will decrease quality and longevity. Dry storage is used only to a limited degree by the flower industry. The main reason for its low level of acceptance has probably been failures due to poor handling of the system.

5. **Preservative solutions:**Fresh flowers deteriorate for one or more reasons. Ten of the most common reasons for early loss of vase life are as follows:

   - Inability of stems to absorb water due to xylem blockage.
   - Wilting (Excessive water loss from the cut flower).
   - Food depletion (short supply of carbohydrate to support respiration).
   - Diseases attacked by bacteria and fungi.

- Accumulation of ethylene.
- Normal maturation and aging.
- Bruising and crushing.
- Fluctuating temperatures during storage and transit.
- Colour change (bluing).
- Poor water quality.

Flower preservatives are chemicals added to water to make flowers last longer. They contain a germicide, a food source, a pH adjuster, water, and sometimes surfactants and hormones. Germicides are used to control bacteria, yeasts and moulds. These microorganisms harm flowers by producing ethylene, blocking the xylem, producing toxins and increasing sensitivity to low temperatures. Bacterial counts of 10 to 100 million per millilitre impairs uptake, while counts of 3 billion per millilitre causes wilting. 8-HQC is the most common one used in commercial floral preservatives. Sucrose is the most common food source used in floral preservatives. It provides energy to sustain flowers longer and to open flowers in the bud stage. One to 2 percent sucrose is the standard amount in preservatives. Sucrose should not be used without a germicide, as it is also primary food source for microorganisms. Acids or acid salts are added to adjust the pH of the water to 3.5 to 5.0. At this pH, fewer microbes can grow and water is taken up by the flowers more easily. Surfactants and wetting agents reduce water tension so water is taken up more easily. Tween 20 and Triton are examples of surfactants.

Water is the most important component of floral preservatives. It should be pure and healthy for flowers, but that is not always the case. There are some properties or characteristics of water that should be understood because of their effect on flowers. They are pH, temperature, soluble salts, alkalinity and hardness. Total soluble salts or total dissolved salts number is expressed as ppm or mg/ liter or microsiemans. Low soluble salt content is best. They can interfere with water uptake because they change the osmotic potential of the water. They can burn leaves and petals because they can accumulate in the tips. Alkalinity is the measure of water's ability to neutralize acid or buffering capacity. It is a measure of the negative ion salts present in the water. Hardness often is used interchangeably with alkalinity but it measures the positive ion salts in the water. These are usually magnesium (Mg^{++}) and calcium (Ca^{++}). The standard practice to decrease hardness in water is to soften it by exchanging the magnesium and calcium ions with sodium ions. This method of decreasing water hardness should not be used for cut flowers. Fluoride is one specific ion that causes many problems. Flowers in the monocots such as orchids are more sensitive to fluoride than others. Fluoride toxicity is more of a problem at a lower pH, which is best for holding flowers.

Water should be tested to determine what the problem is and to determine the best remedy. Besides the standard floral preservative solutions for holding flowers there are some specific solutions and treatments that serve different needs.

- A harvesting solution often will simply be water acidified with citric acid to a pH of 3.5 to 5.0.
- A conditioning, hardening, or hydrating solution is used to restore the turgor of wilted flowers and dry packed flowers.
- Impregnation is a treatment that protects stems against the blockage of water vessels by microbes. Stems are dipped in 1000 ppm silver nitrate solution for 10 minutes. The stems should not be recut after treatment.
- Pulsing or loading is a type of treatment used to extend the vase life of flowers held in water, stored wet or dry for long periods or shipped long distances. It is called a pulse because it is only done for a short period of time, or called loading because the flowers are loaded up with food for a long storage period. Stems are placed in solutions with germicide and a higher concentration of sugar for specific treatment periods depending on the species. Because the higher concentration of sugar can act like a soluble salt causing petal and leaf injury, the treatment is only a few hours or a day. The temperature should be 18^0 to 24^0C and light intensity should be 2000 lux.
- Bud opening solutions are used to open flowers harvested in a tight bud stage. Flowers harvested in the tight bud stage will keep longer in storage and will ship better. Stems are placed in solutions containing higher concentrations of sugar, plus a germicide and hormonal compounds that facilitate bud growth and development. High light and humidity, and room temperatures are used. The high sugar content can make injure on the flowers and leaves.

Preservatives in the form of tablets or powder are prepared from a mixture of chemicals-sugars, germicides, salts and growth regulators. Various types of conditioners are sugar and biocide, antiethylene compound, and hydrated compound. Antiethylene compounds in preservative solutions reduce the action of ambient ethylene as well as autocatalytic production of ethylene by fresh cut flowers. Floral preservatives perform three functions:

- Provide sugar (carbohydrate).
- Supply a bactericide to prevent microbial growth and blockage of the water-conductive cells in the stem.
- Acidify the solution.

There are a number of commercial preservatives on the market, including such products as Floralife®, Petalife®, Oasis®, Rogard®, and Everbloom®. One can also purchase 8-HQC under the name oxine citrate from florist supply companies and add sucrose to make the preservatives. Air embolism, bacterial plugging, and poor water quality are factors that can reduce solution uptake.

Air embolism

Air embolism occurs when small bubbles of air (emboli) are developing inside the stem at the time of cutting. These bubbles cannot move far up the stem, so the upward movement of solution to the flower is restricted. Emboli can be removed by recutting the stems under water (removing about 2 cm), ensuring that the solution is acidic (pH 3 or 4), by placing the stems in a vase solution heated to 40°C by placing the stems in and ice-cold vase solution, by 10 seconds to 10 minutes dipping the stems in a weak detergent solution or by deeply immersing the stems in solution

Bacterial plugging

The cut surface of a flower stem releases the contents of the cut cells, proteins, amino acids, sugars, and minerals into the vase water. These are ideal food for bacteria, and these minute organisms grow rapidly in the anaerobic environment of the vase. Slime produced by the bacteria and the bacteria themselves can plug the water-conducting system. This problem must be addressed at every step of the postharvest chain.

- Clean water should be used for making postharvest solutions (dirty water contains millions of the bacteria that will proliferate at the base of the flower stem).
- Buckets should be cleaned and disinfected regularly (dirt harbors bacteria and may protect them from germicides.

- White buckets should be used (dirt is easier to see in a white bucket).
- Bucket and vase solutions should always contain a biocide. Common bucket biocides are calcium or sodium hypochlorite, aluminium sulphate and salts of 8-hydroxyquinoline. An acidic solution also inhibits bacterial growth.
- Sugar-containing solutions for pulsing, bud-opening or display should contain an adequate germicide.

Mechanical damage

Bruising and breakage of cut flowers should be avoided. Flowers turn with petals, broken stems or other obvious injuries are undesirable for aesthetic reasons. Disease organisms can more easily infect plants through injured areas. In fact, many disease organisms can only enter a plant through an injury point. Additionally, respiration and ethylene evolution is generally higher in injured plants, further reducing storage and vase life.

Diseases

Flowers are very susceptible to disease, not only because their petals are fragile, but also because the secretions of their nectaries often provide an excellent nutrient supply for even mild pathogens. To make matters worse, transfer from cold storage to warmer handling areas can result in condensation of water on the harvested flowers. The most commonly encountered disease organism, gray mold (*Botrytis cinerea*), can germinate wherever free moisture is present. In the humid environment of the flower head, it can even grow (albeit more slowly) at temperatures near freezing. Proper management of greenhouse hygiene, temperature control, and the minimizing of condensation on the harvested flowers all reduce losses caused by this disease. Some fungicides, such as Ronalin, Rovral (Iprodione) and the copper-based Phyton-27 have been approved for use on cut flowers and are very effective against gray mould.

B. Biotic Factors

1. **Genetic or inherent makeup:** Postharvest lasting quality of flower species and cultivars vary considerably due to differences in their genetic make-up. Genetic modification for improved postharvest quality is a realistic goal, as much is

known of the biochemistry and molecular biology of senescence-related processes in plants. Genetic modification has been used to obtain varieties with improved vase-life in several species.

2. **Water relations:** Extended life for cut and potted orchids depends absolutely on a continuing and adequate supply of water. Failure of water supply, whether through obstruction of the cut stems, or through inadequate watering of pots, results in rapid wilting of shoot tips, leaves, and petals. The termination of life of the harvested flowers depends on water uptake and transport, water loss and the capacity of the flower tissue to retain its water. A water deficit and wilting develop, when the transpiration exceeds absorption of water. The rate of water uptake of cut flowers depends on transpiration pull, temperature and composition of solutes. Disruption of water columns in stem vessels by air embolism and resistance to water flow in stems, also develop water deficit. Acidification of water and addition of wetting agent and flower food in the holding solution markedly improve water uptake of cut flowers. Inability to absorb water is a very common reason for premature wilting of cut flowers. The water-conducting tubes in the stem (xylem) become plugged. Bacteria, yeast, and/or fungi living in the water or on the flower or foliage proliferate in the containers holding the flowers. These microorganisms and their chemical products plug the stem ends, restricting water absorption. They continue to multiply inside and eventually block the xylem tubes. Chemical blockage also can occur. Chemicals present in some stems, upon cutting, change into a gum like material which blocks the end of the stem. Excessive water loss from flowers can lead to wilting and reduction in quality and vase life. After harvest, flowers should be removed from the field or greenhouse and refrigerated as soon as possible. Leaving the flowers out of water, in warm air or in warm drafts such as from a heater, causes considerable damage. Flowers should be in water and under cool temperatures as much as possible from the time they are cut until they reach the final customer.

Chapter - 5

Protection Technology

Duff and Daly (2002) briefly describes the commonly found orchid diseases. These include the Bacterial soft rot caused by Erwinia carotovora, black leg or slow decline caused by Fusarium spp., petal blight caused by Curvularia eragrostidis [Pseudocochlio*Bolus* eragrostidis] and Alternaria alternata, top rot/shoot rot caused by Phytophthora parasitica, and stem rot caused by Sclerotium rolfsii [Athelia rolfsii].

The facultative pathogens like *Alternaria, Botrytis, Colletotrichum, Circospora, Curvularia, Gloeosporium, Phyllosticta, Phoma* and *Bacillus* are generally causing leaf spots. However, root rot and wilting are mostly concern with many soil-born pathogens (*Pythium, Phytphthora, Rhizoctonia* and *Sclerotium*). Similarly, the viral diseases are also common in various orchids. They produce a variety of symptoms on leaves and petals of the plant in form of leaf necrosis, blossom necrosis, colour variegation, stripes, streaks, flower break *etc.* Mechanical injuries by tools during cultural operations and presence of insect vectors in polyhouses are the major sources of transmission of pathogens from infected to healthy plants.

The disease management in enclosed environment can better achieve by technical, cultural, physical, biological and chemical methods if they opt in integration. Regulation in temperature and humidity inside the polyhouses, application of balance quantity of fertilizers, use pathogen-free planting maty*Eria*ls, sterilization of tools, sanitation, avoid mechanical transmission of pathogens, eliminate pathogen reservoirs, use resistant varieties, hot water treatment of vegetative propagules and use of suitable beneficial microorganisms, fungicides and insecticides are the major components of integrated disease management in commercial orchid crops.

Fungal Diseases of Orchids

A. Dry Rot and Wilt Disease

Dry rot and wilt diseases of orchids are highly destructive and limiting factor in quality flower production of orchids. Important genera such as *Phalaenopsis* , *Cattleyas, Oncidium, Cymbidium* and *Dendrobium* are more vulnerable to *Fusarium* species. The diseased plants also show other symptoms in form of leaf spots, leaf blight, roots rot, seedling rot and blackening pseudobulbs. *Fusarium subglutinans* and *F. proliferatum* cause leaf spot and root rot of *Cymbidium*. *F. moniliforme* associated with seedling wilt. *F. oxysporum* and *F. solani* are associated with root necrosis of *Cymbidium* orchids. Thirteen isolates of Sclerotium rolfsii [Corticium rolfsii] causing dry rot in orchids, as well as 3 fungal antagonists were isolated from *Dendrobium, Oncidium,* Doritis, *Vanda,* Mokara, *Phalaenopsis,* Papiopedilum and *Rhynchostylis* (Vithoonjit *et al.,* 2005).

During a routine monitoring of orchid diseases in Pakyong, Sikkim, India, a destructive disease was noticed by Bag (2003). The disease was first found to infect pseudobulbs of *Coelogyne* corymbosa, causing pseudobulb rot and death of entire plant in the community pot in June 2001. The fungus was found later to infect several other orchid hosts as well. Some of the orchid hosts found to be affected by the disease during 2001-02 were C. corymbosa, *Eria coronaria* , Spathoglottis sp., *Habenaria* sp., *Aerides* sp., *Dendrobium* sp. and *Cattleya* sp. In all cases, symptoms included basal rot of pseudobulbs, and leaves become yellow and detached from the pseudobulbs. Eventually, the entire plant would become dry and senescent. Numerous small, brown-coloured sclerotia were observed on affected pseudobulbs and white mycelial growth was found associated with leaf bases. The disease-causing organism was identified as Sclerotium rolfsii [*Corticium rolfsii*] based on morphological and cultural observations, as well as pathogenicity test. This is thought to be the first report of orchid disease caused by *S. rolfsii* on *C. corymbosa, Eria coronaria* , *Spathoglottis* sp., *Habenaria* sp., *Aerides* sp., *Dendrobium* sp. and *Cattleya* sp. in India.

Sclerotium rot was observed on greenhouse-grown *V. coerulea* and V. stangeana during June-August 2003 in Pakyong, Sikkim, India. On both species, the disease caused basal rot on pseudostems at the initial stage. The disease gradually spread upward, and the leaves turned yellow and got detached from the stems, ultimately causing the mortality of infected plants. In 2004, the disease was also detected on *V. stangeana, V. coerulea, Acampe papillosa,* Robiquetia spatulata, *Luisia* sp. and *V. coerulea* x *V. roxburghii* grown in the greenhouse. The causal agent was identified as *S. rolfsii* [*Corticium rolfsii*] based on morphological traits. Pathogenicity was also confirmed on the aforementioned hosts under natural conditions. *V. stangeana, V. coerulea, A. papillosa, R. spatulata, Luisia* sp. and *V. coerulea* x *V. roxburghii* are new host records for *S. rolfsii* in India (Bag, 2006).

A total of 29 Fusarium isolates were isolated from *Dendrobium* orchid showing symptoms of root and stem rots. Based on morphological characteristics, three species were identified namely, *F. oxysporum*, F. solani, and *F. proliferatum* which were recovered from root and stem rots of *Dendrobium*. Cluster analysis of RAPD bands clearly separated *F. oxysporum, F. proliferatum,* and *F. solani* into distinct clusters. The present studies showed that three *Fusarium* species were isolated from root and stem rot of *Dendrobium* and cluster analysis of RAPD bands was in agreement with morphological characterization of the *Fusarium* species from root and stem rot of *Dendrobium* (Latiffah *et al.,* 2009).

Cymbidium orchids with blight and rot symptoms were collected, and a total of 63 isolates of *Fusarium* spp. was obtained from pseudobulbs, roots, and leaves of the diseased plants. The isolates were identified based on their morphological characteristics. Out of the 63 isolates of Fusarium spp., 51 isolates were identified as *F. oxysporum*, 10 isolates as F. solani, and the rest as *F. proliferatum*. *F. oxysporum* was isolated from all the *Cymbidium* spp., while F. solani and F. proliferatum were isolated only from *C. ensifolium* and *C. ginatum,* respectively. Isolates of the three *Fusarium* spp. were tested for pathogenicity to their hosts by artificial inoculation. The strongly pathogenic isolates of *Fusarium* spp. induced severe dry rot of pseudobulbs and roots of the host plants. The symptoms progressed up to the basal part of the leaves, which later caused blight of the entire plant. The dry rot symptoms induced on the plants by artificial inoculation with the isolates of *Fusarium* spp. were similar to those observed in the growers' greenhouses. This is the first report of dry rot of *Cymbidium* spp. caused by *F. oxysporum, F. solani,* and *F. proliferatum* in Korea Republic (Lee *et al.,* 2002).

Symptoms: Ichikawa and Saito (1998) proposed 'dry rot' as the common name of rot of *Cymbidium* plants caused by *F. proliferatum, F. oxysporum* and *F. solani*. It was first time reported from Korea in *Cymbidium* spp. (Byung, *et al.,* 2002). The dry rot causes damage on the young plant parts of orchids as they are dark grey or black in colour after infection. In advance stage of disease development, the infected areas are dry, brittle and sharply delimited. Older lesions are markedly decayed, brown blotched and irregular in shape. The basal end of the affected parts most frequently attacked.

The symptoms of the wilt appear first in form of brown-black spots on roots, basal stems and pseudobulbs. In later stage, the physical presence of the pathogen plugs the phloem, thus, blocks the movement of water upside through xylem vessels. The leaves of infected plant are turn yellow, thin, shriveled, wrinkled, wilted and eventually die. The whole plant may show stunting, as successive new shoots tend to get smaller in size with reduced stem thickness. The diagnostic symptom in the plant is a circle or band of purple or pinkish-purple discoloration on the outer layers of the rhizome evident when the rhizome is cut. If the disease is extensive, the entire rhizome may turn purple and the discoloration may extend to the pseudobulbs. Severely infected plants may die in 3-9 weeks, while mildly infected plants gradually decline over a year or so.

Disease Management

Sen *et al.* (2009) showed that antimicrobial activity of the isolate might be linked with the production of siderophores, proteases, chitinases and indole acetic acid (IAA). A talc-based formulation of the antagonist did not only protect *Cymbidium* plants from the disease but improved also its growth.

1. **Chemical Control:** Application of Thiophanate methyl or Iprodione fungicides in pot medium controls soil born inoculum of *Fuasrium*. However, most of the *Fusarium* species show resistance to these fungicides. Benomyl can be applied to seed treatment, foliar spray, drenching of soil and dip of root and propagating material. Chlorothalonil has excellent protective activity while the systemic fungicide Azoxystrobin has protective and curative actions and suppresses the growth of *Fusarium*. Fudioxonil and Triflumizole also provide adequate control of *Fusarium* wilt if applied during the early infection period.

2. **Biological Control:** Biological control is principally based on the adversary nature of living organisms to others. Various species of *Gliocladium, Trichoderma, Bacillus,* and *Pseudomonas* are known as universal biocontrol agents as they reduce a

variety of plant pathogens by the mechanism of antibiosis and competition. The mild strain of *Fusarium* can reduce the incidence of its virulent strain. Biological control can achieve in several ways.

Cross protection

A weak virulent *Fusarium* isolate, strain HPF-1 had antagonistic activity against *Cymbidium* yellow spot caused by the *Fusarium proliferatum* race Y. HPF-1was isolated from an apparently healthy *Cymbidium* leaf in a greenhouse where a severe outbreak of yellow spots had occurred. When a one year old mericlone seedling of *Cymbidium* was spray inoculated with HPF-1 black necrotic spots that were discernible only by microscope were produced on leaves and the development of yellow spots from natural infection was suppressed for about 10 month growth period. (Ichikawa, *et al.,* 2003).

Treatment of vegetative propagating material

Pseudobulbs, rhizome, backbulbs, keiki and areial cuttings are treated before use with the solution of commercial bioformulation of *Trichoderma* (@ 8-10 gram *Trichoderma* powder/litre water) for 30 minutes. *Bacillus subtilis* also successfully control the infection of *Fusarium* if the cuttings and bulbs of orchids are treated with bacterial solution.

Seedling treatment

Prior to transplanting of seedlings, they are dipped in a solution of *Trichoderma harzianum* (@ 8-10 gram / litre water) for about 30 minutes. This treatment avoids the root infection of *Fusarium.*

Foliar spray

This approach involves the establishment of antagonist as epiphytic on aerial surfaces of plants. The innundative foliar sprays of a selected antagonist are essentially protect the plant by making antagonist cover on aerial surface prior to dry rot infection. The curative foliar applications of biocontrol agents may also reduce the infection of *Fusarium.* In both type of foliar applications 8-10 gram bioformulation per litre of water can be used. Four to 6 sprays are required with10-12 days intervals.

Mixing of bioagents in medium

Trichoderma and *Pseudomonas* based bioformulations may reduce the inoculum numbers and their capacity if mixed in the medium @ 1-2 per cent before potting.

3. **Cultural Practices:** Keep benches and floors free of plant debris. Remove all diseased plants or parts of plants, e.g. diseased leaves, pseudo stems, scales, calms, infected roots *etc.* Improve airflow in ployhouses to reduce moist content on foliage and stems. Sterilize cutting tools by dipping them in alcohol or solution of bleaching powder before use.
4. **Resistant Varieties:** Always use healthy propagating materials that might be vigorous and disease-free. Use resistant varieties against *Fusarium.*

B. Black Rot and Seedling Damping Off

Several species of Phytophthora have been reported to cause economic damage on orchids worldwide. Of these, P. cactorum and P. palmivora have the widest host range across orchid genera, and they are the most common species affecting commercial orchid production in Florida. The objective of this guide is to provide a practical diagnostic resource for black rot on orchids in Florida (Cating *et al.,* 2010).

In 1996 and 1997, a severe outbreak of a disease on *Cymbidium* orchid (*Cymbidium hybridium*) occurred in a garden in I-Lan, Taiwan, with numerous plants showing black rot symptoms on leaves, pseudostems and roots and dying. A Phytophthora-like microorganism was constantly isolated from the diseased plants and two of the eleven isolates collected from this garden were used in this study. Inoculation tests were carried out using zoospore suspensions to fulfill Koch's postulates. Results showed that both isolates (P96013 and P96016) were pathogenic on *C. hybridium* with the same black rot symptoms as those developed on the naturally infected plants. The pathogenicity of these two Phytophthora-like isolates was confirmed as the organism reisolated from the inoculated plants was morphologically identical to the culture of the original inoculum. In addition to *C. hybridium,* the two Phytophthora-like isolates were also pathogenic on other Asian *Cymbidium* species, including *C. ensifolium, C. sinense* and *C. rubrigemmum,* causing black rot symptoms and death of plants. Both Phytophthora-like isolates were similar to Phytophthora multivesiculata in most of the morphological characteristics. However, they were different from *P. multivesiculata* in some features such as dimensions of sporangia, oogonia and oospores as well as maximum growth temperature. The ITS sequences of these two Phytophthora-like isolates were identical to that of *P. multivesiculata*. Based on the morphological characteristics and the molecular data, both Phytophthora-like isolates from *C. hybridium* were identified as an aberrant strain of *Phytophthora multivesiculata*. This is the first report of the aberrant strain of Phytophthora multivesiculata in Taiwan. Diseased orchid plants in the garden of Yuan-Shan, I-Lan county, were destroyed in 1998 and this aberrant strain of *P. multivesiculata* was not found in the survey of 1998 and 1999 in I-Lan county or any other orchid production areas of Taiwan(Chern *et al.*, 2011).

Su *et al.* (2010) described that *Fusarium solani* (teleomorph Nectria haematococca) caused sheath and root rot of *Phalaenopsis* in Taiwan. In our field surveys since 2005, a fungal disease widely occurred in *Phalaenopsis* nurseries. The disease incidence of some susceptible varieties of *Phalaenopsis* was as high as 30 to 60% in certain nurseries. The diseased plants showed symptoms firstly on lower leaves which turned yellow with a sheath black rot period. The diseased leaves abscised after a short period of time, and eventually the infected orchid plants died. Usually, several small red granules (perithecia) with white mycelia were observed on the discolored sheath surface. In some cases, the disease lesions extended to the stems and roots which displayed brown to black rot with perithecia on the surface. *Fusarium solani* was isolated from 80% of the diseased tissues. Disease symptoms induced by an isolate FST1 of the suspected candidate pathogen of F. solani in greenhouse were identical to those observed in the field. Meanwhile, F. solani was recovered from the artificially inoculated diseased tissues. The perithecia of *N. haematococca* (sexual stage of *F. solani*) with salmon color were formed when single spores of FST1 were cultured on PDA at 24°C for 2 to 3 weeks.

In 2004-05, diseased leaves and stem parts of *Cymbidium hybridum*, *Dendrobium Phalaenopsis*, *Epidendrum* sp., *Phalaenopsis* lueddemanniana and *Zygopetalum mackaii* were collected from 3 modern greenhouses in Poland, especially during spring and summer time. A total of 11 genera and species were isolated from the diseased parts of the orchids. *Phytophthora palmivora* was isolated from most of the analysed plant parts of the 5 orchid species. The species was recovered from diseased leaf blades and stem parts. *Fusarium oxysporum* was isolated from ~1/8-5/8 of the surveyed plants, especially from D. *Phalaenopsis* and *Epidendrum* sp. Two other Fusarium species (*F. culmorum* and *F. equiseti*) were isolated rarely or sporodically. *Botrytis cinerea* was found in each diseased orchid species. Five isolates of *Phytophthora palmivora* were used for inoculation of leaf blades and roots. All the tested isolates colonized the leaves and roots of the 5 orchid species. On leaf blades, the development of necrotic spots was faster when the isolates from *Epidendrum* sp. and *Phalaenopsis lueddemanniana* were used for inoculation. The isolate from *Epidendrum* sp. caused also faster spread of necrosis on the orchid roots (Orlikowski and Szkuta 2006).

Greenhouse studies were conducted to determine the occurrence of leaf blight and dry rot of 2 *Cymbidium* species (*C. devonianum* and *C. tigrinum*), caused by *Fusarium oxysporum*, in Uttarakhand, India. Pathogenicity tests of the isolated pathogen were conducted on healthy plants of the 2 *Cymbidium* species and the morphological characteristics of the pathogen obtained from the infected leaves, pseudobulbs and roots were recorded. Results showed that all the 18 Fusarium spp. isolated from the 2 *Cymbidium* species were identified as *F. oxysporum* and were strongly pathogenic to *C. devonianum* and *C. tigrinum*. The pathogenic *F. oxysporum* isolates induced severe dry rot of pseudobulbs and roots of the host plants. The symptoms progressed up to the basal part of the leaves, which later on caused blighting of the entire plant (Yadav *et al.*, 2010).

Symptoms: *Cymbidiums, Oncidiums* and *Cattleya* are most susceptible genera. As the name implies, black rot characterized as dark blackish spots or lesions on the affected plant parts. The symptoms of black rot first started from the margin and tip of the leaf during rainy season as small, water soaked translucent spots.

These spots expand quickly and cover the entire leaf surface in moist condition. Affected tissues of the leaf turn brown to black with watery appearance. Black spots further rapidly spread throughout the plant and move to the roots. *Pythium* and *Phytophthora* also attack on juvenile seedlings of orchids, usually at the soil line. The collar portion of the infected seedlings is rotted and translucent in appearance and that causes the little plants to bring down and die which termed as damping-off. In mature plants, the disease can hit to the leaves, pseudobulbs and roots but often starts from new leaves and progresses downward until touch the roots. If infection of the pathogens allowed to reach the crown of a monopodial (single-stemmed) orchid, the rot will kill the plant. Affected leaves may turn yellow around the infected area and the lesions themselves will be soft and exude water when pressure is applied.

Management: Contaminated pots, growing media, propagating materials, tools and water splashing from infected plants to nearby places are the excellent means of spreading the fungus. The continuous moist environment of an extended period favours disease development rapidly. The dense canopy of plants also makes a congenial environment for disease development. The following management practices may apply to the orchid crop.

1. **Preventive measures:** Plants if placed outdoors should be elevated 90 to 120 cm above the ground to shun splash contamination. Early watering in the day is important so that the plants can dry out before dark comes, is the best prevention. Adequate spacing of plants with good air circulation in the growing area is a key of successful management and help plants dry quickly after watering or rains. The addition of airflow device in a growing area improves the air circulation. Use of a high-calcium fertilizer in the spring is recommendation to avoid black rot in new growth of orchids.

2. **Mechanical measures:** During preparation of media, multiplication of propagating materials like back bulbs, keiki, areial cuttings, meristem or tissue culture and seeds, removal of diseased portion of the plants, rouging of infected plants from the beds, potting, repotting, cuttings of flowers be always in practices with sterilized implements. Glass, plastic, cloth, nonmetallic tools, pots, equipments *etc.* can disinfest with a solution of bleaching powder (10 per cent) or by the use of quatenary ammonium compounds such as Physan or Consan. Tools may also be sterilized by flaming with a propane torch or by soaking in a 50 per cent solution of clorox or a saturated solution of tri-sodium phosphate (TSP) for a few minutes. Dead and diseased plant parts, interior portion of pseudobulbs that seems to be healthy on the surface should remove using sterilized knife. In sympodial plants such as a *Cattleya,* it may be necessary to remove infected leaves with some healthy portion bordering the infected area as a safety precaution. Similarly, the pseudobulbs and its infected portions must take out very carefully by sterilized knife to impede further spread of the disease to healthy tissues. If the plant is badly infected, remove the infected part of rhizome growth by growth until get to healthy tissues. In a monopodial orchid such as a *Vanda* or *Phalaenopsis* , remove the infected portion of the leaf or the

entire leaf if necessary to halt the spread of the pathogens before it reaches to the crown.

3. **Chemical treatment:** The surface protectants of propagating materials are most effective means to keep away *Pythium* and *Phytophthora*. The cut portions of the orchid plants should also treated by suitable fungicides to protect the remaining healthy tissues. During the treatment all exposed areas should cover properly with the fungicide. Cinnamon is an excellent fungicide and ground cinnamon may be applied straight from the spice jar directly to the exposed area where infected portions of the plant have been excised. Mixture of cinnamon and casein-based glue such as Elmer's or cooking oil is also recommended in form a thick paste. This paste is fairly waterproof and can use to cover the wound portions. The Bordeaux paste prepared with lanolin oil also protects the open cut ends of the plant. Lanolin seals the wound to keep moisture out from the living tissues. Drenches with protectant fungicides such as Truban or Terrazole are recommended if the disease is diagnosed at early stage. Systemic fungicide such as Aliette, Subdue are more effective against oomycetes. Captan, Dithane M-45 and Physan 20 have also been recommended for control of black rot. Disinfestation of the soil / medium with any one of Captan, Thiram, Copper oxychloride and Carbendazim proficiently controls the disease. These fungicides are mixed in the soil @ 0.2-0.5 percent suspension. Combination of seed and soil treatment is the best approach for management of damping-off. The medium may also be disinfested using 2 per cent formalin solution being keep the medium air tight for 48-72 hrs after disinfestation.

4. **Biological control:** Application of *Trichoderma, Pseudomonas, Bacillus* and *Penicillium* in form of surface coating agents, foliar spray, seedling treatment, soil treatment *etc.* during orchid cultivation reduce the incidence of damping-off and black rot.

C. Top rot / Shoot rot

The top portion of *Vanda* and shoots of *Cattleya* are adversely affected by the pathogen *Phytophthora parasitica.*

Symptoms: A symptom of top rot appears on the new leaves in form of small water soaked lesions in *Vanda* orchid. These lesions turn dark brown later and result the death of new leaves. Discolouration of infected portions may progress downward on the stem. The symptoms of disease may also starts from the base of stem, where it produces dark brown discoloration. Shoot rot was first time detected on *Cattleya* orchids in Darwin. It produces small irregular brown spots on new side shoots in the beginning, which latter collapse each other and infect entire shoot. In the later stage of disease development, infected shoots turn down black in colour. The disease will spread back along the rhizome to the next shoot causing the same symptoms. *Vanda* and *Cattleya* are only the two orchid types in record, which have been attacked by this disease. However, the other orchids may possibly affect in a similar way.

Biology of Pathogen: *Phytophthora parasitica* has a wide host range under the vegetable, fruits and ornamental crops in which it causes variety of symptoms in form of black rot and seedling damping off, stem rot, stem canker and twig blight, fruit rot *etc.* However, parasitic nature of the pathogen differs with genus and species of the plant. *Phytophthora* is hemibiotroph in nature and can survive in dead tissues in absence of living tissues. Fungus produces aseptate mycelium at early stage but it may be septate in old culture. The long and unbranched sporangiophore may come out from stomata or directly by rupturing the epidermis layer. Single, hyaline, ovoid, papillate sporangium develops on the top of sporangiophore. The behaviour of sporangia depends on the available moisture content and temperature at the site of infection. These may germinate directly in dry conditions by producing germ tube and produce many zoospores therein in moist condition. Zoospores are kidney shaped, biflagellate and encysted after liberate from sporangium. The encysted zoospore may germinate by the germ tube. The chlamydospors are also produced by the pathogen characterized as spherical, thick walled, smooth, hyaline or yellowish in colour. Sexual reproduction of the pathogen is not common in nature. However, in certain conditions it takes place with oogonium and antheridia. The oogonium after fertilization produces the oospores. They are resting structures and survive for longer time in the soil.

Disease Cycle: *Phytophthora* is soil born and hemibiotrophic in nature. It survives in living perennial orchid plants, their infected debris, residues and in growing medium in form of oospores, chlamydospores and dormant mycelium. These primary inoculums can directly penetrate to the host tissues if in a right approach to them under dry conditions. In absence of host oospores will produce zoospores through sporangia are also infect to the stem at the soil level and roots of orchids. The dissemination of primary inoculum from infected to near by healthy orchid plants achieved by water splashing. The pathogen is also introduced in to new areas by planting infected propagating materials. The secondary infection is efficiently zoospores and conidia.

Management: Disease free propagating materials should be planted into the disease free growing medium. Orchid growers are advised to have good drainage system in their orchid growing areas / polyhouses. The fumigation of compost mixture with methyl bromide or chloropicrin is also recommended. The *Trichoderma* and *Pseudomonas* based bioformulation is effective to control the foliar, stem and root symptoms of the pathogen. It can be mix in the medium @ 250 g/ quintal growing medium prior to plantation. The protective and curatives innundative foliar sprays of bioformulation of *Trichoderma* and *Pseudomonas* @ 6-8g/ litre of water reduce the incidence of top and stem rot of orchids. The excellent control of disease further achieved by fungicides. Metalexal, Fosetyl Al and Propamocarb are selective fungicides, control the incidence of *Phytophthora*. These fungicides can be used in the treatment of soil, seedling, propagating materials. Foliar spray of the fungicides is equally important to control incidence of disease.

D. Stem Rot of *Dendrobium* and *Vanda*

During June 2003, near Erzurum, Turkey, a decline of *Orchis* palustris was observed in several meadows that had been irrigated for forage production. Stems were chlorotic, wilted and collapsed. There was a soft, watery rot at the crowns and lower stems. White mycelium and black sclerotia formed on necrotic stem and crown tissues. The fungus was isolated from sclerotia on potato dextrose agar and identified as *Sclerotinia minor* based on morphological and cultural characterization as well as pathogenicity tests. This is thought to be the first report of *S. minor* infecting *O. Palustris* (Eken *et al.*, 2005)

Three Fusarium species, *F. oxysporum*, *F. proliferatum* and *F. solani* were isolated from root and stem rot of *Dendrobium* orchid. Their pathogenicity on *Dendrobium* was studied and the inoculation test showed that the three Fusarium species were pathogenic, causing root and stem rot on the orchid. Molecular characterization using PCR-RFLP of ITS+5.8S regions showed that the isolates from the same species produced similar patterns and UPGMA cluster analysis of PCR-RFLP of ITS+5.8S clearly grouped *F. oxysporum*, *F. proliferatum* and *F. solani* into separate clusters. The present study showed that three Fusarium species, *F. oxysporum*, *F. proliferatum* and *F. solani* were associated with root and stem rot of *Dendrobium* orchid (Latiffah *et al.*, 2010).

Symptoms: White visible mycelial growth of the pathogen appears on basal portion of stem, roots and other ground parts of the orchid plants during moist condition. Infected tissues rapidly collapse and rotted. The internal browning of infected pseudostem also observed. In the older infected regions the disease is characterize with the presence of small typical tan to brown mustard seed size sclerotia embedded in a dense white fluffy mat of fungal growth. Mature plants are attacked just below the soil surface and completely girdled by the mycelium. During severe condition, the disease often destroys the entire root system, which results the sudden wilting of plant. Slightly sunken, yellow spots also develop on plant parts that in touch with infected portion at soil level, they rapidly decay, collapse and become covered by a white fungal mass with numerous sclerotia. This disease is also attacking on the top portion of the *Dendrobium* plants where it causing rot followed by rapid collapse of tissues. The disease then advances either up or down the stem.

Management

Cultural management

- Sanitation of infected areas, compost and working places.
- Avoid the excess presence of moisture content in the Polyhouses and pots. The moisture does not adhere on the plant surface. For this purpose well-designed irrigation-drainage system must be use.
- Diseased plants, infected parts of them should be removed, collected and disposed by burying 2 to 3 feet deep or by burning.

Maintain recommended plant density in the orchid growing area. Plants should be staked properly in the way that their branches and leaves do not come in touch with soil / compost surface.

Biological control

The soil application of *Trichoderma, Pseudomonas,* and *Bacillus* reduces the capacity of primary inoculum of *Sclerotium rolfsii*. However, patch application and foliar spray of biocontrol agents are also beneficial in the management of *S. rolfsii.*

Chemical control

The efficacy of mancozeb (0.25%), copper oxychloride (0.3%), difenoconazole (0.1%), carbendazim (0.3%), hexaconazole (0.1%), metalaxyl (8.0%) + mancozeb (0.1%) and copper hydroxide (0.25%), and of the extracts of *Ocimum sanctum* [*O. tenuiflorum*], *Catharanthus roseus, Allium cepa, Azadirachta indica, Bougainvillea spectabilis,* Allium sativum and Gliricidia maculata at 20% each against *S. rolfsii* [*Corticium rolfsii*] was evaluated. The pathogen was incubated with the extract at 24+or-1°C, then radial growth was recorded after 24 h. The inhibitory effect of the fungicides was evaluated by poisoned food technique. The total inhibition of *S. rolfsii* was obtained with the garlic extract. Azadirachta indica and Catharanthus roseus inhibited *S. rolfsii* growth by 79.3 and 70.4%, respectively. Among the fungicides, mancozeb, difenoconazole and hexaconazole resulted in the total inhibition of radial growth (Kolte and Raut 2007).

- Vapam (Sodium methyl dithiocarbamate) is available in liquid form and can be use for compost / medium treatment.
- Application of the PCNB to sterilize the compost / medium before potting of the plants.
- Increase the pH of the medium by adding the calcium components.
- Spray of Zineb @ 0.2 per cent controls the infection

E. Collar Rot and Root Rot

Sui *et al.* (2010) described that *Fusarium solani* (teleomorph *Nectria haematococca*) caused sheath and root rot of *Phalaenopsis* in Taiwan. In our field surveys since 2005, a fungal disease widely occurred in *Phalaenopsis* nurseries. The disease incidence of some susceptible varieties of *Phalaenopsis* was as high as 30 to 60% in certain nurseries. The diseased plants showed symptoms firstly on lower leaves which turned yellow with a sheath black rot period. The diseased leaves abscised after a short period of time, and eventually the infected orchid plants died. Usually, several small red granules (perithecia) with white mycelia were observed on the discolored sheath surface. In some cases, the disease lesions extended to the stems and roots which displayed brown to black rot with perithecia on the surface. *Fusarium solani* was isolated from 80% of the diseased tissues. Disease symptoms induced by an isolate FST1 of the suspected candidate pathogen of *F. solani* in greenhouse were identical to those observed in the field. Meanwhile,

F. solani was recovered from the artificially inoculated diseased tissues. The perithecia of *N. haematococca* (sexual stage of *F. solani*) with salmon color were formed when single spores of FST1 were cultured on PDA at 24°C for 2 to 3 weeks.

Symptoms: The collar region of stem is most susceptible place to the disease than the other parts of the plant for soil-born pathogens as it always in touch with moist soil surface. The narrow and depressed lesions are developed on the stem near the soil line. They are dark brown in color. The growing lesions unite and form a big lesion that latter result collar rot. Orchids will also manifest a root rot problem by showing a decline in health and the roots will turn brown. The disease is also confined mostly to cuttings that have been planted in an infected soil / compost medium. The nonpathogenic strain of *Rhizoctonia* may also act an endophytic mycorrhizal fungus on orchid seedlings. Strains parasitic to other plants may be non injurious to the orchids (Downie, D.G 1957 and Harley, 1965)

Management: Precaution must be taken to avoid over-watering in orchid growing areas. Among the fungicides Carbendazim, Brassicol (PCNB), Flutolanil are very effective against *Rhizoctonia* . Application of PCNB into the medium/compost reduces primary inoculum. Spray of Carbendazim at the rate of 0.7 g / litre water gives a good control of the pathogen. The same fungicide can also be used for treatment of propagating materials and seedlings. Thiophanate methyl, Chlorothanoil, Iprodione and Propiconazole are other fungicides useful in the management of *Rhizoctonia*. Application of *Trichoderma*, *Pseudomonas* and *Bacillus* in the growing medium reduces the incidence of pathogen. The binucleate and hypovirulent isolates of *Rhizoctonia* are effective biocontrol agents of several pathogenic species *R. solani*. They induce systemic acquire resistance into the host plant through cross protection and have a good colonizing ability to host tissues (outer tissues of roots, crowns, hypocotyls, stems or petioles). However, other mechanisms like mycoparasitism and antibiosis are absent during this biological control. Heat treatment of vegetative parts, cutting tools and compost, removal of infected plant debris from the compost, repotting of orchids in a clean pot with fresh sterilized soil are other physical methods for managing the *Rhizoctonia solani*.

F. Petal Blight of *Dendrobium* Hybrids

Symptoms: *A. alternata, C. eragrostidis* cause small brown spots on the petals (Fig. 10.2). These spots are well define on white and light coloured flowers. However, *B. cinerea* develops small to large watery brown marks on flower parts. *Dendrobium* flowers are most susceptible to the petal blight. Disease is very difficult to control and is most severe during rainy weather.

The genus *Alternaria* infects more than 380 plant species as per the record of USDA Systematic Botany and Mycology Fungus, Host Distribution Database. Fungus distinctively stands for its transversely and longitudinally septate muriform conidia. The conidiophores are dark and septate which arise directly from infection site. The top portion of conidiophore looks like bushy due to appearance of 4-8 large catenates conidia chains. The secondary conidiophores are generally short and 1-celled. Typical mature conidia are 10-30 x 5-12 μm in size, bear short conical beak or beakless, narrowly ellipsoid to ovoid and elongated on branching chains. Juvenile conidia are narrowly elliptical and punctuate with rough ornamentation at older stage. Conidia are dull olive in color and possess 3-7 transverse and 1-5 longitudinal septa along with individual chains of 5-15 conidia. Complex of branching chains contain 50-60 conidia.

Botrytis cinerea (perfect stage: *Sclerotinia fuckeliana*) is commonly known as grey mold. The generic name of the fungus is derived from Greek word *botrys* that stands for a bunch of grapes. This genus was first described by de Bary from vine leaves. The mycelium of *Botrytis cinerea* is septate and brown in colour. The young hypha is thin and hyaline, conidiophores are light brown in colour, septate and irregularly branched at the top. The top portions of conidiophores and lateral branches attached therewith are slightly enlarged and have minute sterigmata. Conidia are hyaline, aseptate, oval, globose or short cylindric and born on a sterigmata in cluster which seem to be bunch of grapes. Conidiophores and conidia stand freely on the hyphae. Fungus has a layer of sporodochia (sclerotia) comprises by interwoven brown septate hyphae and produces round to oval aseptate unfunctional conidia. Sporodochia turn to be hard soon and act as a primary source of inoculum.

The many species of *Curvularia* (perfect stage: *CochlioBolus*) are commonly known as leaf spot fungi. The mycelium and conidia of the pathogen are dark brown in colour and septate. The third cell of conidia from its basal end is curved, large and dark than remaining cells of the spore.

Management: Pot and field sanitation is essential practice to reduce the primary and secondary infection of the pathogens from orchid growing areas. Infected residues of orchid plants particularly fallen petals are collected and damp or burn away from the orchid growing area otherwise pathogen may survive on them as saprophyte. The infected petals pruned just prior to beginning of sporulation of the pathogens on to the infected regions. The petal should be free from moisture. The chemical control of the pathogens can be achieved by several fungicides. Spray of Captan, Thiram and Benomyl is recommended against the *Botrytis cinerea*. A preventive spray of Captan @ 0.2 per cent is giving before emergence of flowers. However, more spray should be given on to the flowers if needed. The contact fungicides Iprodione and Vinclozolin are also effective to the *B. cinerea*. The *Alternaria* and *Curvularia* can better control by Dithan M45 (0.2 per cent), Blitox (0.25 per cent), Difolatan, Deconil, Brestan, Antracol and Captan.

G. Leaf Spots of *Vanda* and *Dendrobium*

Several fungal pathogens cause the leaf spots on orchids. The details of the disease are as follows.

Symptoms: The leaf spot is a serious disease of *Vanda* orchids and related hybrids. Symptoms may occur on either side of the leaf, which start as small black raised lesions giving a rough texture of the leaf surface. As spots are getting maturity, they enlarge in size and form pustules, which contain the spores for new infections.

Disease Cycle: Pathogen survives in form of perithecia and pycnidia in infected compost, pot, plant parts and infected debris of orchids. During favorable conditions these structures release ascospores and conidia after got wet. The ascospores liberate from ascus forcibly and continue rise through in spring and summer season. The conidia are exuded in viscid mass from which splashed away by rain or physical and mechanical means. Thus, ascospores and conidia may reach to the new sites of the healthy plants where they initiate the primary infection. Ascospores produce single germ tube after germination. The germ tube has an appressorium which directly penetrates host epidermis and grow inside the host inter and intracellularly. The incubation period varies from few to several days. The temperature, humidity, plant type *etc.* affect the incubation period. Pycnidia are formed in these spots, which provide the conidia for secondary infection. The conidia get germinate within 10-12 hours in presence of warm and humid conditions.

Symptoms: *Cercospora* leaf spot is resulted in a range of symptoms. Hybrids of *Dendrobium* infected with *Cercospora* showed large areas of very pale green with slightly sunken purple coloured spots on lower and upper surface of leaves. These spots turn into purple black with age. This is a most common symptom of the disease.

Management: Leaf spot disease of orchids caused by the above pathogens is difficult to manage by a single method in warm and humid enclosed areas, reason being the pathogens can sustain a wide range of temperature with good saprophytic survival ability in absence of living tissues and produce various enzymes and toxins against host tissues. These all together make pathogens very vulnerable to the plants. The better management can be achieved by adopting legal, cultural, biological, physical, chemical and mechanical methods. The seed and vegetative propagating materials if brought from outside must be disease free and certify legally by any plant protection agency under the quarantine. Cultural methods involve a preparation of disease free planting stock, adjustment of date of sowing, density of the orchid plants, management in irrigation and fertigation system, preparation method of growing medium and minimizing the influx of inoculum from neighboring crops. The sanitation engrosses as the collection of infected residues, training and pruning, removal of infected and unwanted plants and weeds (collateral, alternate and volunteer host plants) from orchid nursery, field and polyhouses. These practices are cost effective and easy to opt by the orchid growers. If orchids are growing in polyhouses, the regulation in the ventilation, temperature and humidity seem to be important in diminish the disease infection. The crop canopy is not to be dense. Intermittent application of antagonistic organisms to the medium / compost, seeds, seedlings, vegetative propagating materials and standing crops of orchids discourage the development of disease. Management of the leaf spots however, much possible by fungicides. The Captan, Folpet, Benomyl, Copper oxychloride, Triadimefon, Ferbam *etc.* can be used at the rate of 140 g a.i./ha. Spray of Benomyl on standing crop reduces the secondary infection. Azoxystribin is safe and extremely useful fungicide against this disease. *Bacillus* spp. and *Trichoderma* are known as good biocontrol agents of *Cercospora* leaf spot of orchids.

H. Anthracnose of Orchids

Various types of anthracnose are found on a variety of orchids. The details of anthracnose are giving below.

1. **Francy Anthracnose:** *Gloeosporium macrospurs* Sacc causes anthracnose on orchids. The same stage also produces a similar disease on foliage of *Haya carnosa, Citrus aurantium* and Aloes. Mangin noticed the serious incidence of disease on greenhouse orchids in France. *Physalospora Cattleyae* Maub.and Las. is a perfect stage of this pathogen.

Symptoms: Initially infected tissues have light yellowish soft spots. When the epidermis is torn away from one of the spots, a clear liquid will ooze out. At this stage of the woe, infected leaves lose their normal color, collapse and drop off. Ordinarily there is no fruit of any fungus formed on the spots but under moist conditions, acervuli appear. The disease, although attacking the foliage, does its greatest damage to the stems. Infection can take place only through a wound made on the epidermis.

2. **American Anthracnose:** Caused by *Glomerella cincta* is a perfect stage (winter stage) which described by Stoneman. *Glomerella cincta* produce flask-shaped perithecia. The asci placed within perithecia are clavate in shape and ascospores vary from elliptic to curve in their shape. The summer stage of this fungus was described by Halsted in form of conidial stage, perhaps *Colletotrichum*. The conidia of *Colletotrichum* are elliptic and guttulate. Setae may often be present but they are generally obscured by the great masses of spores formed in the acervuli.

Symptoms: This anthracnose is very prevalent on polyhouses of orchids. The *Sobralia macramtha* is most susceptible variety to this disease. The disease is first noticed as a patch discoloration on the stems, which soon become brown to almost black while the tender interior tissues become soft and decayed. Later the spore bearing pustules appear in great abundance on the dead parts. The symptoms of disease appear on leaves are in a way similar to that of stems. The disease, however, always starts from the tip and moves to downwards. There is usually a distinct line of demarcation between the healthy and the diseased tissues.

3. ***Bletia* Anthracnose:** *Colletotrichum Bletiae* Hals. is main pathogen of *Bletia* anthracnose. The acervuli of the pathogen are light brown in color and possess numerous dark setaes.

Symptoms: The beautiful *Bletia* orchids often subject to the attacks of an anthracnose, the cause of which is closely associated with Sobralia anthracnose. The disease usually begins at the tender tips and affects the foliage showed a ragged appearance. On the *Bletia*, the anthracnose is manifested as a spotting that disfigures the leaves. The spots are almost black and very soft. As disease progresses, the soft tissues rot and break away the fibrous portions. During severe conditions the infected areas increases in size and colease to form a large infected area, which results in the reduction of photosynthesis of foliage.

4. **European Anthracnose:** Eup*Eria*n antracnose has been reported by Sorauer as a very serious problem on cultivated

orchids. The disease is prevalent on those plants, which grow in excessively fertilized medium in overheated polyhouses. *Glaeosporium affinae* Sacc. causes European anthracnose. Pathogen differs very little from other *Gleosporium* species. The fungus produces pseudostroma in the pustules located underneath of epidermis, which upon the bursting release the spore mass outside the epidermis. The spores are one-celled, hyaline, cylindrical and often slightly curved. The spores germinate quickly, usually after forty eight hours. *Glaeosporium affinae* is serious problem on orchids in Europe but has not yet proved very serious in the United States.

Symptoms: Symptoms of anthracnose appear first on youngest foliage on the tip of the plant in form of brown discoloration. This explains the disappearance of the green coloring matter in the affected parts. The periphery or border of the leaf also affected. Consequently, the affected portions wither and dry up. In severe cases, the older leaves and even the bulbs become diseased. The disease is carried with infected bulbs. The same disease also harms to other orchids such as *Cattleya, Mendelii* and *Cypripedium laevegatum*.

5. **Anthracnose and leaf blight:** This disease infects to *Cattleya, Phalaenopsis* , *Dendrobium* and *Vanda*.

Symptoms: Leaf spot disease produces small brown spots with a distinct yellow/green halo around it on orchid leaves. The infected area of the spots quickly enlarges in size under favorable conditions. Consequently colease to each other and form big spots. In the later stage of disease development, these spots turn into black in colour and die. Most orchid species are susceptible to this disease. More often brown dead tissues have a appearance of concentric ring.

Management: Care and vigilance should always be exercised to prevent the introduction of the disease. The source of infection should be removed carefully. All infected plants should be destroyed by fire. The biological control of *Colletotrichum* and its related pathogens can achieve by the use of *Gliocladium catenulatum, Trichoderma harzianum, BurkholdEria cepacia* at pot, greenhouse and field level. However, *BurkholdEria cepaciaii* has mush antagonistic potential against *Colletotrichum gloeosporioides*. It reduces mycelial growth and spore germination of pathogen by producing secondary metabolites such as pyrrolnitrin, phenazine and cepabactin. Pyrrolnitrin is a broad-spectrum antifungal antibiotic. Use of copper-based fungicides is very effective against the disease. Spraying of 4-4-50 Bordeaux mixture, 0.15 per cent

Cuprocide, Fytolon or Blitox-50 will reduce the infection of *Colletotrichum*. Captan is a widely use fungicides also found effective against the pathogen. Benomyl will also help to protect the plants from infection. Manganese ethylenebisdithiocarbamate (Maneb) and Strobilurin fungicides like Azoxystrobin (Quadris), Trifloxystrobin (Flint), and Pyraclostrobin (Cabrio) have recently been labeled against anthracnose.

Other Related Species of *Gloeosporium* Causing Anthracnose and Leaf Spots in Orchids

1. *Glceosporium oncidii* Oud. (*G. maxillaria*). This pathogen is confined to leaves of *Maxillaria infestans.*
2. *Gloeosporium epidendri.*Henn. This pathogen attacks on stems of *Epidendrum* sp.
3. *Gloeosporium stanhope.* Allesch. is found on leaves of Stanhopee.
4. *Gloeosporium Laeliae* Henn. is found on leaves of Lalia.
5. *Gloeosporium pallidum* Karst. and Han occurs on leaves of *Liparis longipes.*
6. *Colletotrichum orchidearum* K. and H. appears on leaves of *Bolbophyllum labbi,* B. *longiflorum, Cymbidium* sp., *Physiophon loddigesii, Ezia stelleta, Coelogyne mayEriana, Pleurothallis tribuloides, Sarcanthus pugioniformis, pseudobulbs of Eulophia saundersiana* and *Oncidium pulvinatum.*
7. *Colletotrichum dicheae* Henn. grows on foliage of *Dichaea vaginata.*
8. *Colletotrichum roseolum* Henn. develops on the pseudobulbs of *Stanhopea oculata.*

Bacterial Disease of Orchids

In August 2008, approximately 50 *Oncidium* 'Gower Ramsey' orchids were discovered at a commercial orchid nursery in South Florida (USA) with brown, macerated leaves typical of soft rot disease reported in other orchids (Cating *et al.*, 2009, Cating and Palmateer, 2011). Based on biochemical characteristics, nucleotide sequence analysis, and pathogenicity tests, the causal agent was identified as *P. chrysanthemi.* An *Erwinia* sp. has been reported to cause soft rot symptoms on *Oncidium* aureum. This is thought to be the first report of a *Dickeya* sp. (=P. chrysanthemi) causing soft rot symptoms on *Oncidium* orchids grown in large-scale commercial production in the United States.

A causal agent of Bacterial soft rot in moth orchid was isolated and identified which occurred in Hangzhou, Zhejiang Province(Wu *et al.*, 2010). The pathogenicity test, Biolog test and fatty acid methyl esters (FAME) test were undertaken to elucidate the precise position of the causal bacterium. The results indicated that the causal agent's characteristics were same to *Erwinia chrysanthemi*. And, 16S rRNA gene sequence analysis also showed that the pathogenic bacterium was closely related to *E. chrysanthemi* and well separated from all other *Erwinia* species. Therefore, it could be concluded that the pathogenic bacterium causing Bacterial soft rot of moth orchid was identified as *E. chrysanthemi*.

During a regular survey of plant diseases, a bacterial disease was found in hybrid orchid *Dendrobium* species (cv. Aranda Hee Nui) in Guwahati, Assam, India. Initially, the disease was observed on succulent leaves showing small, pale brown water-soaked spots on the upper surface of the leaf, which later enlarged, became soft and finally rot. The causal bacterium was isolated in pure form and its pathogenicity was confirmed by using both injection-infiltration and needle-prick inoculation techniques. Based on the disease symptoms and Koch's postulation studies, in addition to morphological, cultural, biochemical characters and electron microscopy of the bacteria, the isolate was tentatively identified as non-fluorescent Pseudomonas *Cattleya*e [*Acidovorax avenae* subsp. *Cattleya*e]. This is the first report of occurrence of the bacterial disease of hybrid *Dendrobium* orchid (cv. Aranda Hee Nui) from India (Borah *et al.*, 2002).

The incidence of Bacterial pathogens is contributing large economic losses in several orchid nurseries. Loss of leaves and death of complete plant are especially severe during prolonged rainy periods. Today, Bacterial diseases are found in all orchid growing areas worldwide and become a serious threat to orchid industry. Though, the Bacterial diseases are still poor recognized by growers while, they destroy entire crop. Thus, their importance cannot be over emphasized. The common bacterial diseases are as follows.

A. Leaf Spot / Brown Spot / Brown Rot

Ornamental orchids of the genera *Phalaenopsis* and *Doritaenopsis* were seen to be affected by a previously unknown Bacterial disease in Baden-Wurttemberg, Germany. Symptoms are illustrated: they include dark brown, firm, circular lesions 1-2 mm in diam., particularly on the underside of leaves, some surrounded by a chlorotic halo or a water soaked margin, leading to dieback of the affected leaf or peduncle, or occasionally the whole plant. The disease was shown to be caused by a bacterium belonging to the group of non-fluorescent oxidase-positive Pseudomonas spp.(Moltmann, 1995).

Bacterial black spot disease was observed on *Odontoglossum*, *Odontioda* , Odontocidium, and *Vuylstekeara* orchids in Japan. Typical symptoms on the leaves were dark or black spots (or both) with a yellow halo. The causal agent was identified as *Burkholderia* and ropogonis (Smith 1911, Gillis, Van Van, Bardin, Goor, Hebbar, Willems, Segers, Kersters, Heulin and Fernandez 1995). The isolates were pathogenic on four original host orchids, *Phalaenopsis* orchid, and tulip; they were not pathogenic on white clover or corn after needle stab inoculation. An antibiotic bactericide (oxytetracycline/streptomycin mixture WP) was most effective for controlling the disease (Takahashi, *et al.*, 2004).

An unknown disease causing leaf rot on leaves of *Odontioda* sp., which is an intergeneric hybrid of *Odontoglossum*, occurred in Tochigi Prefecture, Japan, during the summer of 1995. One bacterial species was almost purely isolated from the infected leaves. The isolated strains were pathogenic to *Odontioda* cv. Baiser orchids in the dark with high temperature stress and high humidity. Based on the results of API 20E strips and other physiological and biochemical tests, the strains were identified as E. cloacae. They were also pathogenic to other orchids such as Calanthe discolor, *C. furcata*, *Habenaria* radiata, *Miltonia* sp. and *Phaius* minor *F. punctaus*, after inoculation and incubation for 2 days at 35°C. This is the first report of an orchid disease caused by *E. cloacae* and the name "Bacterial leaf rot disease of *Odontioda* " was proposed(Takahashi *et al.*, 1997).

Small to large leaf spots with or without water-soaking or soft rots were observed on various orchid genera, including *Dendrobium*, *Oncidium*, and *Miltonia* spp. and hybrids. Bacteria isolated and cultured from the lesions were tentatively identified using analytical profile index (API) strips and standard physiological and biochemical tests, and confirmed by species-specific polymerase chain reaction and sequencing of the 16S rRNA gene. The variation in pathogenic, morphological, cultural, and molecular characteristics of the orchid isolates also was evaluated. In our studies, a gram-negative, aerobic, rod-shaped bacterium that produced pale yellow, opaque, round colonies with entire margins on nutrient broth yeast extract agar (NBY) was isolated consistently from diseases orchid plants. On yeast dextrose calcium carbonate agar, the isolates produced brownish-yellow, nonmucoid colonies, with the majority of the strains secreting a diffusible yellow or tan pigment into the media. The bacterium was identified as

BurkholdEria gladioli. Molecular analysis indicated very little diversity in the 16S rDNA gene. Testing *B. gladioli* isolates using media containing copper or streptomycin indicated varying levels of resistance (copper resistant=Cur; streptomycin resistant, Smr), with approximately 75% of the strains resistant to copper and 94% of the strains resistant to streptomycin. The minimum inhibitory concentration (MIC) of cupric sulfate among Cur strains ranged from 50 to 1,000 micro g/ml and the MIC of streptomycin was 50 to 100 micro g/ml for all Smr *B. gladioli* strains tested. Field and laboratory data suggest the frequent use of these chemicals in nurseries may have inadvertently resulted in the development of copper and streptomycin resistant in *B. gladioli* from orchids(Keith *et al.*, 2005)

Symptoms: It is complex disease caused by many bacteria. The bacterium attacks to the most valuable orchids and ruins them in a very short time if moist condition prevails. Bacterial leaf spot is very common in Japan. Symptoms of the disease started as light amber-colored spots on the leaf blades. The spots quickly enlarge and in a few days, the entire leaf becomes invaded by bacterium and discolored due infection. As disease progresses the foliage turns brown then deep chestnut brown in colour. The upper surface of leaf becomes wrinkled with loss of luster. The lower surface of the leaf, just underneath the spots, rapidly takes faintly pale color and gradually turns into the same color as that of the upper part. If infection takes place on the lower portion of the leaf, the upper half soon becomes yellowish and dies in lack of food. The rot from the leaves goes down to the stem consequently the entire plant is damaged.

The brown spot is caused by the *Xanthomonas Cattleyae,* which may severe in full grown plants and seedlings. It develops small water soaked soft spots on the leaves which turn into brown spots after dry. The whole plant of orchid may die when the infection reaches to the crown. The numbers and size of brown spots progress quickly during cold conditions when orchid leaves are allowed to remain wet forever. *Phalaenopsis schillEriana* and *Cypripedium phillipinense* are seem to be more susceptible varieties of orchid to the brown spot

The brown rot is caused by *Erwinia cypripedii* is extremely harmful to *Paphiopedilum.* This pathogen causes water soaked irregular transparent spots followed by yellowing and browning of leaves and ultimately death of plant occur in a severe conditions. Brown rot infected orchids emit the spice odour. The bacteria also attack on fleshy and succulent leaves of *Phalaenopsis amabilis, P. schilEriana Cypripedium haynaldium, C. philippinense, C. laevigaturn, C. godefroyae.*

B. Acidovorax Leaf Spot

This disease causes huge losses in *Dendrobium, Oncidium, Rhynchostylis, Cattleya, Cypripedium* and *Vanda.. Acidovorax avanae* subsp. *Cattleyae* is a major causal organism of this disease. \

A leaf spot disease of *Cattleya* sp., that caused considerable damage in an orchid house in Aragua state, Venezuela, was studied. A bacterium isolated from the advancing margins of the lesions, proved pathogenic upon inoculation, and was readily isolated again. On the basis of biochemical and physiological tests, the bacteria was identified as a non fluorescent, oxidase positive Pseudomonas. Pseudomonas *Cattleya*e [Acidovorax avenae subsp. *Cattleya*e] is the only pathogen known from this genera that affect *Cattleya* worldwide. The bacteria studied in this work agreed with the genus but the results of some tests indicated a different species. This is the first report of such a pathogen affecting orchids in Venezuela(Trujillo and Hernandez,1999)

Symptoms: Bacterial leaf spot of orchid is a destructive disease caused by *Acidovorax avanae* subsp. *Cattleyae*. The first symptoms on orchids are soft, water-soaked lesions, which eventually turn brown or black. The disease may cause death of seedlings. Infection may occur anywhere on the leaf of older plants. The entire plant may die if infection reaches to the growing point. Diseased areas always show a considerable amount of exudate, which contains infectious bacteria. The transmission of bacteria is possible to other plants by splashing water. On *Cattleya*-type orchids, the disease appears on the leaves as sunken, black, water-soaked spots.

C. Bacterial Soft Rot of *Dendrobium* and *Oncidium*

Erwinia chrysanthemi is a devastating Bacterial pathogen in *Phalaenopsis* amabilis and causes soft-rotting disease by secretion of cell wall-degrading enzymes. However, the molecular mechanisms underlying the interaction of *P. amabilis* with *E. chrysanthemi* remain elusive. In this study, early molecular events of the plant in response to the pathogen attack were investigated. The alteration in reactive oxygen species accumulation and peroxidase activity occurred at the site of infection. Subsequently, a systematic sequencing of expressed sequence tags (ESTs) using suppression subtractive hybridization (SSH) was performed to obtain the first global picture of the assembly of genes involved in the pathogenesis. The majority of the SSH clones showed a high identity with genes coding for proteins that have known roles in redox homeostasis, responses to pathogens and metabolism. A notable number of the SSH clones were those encoding WRKY, MYB and basic leucine zipper transcription factors, indicating the stimulation of intracellular signal transduction. An orchid gene encoding trans-2-enoyl-CoA reductase (ECR) was the most abundant transcripts in the EST library. ECR is an enzyme catalyzing the very long chain fatty acids (VLCFAs) biosynthesis, and the full-length cDNA of the ECR gene (PaECR1) was obtained. Functional analysis of PaECR1 was conducted by virus-induced gene silencing to knock down the gene expression in *P. amabilis*. The PaECR1-silenced plants were more susceptible to E. chrysanthemi infection, implying potential roles for VLCFAs in the pathogenesis. In summary, the pathogen-responsive gene expression profiles facilitated a more comprehensive view of the molecular events that underlie this economically important plant-pathogen interaction(Fu *et al.*, 2012).

Symptoms: The bacteria can affect any part of a plant but symptoms on leaf are likely to be noticed first. Faint, irregular, olive-green areas appear on the leaf at the early stages of infection. As the disease progresses, these areas develop into larger dark green to brown water-soaked spots and eventually the entire leaf rots. Yellowing of leaves and yellow margins around spots on the leaf followed by leaf drop are common symptoms. The translucent tan leaves also seen in heavily infected orchid crop. The symptoms further extend into the canes from infected leaves. The exterior portions of infected canes are yellow to brown, while insides area of canes is greenish yellow to brown. The infected canes are often soft and mushy. Diseased canes become weakened, brittle and hollow, often breaking or bending at various points along the cane in the later stage of disease. Bacteria may also enter into the lower parts of canes through wounds and move upward inside the cane to healthy leaves. Watery rots of leaves begins at the stem-end of the leaf is very good indication of Bacterial infection. Severely infected canes die and results in stunted plants with poor vigor. Young shoots may destroy within 2 weeks and many young plants die in a few months. Three to 4 year old plants decline gradually and die in 1 to 2 years. Diseased plants have a foul odor.

The sudden outbreak of the orchid plants in fields includes rapid and severe damage of commercial *Dendrobium*. However, slow damage is less dramatic. Poor growth of the plants, stunted plants, few new shoots, small shoots; few, small, or incomplete spikes (floral sprays with missing flowers) and early decline in productivity of the crop are the main symptoms of slow damage. Usually bacteria do not affect the floral sprays in field. The cut ends of spikes are quickly invaded by pathogenic bacteria, kept in the vase or bucket water. These bacteria multiply and move throughout the flower stem causing rancid rots and early wilts of the lower flowers on the spike. In some cases, the entire spike rots. This is a major post-harvest problem that reduces the quality of *Dendrobium*.

Management of Bacterial Diseases: Orchids are vulnerable to Bacterial infection if plants are over fertilized under insufficient light and ventilation and facing continuous wetting conditions. Bacterial population is extremely difficult to control due to its rapid reproduction rates, massive population and good tolerance capacity against many chemicals and antibiotics. The following principles for Bacterial disease control are as under.

1. **Exclusion:** Growers must start their fieldwork using disease free propagating materials. The seedlings and plants of orchids should obtain from disease free places and environments. These plants must be carefully grown in already disinfested area following strict guidelines, which can prevent Bacterial contamination. If the disease-causing bacteria are keeping out from nursery, disease will not occur. Healthy seedling must be planted in pasteurized or sterilized potting media. The soil and composting material can sterilize by 2-3 per cent formalin if the field is prepared in areas that had heavy vegetation. Gravel, cinders or other bedding materials must not have been previously used and must be free of organic matter and soil. If soil or small bits of plant debris are present, the medium must be fumigated or pasteurized to reduce common pathogens that are likely to cause decline problems as the crop matures. New crop of orchid should not be planted in the same nursery have Bacterial infection earlier.
2. **Elimination:** Once a field is contaminated, it is very difficult to eliminate the bacteria. Therefore, as soon as a diseased plant observed, the entire plant and those around it must remove and destroy. All leaves, stems, roots and flowers from diseased plants remove and gather from the nursery site. Avoid the replantation in those spots from which the diseased plant(s) had been removed for at least 8 months.
3. **Moisture:** Orchids grow well in wet environments that must free from Bacterial contamination. However, Shady area if endure with Bacterial population along with high moisture content and temperature can describe better for sudden outbreak of the Bacterial diseases. Thus, moisture content seem

to be a crucial factor in the disease development. This can easily manage in Polyhouses.

4. **Sanitation:** Sanitation is a simple and effective method to reduce the primary inoculum of bacteria from the polyhouses. Collect the infected leaves, branches, flower, pseudobulb, roots *etc.* and burned them. Diseased plants should immediately remove from the field and destroy. Better ventilation, removal and burning of infected tissues, balance use of fertilizers being reduce the nitrogen and increase the potassium component in potting medium, excise of small isolated leaf spots are the possible practices must take up during cultivation.
5. **Pests:** Several insects and pests are carriers of pathogenic bacteria, necessarily eliminated from the source. The entry of birds, large lizards and toads also prevented, they otherwise can spread the bacteria into the orchid growing areas. If the cultivation of orchids is doing in Polyhouses, the Bacterial problem can easily control by adopting sanitation practices and use of various insecticides.
6. **Resistant varieties:** For many crops, breeding and biotechnological approaches have provided resistance to Bacterial diseases. However, time may be consume to develop resistance cultivar of commercial orchid against the pathogenic bacteria
7. **Chemical control:** Few agricultural chemicals are available for Bacterial disease control, which are as follows.
   - Agribrom is effective as a post-harvest treatment in controlling *Erwinia* and *Pseudomonas* on flower stems.
   - Streptomycin sulfate I.P. 90 % at the rate of 0.01 per cent effectively eliminates the Bacterial infection from vegetative propagating materials.
   - One percent solution of bleaching powder or salts (seawater) controls the Bacterial infection if spray on the diseased plants.
   - Sprays of Coperoxychloride (Blitox-50) in combination with Streptomycene suplphate I.P. 90 % on the diseased plants reduce the incidence of Bacterial infection.

Viral Diseases of Orchids

Impatiens necrotic spot virus (INSV) had infected *Oncidium* varicosum. Infected *Oncidium* showed symptoms typical of a virus disease, with leaves having various types of yellow mottling or streaks, irregularly shaped necrotic (brown to black) spots, and yellow to necrotic concentric ringspots. The virus belongs to the Tospovirus group which includes the tomato spotted wilt virus. Methods for reducing the risk of the disease include using uninfected plant material, removing weeds which act as hosts of the insect vector (*Thrips* sp) or using insecticides to prevent thrips from transmitting the pathogen (Koike and Mayhew, 2001).

Yi *et al. (*2008) conducted a surveys and pathogen identification of viral diseases on *Oncidium, Phalaenopsis* and *Cymbidium* hybrids were conducted in Yunnan, China. Black and brown necrotic spots were the common symptoms of the infected orchids. The T-series were the most susceptive cultivars in different *Oncidium* samples. It was confirmed that *Cymbidium* mosaic virus (CyMV) and *Odontoglossum* ringspot virus (ORSV) were the main causal agents by the detection methods of electron microscopy, serology and RT-PCR, and CyMV was the common species. CyMV and ORSV co-infection were detected from some diseased orchids.

Cymbidium mosaic potexvirus [*Cymbidium* mosaic virus] (CymMV) and *Odontoglossum* ringspot tobamovirus [*Odontoglossum* ringspot virus] (ORSV) were isolated from *C. hybrida* plants in Lithuania. The viruses were identified using test plants and electron microscopy. The optimum conditions for the purification of the viruses were determined. The yields of purified preparations were 7.5 mg/100 g of infected tissue for CymMV and 12.4 mg/100 g of infected tissue for ORSV. The titres of the specific diagnostic antisera in the DAS [double antibody sandwich]-ELISA test were 1:2560 and 1:40 960 for CymMV and ORSV, respectively. The immunoglobulins from the antisera were precipitated and purified. Specific enzyme-labelled ORSV antibodies with horseradish peroxidase and inorganic pyrophosphatase of Escherichia coli were prepared. The optimum conditions for the DAS-ELISA test consisted of an antigen concentration for adsorption on microplates of 3.3-10.0 micro g/ml, immunoglobulin concentration of 10 micro g/ml, and conjugate dilution of 1:400-1:800. With inorganic pyrophosphatase, the background was very low, lower by approximately 4-fold than with horseradish peroxidase. Inorganic pyrophosphatase gave good results and high green colour intensities. This method allowed the reliable detection of

ORSV directly in the crude sap from *Cymbidium* leaves (1:1000 dilution). The results indicated the possibility of using inorganic pyrophosphatase from *E. coli* for labelling immunoglobulin instead of horseradish peroxidase or alkaline phosphatase, which are usually used in the detection of plant viruses by ELISA (Navalinskiene *et al.*, 2005).

In 2010, a survey for viral diseases in commercial, orchid-producing greenhouses was carried out in Morelos, Mexico by Soto *et al.* (2012). Many symptomatic plants were observed. The most common leaf symptoms were yellow mottle, yellow streaks, and chlorotic and necrotic ringspots. Leaf samples were collected from symptomatic plants (*Encyclia, Oncidium*, Schomburghia, Brassia, Guarianthe, *Cattleya, Epidendrum, Vanilla*, Xilobium, *Laelia* and Brasso *Cattleya*). The causal organisms were identified as *Cymbidium* mosaic virus (CymMV) and *Odontoglossum* ringspot virus (ORSV) based on genetic analyses (RT-PCR and DAS-ELISA). This is thought to be the first report of CymMV and ORSV in orchids in Mexico.

A putative virus-induced disease showing chlorotic spots on leaves of *Phalaenopsis* orchids was observed in central Taiwan. A virus culture, *Phalaenopsis* isolate 7-2, was isolated from a diseased *Phalaenopsis* orchid and established in *Chenopodium quinoa* and *Nicotiana benthamiana*. The virus reacted with the monoclonal antibody (POTY) against the potyvirus group. Potyvirus-like long flexuous filament particles around 12-15x750-800 nm were observed in the crude sap and purified virus preparations, and pinwheel inclusion bodies were observed in the infected cells. The conserved region of the viral RNA was amplified using the degenerate primers for the potyviruses and sequence analysis of the virus isolate 7-2 showed 56.6-63.1% nucleotide and 44.8-65.1% amino acid identities with those of Bean yellow mosaic virus (BYMV), Beet mosaic virus (BtMV), Turnip mosaic virus (TuMV) and Bean common mosaic virus (BCMV). The coat protein (CP) gene of isolate 7-2 was amplified, sequenced and found to have 280 amino acids. A homology search in GenBank indicated that the virus is a potyvirus but no highly homologous sequence was found. The virus was designated as *Phalaenopsis* chlorotic spot virus (PhCSV) in early 2006. Subsequently, a potyvirus, named Basella rugose mosaic virus isolated from malabar spinach was reported in December 2006. It was found to share 96.8% amino acid identity with the CP of PhCSV. Back-inoculation with the isolated virus was conducted to confirm that PhCSV is the causal agent of chlorotic spot disease of *Phalaenopsis* orchids in Taiwan. This is the first report of a potyvirus causing a disease on *Phalaenopsis* orchids(Zheng *et al.*, 2008).

Lakani *et al.* (2010) found that most of the commercially orchids sampled from Bogor was infected by virus-like disease. The symptoms depend on the orchid species. Thus, the symptoms varied such as mottle, mosaic and necrotic flecks either on the leaves or on the flowers. Similiar symptoms were not found in samples obtained from Central Java and East Java. A Tobamo-like virus was inferred to be possible cause of the viral disease-like symptoms. Serological test of the samples by ELISA showed positive against Odontoglosum Ringspot Virus (ORSV) antibody and was negative against *Cymbidium* Mosaic Virus (CyMV) antibody. Total RNA was extracted from symptomatic plants and RT-PCR was carried out by using a pair of coat protein (CP) gene primer of ORSV. It was successfully amplified a 500 bp of DNA fragment and it was directly sequenced. The nucleotide sequence of CP gene had confirmed the identity of ORSV. Phylogenetic analysis based on the CP nucleotide sequences were grouped into only one major cluster. The ORSV isolate from Bogor (Sd 21) and the other isolates were clustered in the same group and had highest nucleotide homology (99%). These results provide first evidence of ORSV infecting orchids in Bogor, Indonesia.

Tobacco rattle virus (TRV) was identified and isolated from *Cymbidium* hybridum and *Phalaenopsis* sp. grown in Lithuania. TRV was found infecting orchids in mixed infection with *Odontoglossum* ringspot virus (ORSV) and *Cymbidium* mosaic virus (CymMV). Viruses have been identified by the mechanical inoculation of differential host plants, electron microscopy and DAS-ELISA. TRV identification was confirmed in RT-PCR. ORSV was identified in all examined orchid plants grown under greenhouse conditions in Ukraine. CymMV mixed infection with ORSV was found in three plants. The survey of orchid genera (20 species) growing naturally in Ukraine revealed the presence of Arabis mosaic virus (ArMV), TRV and Tomato aspermy virus (TAV). The most sensitive to viral infection were plants of Listera ovata, Neotia nidus-avis and *Orchis* purpurea species (Navalinskiene *et al.*, 2008).

Orchid genera like *Vanda, Dendrobium, Oncidium, Phalaenopsis* and *Cattleya* are most susceptible to viral diseases. With the increasing importation and rapid exchange of orchid plants worldwide, the introduction and spread of viruses among orchid plants in a greenhouse is highly probable. The infection, symptoms and disease severity caused by viruses depend on the viral strains, hybrids, species and genera of orchids. The age of the leaves, time of year, environmental conditions and presence of insect vectors in orchid field also affect viral symptoms. Plants infected with more than one virus express variable symptoms.

The same virus may cause widely different symptoms on different genera e.g., a flower breaking on the blossoms of one genus, a bar mottle or other symptoms on other genus. Virus-infected plants bloom less efficiently, inferior in vigor and produce flowers of lesser quality than healthy plants. The introduction of virus-infected stock may act as virus reservoir that further contaminates other orchid plants and seedlings. Viral symptom are difficult to diagnose as they are resemble with the symptoms of nutrient deficiency, fungal infection, damage caused by the aphids and thrips. Sometimes virus-infected plants do not show any symptoms. Such silent plants are excellent carriers of virus; they rather express strong symptoms on susceptible orchids. Thus, before establish efficient control measures against viral diseases, the rapid and accurate diagnosis is very important.

Viruses are very small obligate parasites composed of nucleoprotein. The genetic material either in form of RNA or DNA is enclosed by a protein coat. The viruses always replicate only in living cells by utilize their metabolism, where they produce many new copies of daughter virus particles. New synthesized virus particles may transfer to the healthy plants in various ways, but insect and mechanically transmission are more pronounced in orchid crops. The infected cell sap transmitted during feeding of healthy tissues by aphids, training and pruning by hands and tools during vegetative propagation, mericloning and division. Some common viral diseases are as follows.

A. Leaf Necrosis and Black Streak : This disease is caused by *Cymbidium* mosaic virus (CyMv). The infected plants have irregular, brown to black, elongated spots and streaks of dead tissues on both surfaces of older leaves. Infected leaves that show symptoms tend to age quickly and dry up.

Flowers from such plants are usually symptomless but they may open in an unthrifty manner. If the leaves die prematurely, the flowers are usually fewer in number and smaller. Diseased plants are usually less vigorous; however, not all infected plants show symptoms. The infected cell sap carrying virus particle is strongly transmitted mechanically by hands and tools. No insect vector is known to transfer the CyMV. Leaf necrosis affects *Cattleya* and its hybrids as well as many species of *Angraceum, Cymbidium, Epidendrum, Laelia, Oncidium, Spathoglottis* and *Zygopetalum*. Certain strains of the tobacco mosaic virus also produce leaf necrosis on *Cattleya, Cymbidium* and other orchids that closely resemble with the symptoms produced by the CyMV.

B. Blossom necrotic streak : A strain of *Cymbidium* mosaic virus causes latent streaks or spots on orchid blossoms. The blossoms first appear symptomless but brown necrotic spots and streaks become visible one or more weeks after infection. Delay appearance of flowers after the blossoms open, may creates a problem in marketing. Infected flowers may develop necrotic symptoms 3 to 5 days after cutting from the plant. Long, yellowish, irregular streaks may develop in the leaves. Transmission of this strain of *Cymbidium* mosaic virus occurs with contaminated pruning tools. *Cattleya* and its hybrids are susceptible along the species of *Angraceum, Cymbidium, Epidendrum, Laelia, Oncidium, Spathoglottis* and *Zygopetalum*.

C. Mild flower break: This symptom is caused by tobacco mosaic virus. The virus can produce symptoms of mild and sever flower breaking. The first one is much more common in *Cattleya* than second one. The plants have symptoms of mild flower breaking chrarcterized as much less variegation in the flowers than plants infected with severe flower break. The flowers are not malformed and the leaves have only mild mosaic symptoms. The transmission of tobacco mosaic virus is unknown in orchids. The symptoms of mild flower breaking have been noticed on about 30 species and hybrids of *Cattleya*.

This virus also produces spots and rings of dead tissues on the leaves of *Odontoglossum* and diamond mottle on *Cattleya, Cymbidium, Odontoglossum* and *Phalaenopsis*. Diamond mottle is distinguished by elongate chlorotic areas that are often diamond-shaped. Older leaves sometimes develop brown to black flecks and streaks.

D. Severe flower break: This is a conspicuous virus disease of *Cattleya* and its hybrids, which is characterized by rolling and twisting of the sepals and petals and variegation in the flowers.

The normal pigment of the petals, sepals and petioles is replaced by irregular streaks, blotches, ring and line patterns which are either more or less intense than the normal color. Leaves develop after infection are sometime twisted and mottled with streaks of light and dark green. Leaves formed before viral infection is symptomless. The dark-green areas are somewhat raised, producing ridges and bumps. The severe flower-breaking virus is transmitted by the green peach aphid (*Myzus persicae*), grafting or mechanical sap inoculation. The incubation period of virus may vary from 12 to 19 days. However, if virus is inoculated through flower buds 3 to 5 months are required for leaf infection. In addition to *Cattleya* and its hybrids, the severe flower break virus also infects *Cymbidium* and *Oncidium* and possibly other genera. In *Cymbidium*, the symptoms consist of numerous severe chlorotic and rectangular areas on the leaves, denoted the disease as Bar mottle.

E. Symmetrical flower break: In addition to severe and mild flower streaking, where the variegation is irregular, a symmetrical variegation in *Cattleya* also occurs, which is characterized by pigmentation on the sepal margins and over most of the petals except in middle areas that have little or no pigment. The infected leaves also develop an inconspicuous mottling. The means by which the virus causes symmetrical flower break disease is transmitted in the greenhouse is unknown, but it can be experimentally transmitted by juice inoculation.

F. Ring spot: *Cymbidium* rings spot virus causes single or concentric ring spot patterns on old and young leaves. The rings may turn brown to purple-black and enclose a central light green or yellow "Island". Rings may overlap or merge to form large compound patterns. Ring spot may cause death of young shoots and often the entire plant. Species of *Cattleya, Cymbidium, Spathoglottis,* and *Trichosma* are affected. Ring spot symptoms may also a result of infection caused by strain of tobacco mosaic virus and *Cymbidium* mosaic virus together.

Management of Viral Diseases

Virus indexing

Under certain conditions, some infected orchid plants do not show any distinct symptoms through which pathogen presumed to be virus can be identified. In such cases, a indexing is required to determine whether a plant is infected with a virus or symptoms are due to other causes. The indexing can be performed on certain plant species or varieties called indicators, outside from the Orchidaceae family. Useful indicator plants include Coffee senna (*Cassia occidentalis*), Jimson-weed (*Datura stramonium*), Globe-amaranth (*Gomphrena globosa*) and *Chenopodium amaranticolor*. The indicators are sensitive to specific viruses and on inoculation these viruses develop characteristic symptoms on them. Along with symptoms produce by indicator the physical presence of inclusion bodies of viruses may also be confirm by electron microscope and fluorescent antibody microscopy. Serological test such as Enzyme-Linked Immunosorbent Assay (ELISA), Dot-blot Immunobinding Assays, Serologically Specific Electron microscopy (SSEM) are other useful methods for detection of virus particle in the sample.

Cultural management

- Exclusion and eradication of diseased plants. Prompt removal and burning (or burying) of all infected plants is an excellent cultural practice. Once plants infected with virus, never recover although symptoms may vanish at least temporarily after manipulation in prevailing environmental conditions and chemical control. Thus, it is necessary to destroy suspected unhealthy plants and those showing flower breaking at early stage of the growth. Otherwise, infected plants may serve and act as a source of infection to other plants. Seedlings from older or newly traded and purchased plants should be isolated until known to be virus-free.

- The plantation of orchids should begin with known virus-free propagating stock and seedlings. The mechanical injury should be avoided during plantation. The nursery prepared from virus-free meristematic tissues of orchid can produce healthy seedling. Though, this technique is time-consuming and expensive to set up but can be used to save valuable stocks.
- Rigorous sanitary measures during pruning, dividing and harvesting flowers are necessary. Tools must sterilize before every use. They can be dipped in 70 per cent rubbing alcohol and then flamed or soaked for 20 to 30 seconds in the solution of bleaching powder (1 part bleaching powder : 5 parts water). All fragments of plant materials should be wiped-off with sterilized knife. When the bleaching solution turns green from plant juices, discard it and make up a fresh solution for further use. Use clean pots and new potting media. Avoid handling of plants unnecessary. Carefully wash hands with soap and hot running water after handling diseased plants and tools every time. Keep down all weeds around growing orchids. They may harbor to one or more viruses.

1. **Chemical control:** Fumigation of polyhouses with suitable chemicals can down the population of virus-carrying insects. Timely applications of suitable insecticides against the aphids, thrips, scales and other insects can reduce the incidence of virus in greenhouses.

Harmful Insects of Orchids and their Management

1. **Aphids:** The aphids are probably the most ubiquitous insect. Both the nymph and adult stages are harmful to the plant. They have piercing and sucking type of mouth parts. These stages are usually found on the new succulent growth including flower buds. Affected parts are deformed, curled

and become pale. Severely infested plants are failed to form flowers. Aphids are also effective vectors of orchid viruses.

Control: Spray of systemic organophosphorus insecticides @ 0.05% dimethoate or 0.025% Imidacloprid 17.8% SL.

2. **Mealybugs:** Mealybugs look like white cottony masses. The growing tips, buds and flower stems are affected by mealy bugs. One type is also found on the roots. It's found in similar areas as aphids. This is a bothersome pest that usually needs repeated treatment of insecticide to get rid of it.

   Control: Spray of systemic insecticides @ 0.05% dimethoate or 0.025% Imidacloprid 17.8% SL and or 0.05% diazinon can effectively control mealybug.

3. **Thrips:** Thrips are very destructive, especially to flower buds, maturing flowers and young leaves. They are iniscule buggers that look something like long gnats and are very difficult to see with the naked eye. Their damage is easier to detect as it shows up as light streaks on the flowers or stippling on the leaves. The flower buds are also usually deformed.

   Control: Spray of organophosphates like 0.3 % mono corotophos or 0.03 % dimethoate or 0.02% diazinon provides good control.

4. **Scale:** Scales are very common insects of orchids. Various forms of scale occur in nature but most of them characterized by a shell that serves as a type of armour to protect the insect body from biotic and abiotic stresses. They are frequently found on the undersides of the leaves near the middle vein or on the edges of the leaf, where they suck the plant sap. The scales commonly hang out on the flower stems. These insects can better control before shell development.

   Control: Use one or two spray of 0.05% metasystox and 0.025% imidacloprid. Nursery should fumigate with 2% formalin solution.

5. **Spider mites:** Spider mites are very small in size and difficult to see. They are very destructive during hot and dry growing conditions in poly house. All stages of mites confine to the under surface of leave. In severe infestations fine webbing on the leaves appeared where dust particle get entangled later on. Continuous sucking of cell sap produces yellow spots on them which ultimately turn brown.

Control: Spray of Kelthane 18.5 per cent WT and 20 per cent emulsive concentrate and is active against all stages of spider mites.

6. **Slugs and snails:** Snails and slugs head most people's list as being one of the most revolting of all orchid pests. They usually come out during night and find on the bottom of the flower pots. They love cool, damp spots. If they travel across dry surfaces, they'll leave a telltale slime trail. They cause extensive damage to young orchid roots and stems and developing and maturing flowers.

   Control: Snails could be handpicked and killed. Metaldehyde (5%), Copper sulphate solution (1%) and insecticides (arsenates, persistant organochlorines *etc.*) are employed as molluscides. A bait of 30 % bran + 2% metaldehyde proves a good killer for slugs.

7. **Roaches**: Cockroach munches on flowers and flower buds during night.

Safe use of pesticides: The followings guidelines should always under consideration while using the pesticides into orchid house.

1. Always read the instructions about the doses and application methods of pesticide .
2. The recommended insecticide and its specific dose always should be used against identified insects. The higher dose of the chemical may damage the plant.
3. Spray of the chemical should be done in early morning.
4. Prior to spray of insecticide the orchid pots must have proper moisture content.
5. Proper interval between two or more spray should be maintain.

Glossary

Abaxial: Underside of leaf or side facing away from the stem is known as abaxial.

Abscission: The separation of a leaf, flower, or fruit from a plant as a result of natural structural and chemical changes.

Abscission zone: is a layer of weak, thin-walled cells that form across the base of the plant part where the break eventually occurs.

Absorption: Intake of water and mineral elements by the root system.

Acervulus: A mass of closely clustered conidiophores and conidia not covered by fungal tissue, initially subcuticular or subepidermal but eventually exposed.

Actinomycete: Any member of the order Actinomycetales; the name is often used to refer specifically to those species which form mycelium.

Acute: 1. Sudden development of disease symptoms. 2. Denoted to less than 90 degrees.

Adaxial: The upper side of a leaf is known as the adaxial surface.

Adjuvants: additives, as chemicals added to a preservative to enhance its effectiveness.

Adsorption: It is the process of attraction of atoms or molecules from an adjacent gas or liquid to an exposed solid surface. Such attraction forces (adhesion or cohesion) align the molecules into layers onto the existent surface.

Aerial roots: Roots that arise above the ground, from either the stem or the base of the plant.

Alliance: A tribe or group of related genera. Among orchids many of these genera can crossbreed with other genera.

Alba: White form of a flower.

Alternate host: One of two kinds of plants on which a parasitic rust fungus must develop to complete its life cycle.

Amphistomaty: A leaf having stomata on both surfaces.

Amyloid: Spore walls, spore ornamentation, hyphal walls, ascus tips, etc. are blue or grayish to blackish-violet in colour followed by staining in Melzer's reagent, presumably because of the presence of starch or a starch-like compound.

Anamorph: An asexual (usually conidial) stage in the life cycle of a fungus.

Anastomosis: Fusion of somatic hyphae.

Annuals: Plants that complete their entire life cycle within one growing season. The plant germinates from seed, grows blooms, and then produces seed and dies.

Antagonist: An agent or substance that counteracts the action of another.

Anther cap: In orchids, a cap of tissue that covers the clusters of pollen.

Anther: The portion of the flower that bears pollen.

Antheridium: A male gametangium.

Anthracnose: Any of various plant diseases, particularly those caused by fungi of the Melanconiales, in which discrete, dark-colored, necrotic lesions develop on the leaves, stems, and/or fruits.

Antibiosis: An association between two organisms that is detrimental to the vital activities of one of them.

Antibiotic: Substance used to inhibit the growth of micro-organisms, including bacteria and fungi.

Apomorphy:A novel evolutionary trait that is unique to a particular species and all its descendants and which can be used as a defining character for a species or group in phylogenetic terms.

Apothecium: An ascus-bearing structure (Ascocarp) in which the ascus-producing layer (hymenium) is not covered by fungal tissue at maturity.

Appressorium: An enlargement on a hypha or germ tube that attaches itself to the host before penetration takes place.

Aropetal: Development of structures (such as spores) in succession from the base to apex.

Ascocarp: Ascospore-bearing, multicellular sporocarp formed by a member of the Ascomycotina.

Ascomycete: It is a class of fungi describe under subdivision Ascomycotina.

Ascostroma: A fructification of the Ascomycetes consisting of an undifferentiated mass of tissue or stroma on or in which the asci are developed.

Ascus: A cell that is the site of meiosis and in which endogenous spores usually meiospores are formed.

Aseptate: Lacking septa (nonseptate or aseptate).

Avirulent: Not exhibiting virulence, lacking virulence or nonpathogenic.

Avoidance: First principle of plant disease control defined as avoiding the disease by planting at a time when or in area where the inoculum is ineffective due to environmental condition or rare or absent.

Axil: the upper angle between an axis and an offshoot such as a branch or leafstalk.

Bacillus: A member of the genus *Bacillus*. 2. Any rod-shaped Bacterial (a cell whose length is two or more times greater than its width).

Back bulb: A pseudobulb that has completed its growth cycle and become dormant. It serves as a storage organ and may bear a dormant bud. Under favorable conditions, this bud may develop into a new plant if the pseudobulb is severed from the parent.

Ballistospore: A spore that when mature is actively projected.

Basidiospore: An exogenous sexual spore (meiospore) borne on a basidium.

Basidium: A cell in which karyogamy and meiosis take place and which bears exogenous spores of sexual origin.

Basipetal: Successive from apex to base.

Biennials: A plant that take two growing seasons to complete its life cycle. It germinates,grows leaves stems in the first year; produces flowers and fruit (seed) in the second, and then dies.

Bifoliate: Having two leaves on a single pseudo bulb.

Biological control: The deliberate use by humans of one species of organism to eliminate or control another.

Blight: A disease characterized by rapid and extensive death of plant foliage. A general term applied to any of a wide range of unrelated plant diseases (e.g., chestnut blight, fire blight, late blight, halo blight).

Bolting: It is the growth of an elongated stalk with flowers grown from within the main stem of a plant.

Bract: A modified leaf that serves to protect a flower, bud, or shoot.

Bud: It is an undeveloped or embryonic shoot and normally occurs in the axil of a leaf or at the tip of the stem. Once formed, a bud may remain for some time in a dormant condition, or it may form a shoot immediately.

Bullheads: Over size buds, normally having more petals than normal.

Calyx: Collectively sepals of flowers are known as calyx.

Cane: A long, narrow pseudobulb, usually having many joints or nodes.

Capsule: The seed pod of an orchid, often containing thousands, even millions, of seeds.

Caudicle: A slender stalk-like appendage of the pollinium or pollen mass.

Central growing point: On a monopodial orchid, this is where the upright vegetative growth will begin.

Chlamydospore: A thick-walled, nonsexual spore; a transformed hyphal cell.

Chlorosis: The loss of chlorophyll from the tissues of a plant, resulting from microbial infection, viral infection, the action of certain phytotoxins, the lack of light, to magnesium or iron deficiency, etc. Chlorotic tissues commonly appear yellowish.

CITES: Convention on International Trade in Endangered Species of Wild Fauna and Flora.

Clamp connection: A curving outgrowth of a cell that, at cell division, acts as a bridge to allow passage of one of the products of nuclear division into the penultimate cell, thereby assuring maintenance of the dikaryotic condition in the members of the Basidiomycotina.

Clavate: Club-shaped.

Cleistogamous: used to describe a flower that self-pollinates, often without even opening the flower fully; relatively rare in orchids.

Cleistothecium: An ascocarp with the asci surrounded by fungal tissue and without regularly formed openings.

Clone: A sexually propagated offspring of a single parent plant. A clone is identical to its parents.

Cold frame: A wooden or concrete block box in which plants can grow or hold dormant during the cold winter months.

Column: The fused sexual organ of an orchid flower, found atop the lip.

Community Pot: Many tiny seedlings planted together in a single container before they are individually repotted.

Competition: A more or less active demand on the part of two organisms for sources of survival that rather inadequate to provide for all organisms present.

Complete fertilizer: Any fertilizer that contains all three of the primary nutrients (nitrogen, phosphorus, potassium).

Conidiophore: A hypha often specialized in structure that bears one or more conidia.

Conidium: A thin-walled, asexual spore that is borne exogenously on a conidiophore and is deciduous at maturity.

Corolla: The corolla is the name for all the petals of a flower, the inner perianth whorl term used when this is not the same in appearance as the outermost whorl and is used to attract pollinators based on its bright color.

Cross Protection: The protection conferred on a host by infection with one mild strain of a virus that prevents infection by a closely-related pathogenic strain.

Cross: The progeny that result from transferring pollen from one plant to the flower of another; the act itself.

Cross pollination: Transfer of pollen from the anther of a flower to the stigma of a flower of another plant.

Crown: The central part of the rosette of leaves in a monopodial orchid from which new growth arises upward.

Cultivar: A specific cultivated plant grown from a single seed.

Damping-off: A disease that results in the collapse and death of seedlings before or after they emerge from the soil (pre-emergence and post-emergence damping-off, respectively). Common causal agents include species of *Pythium* and *Phytophthora*.

Deadheading: The practice of pinching or cutting off spent flowers.

Decay: The gradual decomposition of dead organic matter.

Deciduous: Loss of leaves or other growths upon maturity or at the end of a growing season, with re growth after a dormant rest.

Deciduous: Plant that sheds leaves once a year.

Deuteromycotina: (Fungi Imperfecti) A non-phylogenetic category originally created for fungi with no known sexual stage; but it also includes the asexual stages of various fungi now known to have a sexual stage in the Ascomycotina or Basidiomycotina.

Dextrinoid: When spores of fungi stained yellowish or reddish-brown by Melzer's iodine.

Dieback: Progressive death of shoots, branches and roots generally starting at the tip.

Diploid: Having a normal number of two sets of chromosomes; also known as 2N.

Disease cycle: A chain of events involve in disease development.

Disinfectant: A physical or chemical agent that frees a plant, organ, or tissue from infection.

Disinfest: To kill pathogens that have not yet initiated disease, but that occur in or on such inanimate objects as soil, tools etc. or that occur on the surface of such plant parts as seed.

Disinfestant: An agent that kills or inactivates pathogens in the environment or on the surface of a plant or plant organ before infection takes place.

Dispersal: Spread of a pathogen within an area.

Dormancy: A rest period during which no vegetative growth occur.

Endogenous: Something that originates from within an organism.

Epidemiology: The study of sudden outbreak of disease.

Epinasty: Downward curling of a leaf blade resulting from more rapid cell growth on the upper side of a petiole than on the lower side; often a hyperplastic symptom of plant disease.

Epiphyte: A plant which naturally grows upon another plant or other means of support above the earth's soil but is not parasitic; deriving it's needed moisture from the air.

Eradication: Control of plant disease by eliminating the pathogen after it is established or by eliminating the plants that carry the pathogen.

Evergreen: Evergreen plant is a plant having leaves all year round.

Exclusion: It is the second principle of plant disease control in which the pathogen is prevented from entering or establishing in the field or area where it does not exist earlier.

Exogenous: Something that originates from outside of an organism.

Exudate: A liquid discharges from plant organs followed by infection such as leaf exudate, root exudate, etc.

Facultative parasite: An organism that normally lives as a saprophyte but under certain conditions can live as a parasite.

FAO: Food and Agriculture Organization.

Firing of leaf: Yellowing, followed by rapid death of lower leaves, moving up the plant and giving the same appearance as if someone touched the bottom of the plants.

Floriferous: Term used to describe a plant that flowers freely.

Flowers: Also known as a bloom or blossom is the reproductive structure found in flowering plants. The biological function of a flower is to mediate the union of male sperm with female ovum in order to produce seeds.

Footcandle: A measure of light useful in determining intensity of light for growing orchids; the illumination produced by a candle at a distance of one foot.

Fungicide: A chemical or physical agent that kills the growth of fungi.

Fungus: A eukaryotic, heterotrophic organism that usually has threadlike cells.

Gene: Unit that carries hereditary traits.

Genus: A group of plants that are classified together because of similar traits and an assumed common ancestry.

Germ cell: Reproductive cell which, when united with a germ cell of the opposite sex, develops into a new individual.

Germ tube: Short, juvenile hyphae arise from certain types of spores upon germination.

Germination by repetition: Formation of a secondary spores followed by germination of a spore.

Germination: The spores of fungi and bacteria, give rise to a vegetative cell or hypha.

Gram stain: An important staining procedure of bacteria discovered by Danish scientist Christian Gram empirically in 1884. When bacteria are stained with certain basic dyes, some of them decolorized with organic solvents termed as Gram-negative species. Cells of Gram-positive species resist decolorization.

Grex: A flock or group, applied collectively to the offspring of a given cross.

Habitat: A place with a set of defined environment suitable for the growth of an organism.

Harden off: Process of acclimating plants grown indoors gradually to the brighter light and cooler temperatures of the outside world.

Hardiness: Ability of a plant to survive is called its hardiness.

Haustorium: A specialized branch of a hypha formed inside a host cell by certain plant-parasitic fungi (especially obligate parasites) in order to obtain nutrients.

Host: A plant gets harm while infected with pathogen as it grows and develop in/on to the plant.

House Plants: It is a plant that is grown indoors in places such as residences and offices for decorative and health reasons such as indoor air purification.

Humus: A stable end product of organic-matter decomposition that's believed to increase microbial activity in soil, improve soil structure, and enhance the root development of plants.

Hyaline: Transparent, translucent and or colorless.

Hybrid: Resulting progeny from the union of two different species (known as a primary hybrid), or of a species and a hybrid, or of two hybrids (known as a complex hybrid).

Hymenium: A palisade-like layer of asci and basisia along with any sterile cells, such as basidioles, paraphyses and cystida

Hymenomycete: A member of the class Hymenomycetes.

Hypha: Filamentous part of a fungus.

Imbibitions: Absorption of fluid by a solid or colloid that results in swelling (Uptake of water, or other liquids, by the seed thus enabling germination).

Immunity: The state of being absolute resistance in which plant remains free from disease.

In vitro: In an artificial environment.

In vivo: Within a living organism.

Incubation period: The period between penetration of a host by a pathogen and the first appearance of symptoms on the host.

Indexing: A method to determine whether a given plant is infected by a virus. It involves the indicator plants sensitive to the virus and can produce symptoms.

Indicator host plant: A plant species that gives characteristic symptoms to a specific virus.

Indigenous : Native to a country or region.

Infection: It is process by which pathogen get establish in to the host.

Inflorescence: Arrangements of flowers in plants.

Inoculation: The proper placement of pathogens or infecting propagules at a site where infection is possible.

Intercalary: A position of any thing (chlamydospores) in between apex and base of a given structure.

Intercellular: Between cells.

Intergeneric : Between or among two or more genera.

Internode: The portion of the stem, cane or pseudobulb located between two nodes or joints.

Interveinal chlorosis: Yellowing in between leaf veins, but with the veins themselves remaining green. In grasses, this is called striping.

Intracellular: Within cells.

Intrageneric: Within a genus.

Introgression: The crossing of two species producing a fertile hybrid, which backcrosses for several generations with one of the parent species. In this way, genetic material and characteristics is transferred from one species to another.

Invasion: Spread of a pathogen through tissues of a diseased plant.

Juvenile: A youngest plant, branch, bud and generally refers to the life stage of a nematode between the embryo and the adult; an immature nematode.

Keiki: A plantlet that grows high up on a cane, pseudobulb or flower stem. When it has developed sufficient roots system, it may be detached and grown separately.

Labellum: Third petal of an orchid flower, modified by evolution into a lip often used as an attractive landing platform for pollinators.

Latent infection: Infection unaccompanied by visible symptoms.

Latent period: 1. The period between infection and the appearance of first sign of the symptoms (See incubation period) 2. Elapsed time between phage infection or induction and lysis of Bacterial cells. 3. Period after acquisition of virus by a vector before it becomes infective.

Lateral: In orchids, the lateral, or side, sepals are the lower pair of sepals.

Lateral sepal: Term used to refer to the two lowermost sepals that extend to the sides, versus the top most dorsal sepal.

Leaf spot: A self-limiting lesion on a leaf.

Leaf: Is an above ground plant organ specialized for photosynthesis.For this purpose, a leaf is typically flat and thin.

Lesion: A localized area of diseased or damaged tissue.

Lip: Orchid labellum.

Lipolysis: The breakdown of storage lipids (e.g. fats and oils) in living organisms.

Lithophyte: A plant that grows on rocks taking nourishment form mosses, litter, rain and their own dead tissues.

Lux: The lumen is the metric unit of light intensity and the term lux refers to the number of lumens per square metre of surface area and Klux stands for kilo lux or 1,000 lux.

Mechanical inoculation: It is a method of transmitting the pathogen from infected to healthy plant through mechanical means; juice from diseased plants is rubbed on test-plant leaves that usually have been dusted with carborundum or some other abrasive material.

Medium: The potting material or mix of materials being used inside the orchid pot.

Medulla: Central part of an organ.

Mericlone: A generally exact copy of an original orchid plant made via the laboratory technique of meristem propagation; since it is a specific cultivar, it is designated by single quotes around its name.

Meristem : Undifferentiated tissue that is capable of developing into specialized tissue such as toot, leaf or stem. In orchids, a plant produced by culturing meristematic tissue.

Microconidium: 1. A spermatium. 2. A small conidium.

Micropropagation: Making new orchids by any of the laboratory techniques, including meristem tissue propagation and sterile seed culture.

Monopodial: Growth of orchid wherein a single vegetative shoot grows continually upward, such as in the central rosette of *Phalaenopsis* and the more vine like *Vanda*.

Mosaic: A common symptom induced in leaves by many plant viruses in which there is a pattern of dark green, light green and sometimes chlorotic areas. This pattern is often associated with the distribution of veins in the leaf. In monocotyledonous leaves it shows as stripes.

Mottle: A diffuse form of the mosaic symptom in plant leaves in which the dark and light green are less sharply defined. This term is frequently used interchangeably with mosaic.

Mulch: A protective covering that is spread on the ground around plants to inhibit evaporation of moisture, weed growth, control soil temperature, enrich the soil and prevent the dispersal of pathogens. It may be organic material such as leaves, peat, or wood chips, or inorganic material such as plastic sheeting.

Multifloral: Having more than one flower per inflorescence.

Mutation: A sudden departure from the parent type, due to a change in a gene or chromosome; an individual resulting from such a departure.

Mycelium: A mass of hyphae, often used to denote all hyphae comprising a thallus.

Natural hybrid: A hybrid that occurs in the wild without the help of humans.

Necrosis: The plant tissue browns and dies. This is called necrosis. The tissue which has already died on a still living plant is called necrotic tissue.

Necrotroph: 1. An organism that kills part or all of another organism before deriving nutrients from it (usually applied to plant pathogens). 2. An organism that derives nutrients from dead plant or animal tissues, whether or not it is responsible for the death of those tissues.

Node: A distinct joint or notch on an inflorescence, stem, or pseudobulb from which a flower stem, leaves, or roots can emerge.

Nomenclature: A system of naming.

Nonresupinate: In orchids, those plants whose flower lips are positioned uppermost relative to the inflorescence axis; the vast majority of orchid flowers are resupinate.

Obligate parasite: A parasite that in nature always grows and multiplies only on living organism.

Oblong: The spores are longer than broad with sides nearly parallel and with ends more or less flattened.

Obovate: Egg-shaped, with the wide end outward.

Obovoid: Egg-shaped, with the narrow end outward.

Oogonium: A female gametangium that contains one or more discrete gametes.

Oospore: A thick-walled sexual spore that develops from an oosphere through plasmogamy or parthogenesis.

Organic matter: Once-living stuff like compost, sawdust, animal manure, ground bark, grass clippings, and leaf mold (composted tree leaves). Used to enrich soil and improve soil texture.

Osmotic pressure: Pressure exerted by the water-absorbing property of a solution.

Ovoid: Egg-shaped.

Panicle: A type of flower inflorescence wherein the flowers are loosely arranged on a branching stem and open from the lowest or inner branches to the top.

Parasite: An organism that draw nutrition from another living organism (host) without giving benefit to latter.

Pathogenesis: The ability of pathogen to cause disease.

Pathogenic: Producing disease.

Pathogenicity: The capability of a pathogen to cause disease.

Pedicel: Flowers having short floral stalk called pedicel.

Peloric: In orchids, a term used to describe an unusual and often beautiful (sometimes grotesque) condition where all three petals (instead of just one) attempt to fashion themselves into lip colors and/or shapes.

Penetration peg: In some plant parasitic fungi : The peg-like hypha emerging from an appressorium that penetrates the epidermal cell wall.

Perennials: Any plant with a life cycle of three or more years. Herbaceous (non-woody) perennials include flowering plants and herbs, mainly. Woody perennials include trees and shrubs. Longevity depends on the plant and growing conditions.

Perfect state: The state of a fungus characterized by sexual spores.

Perithecium: A closed ascocarp with a pore at the top, a true ostiole and a wall of its own.

Peritrichous: The flagella distributed more or less uniformly over the cell surface of bacterium.

Pest: A person or thing that causes trouble, annoyance, discomfort, etc.; nuisance; specific, any destructive or troublesome insect, small animal, weed, etc.

Pesticide: A pesticide is any substance or mixture of substance intended for preventing, destroying, repelling or mitigating any pest. A pesticide may be a chemical substance, biological agent (such as a virus or bacterium), antimicrobial, disinfectant or device used against any pest.

Petal: In orchids, one of the three inner segments of the flower that are positioned between the three sepals; one of the petals is modified into a lip.(A petal is one member or part of the corolla of a flower.

Photo morphogenetic: Influences of the light on growth and development of plants.

Pod: Term used to refer to the seed pod or capsule.

Pollinia: Waxy pollen clumps or grains usually found in the anthers of most orchids; often yellow, distinct, and found under the pollen cap of the column.

Predisposition: An action of set of environments prior to penetration and infection that makes plant or plant parts susceptible to the pathogen.

Primary infection: The first infection of the plant caused by a pathogen that emerging from a dormant stage (overwintering and oversummering propagules).

Primary inoculum: The overwintering and oversummering propagules (resting structures) of pathogen that can cause primary infection on the host.

Proliferation: A rapid and repeated production of new cells, tissues or organs specifically, a hyperplastic symptom of plant disease in which organs continue to develop after they have reached the point beyond which they normally do not grow.

Protection: Fourth principle of plant disease control in which a chemical, physical or biological barrier is created between host surface and pathogen.

Pseudobulb: A thickened portion of a stem, resembling a bulb but not being a true bulb, which is an underground modified bud.

Pyriform: Pear-shaped.

Quarantine: Legal restriction on the movement of agricultural commodities for the purpose of exclusion, prevention or delay in the establishment of plant pests and diseases in area where they are not known to occur.

Raceme: A raceme is a type of inflorescence that is unbranched and indeterminate and bears pedicellate flowers along the axis.

Raceme: A simple type of flower inflorescence that looks like a long stem with flowers arising along it.

Resistance: The ability of an organism to exclude or overcome, completely or in some degree, the effect of a pathogen or other damaging factor.

Resting spore: A thick-walled spore, formed after a sexual process, germinating only after an extended period of dormancy (overwinter and oversummer).

Resupinate : In orchids, those plants whose flower lips are positioned lowermost relative to the inflorescence axis; the vast majority of orchid flowers are resupinate(inverted, upside-down).

Rhizome: A stem that produces roots and above-ground organs such as stems, pseudobulbs and flowers. In epiphytic orchids it is usually found on the surface; in terrestrial orchids it may be underground.

Ringspot: A type of local lesion consisting of single or concentric rings of discoloration or necrosis, the regions between the concentric rings being green. The center of the lesion may be chlorotic or necrotic.

Rot: The softening, discoloration, and often disintegration of a succulent plant tissue as a result of fungal or Bacterial infection.

Sanitation: Principle of plant disease control involving removal and burning of infected plant parts and decontamination of tools, equipment, hands, etc.

Scape: A simple flower inflorescence that is topped by a solitary flowex, such as in many *Papbiopedilum*.

Sclerotium: Hard, resistant, multicellular resting structure usually differentiated in a cortex, rind and medulla that under favorable conditions can germinate to produce mycelium or sexual or asexual fruiting bodies.

Secondary infection: Any infection caused by inoculum produced as a result of a primary or a subsequent infection; an infection caused by secondary inoculum.

Secondary inoculum: Inoculum produced by infections that took place during the same growing season.

Seed pod: The capsule bearing the seeds of an orchid.

Self-pollination: Transfer of pollen from the anther of a flower to the stigma of the same flower or of another flower on the same plant.

Semialba: A white flower with a colored lip.

Sepal: Sepals in most flowers are green and lie under the more conspicuous petals (One of the three outer parts of an orchid flower, one of which is usually topmost and known as the dorsal, the other two lower sepals being known as the laterals).

Sepal: One of the separate parts of the flower forming the calyx.

Septate: Having cross walls.

Septum: A cross wall in a hypha or spore.

Serology: Branch of science dealing with properties and reactions of sera, particularly the use of antibodies in the sera to examine the properties of antigens.

Sheath: A tubular structure surrounding some parts of the plant.

Sib cross: Method of seed propagation of an orchid wherein the pollen of one orchid is placed on the stigma of another orchid that was originally grown in the same seed pod as the first orchid, therefore a cross pollination of siblings.

Sibling: An orchid that is related to another orchid by virtue of having been produced from the same seed pod.

Species: A group of plants showing intergradation among its individuals and having in common one or more characteristics which definitely separate it from any other group; a kind of plant distinct from other plants.

Specimen: Term usually used to refer to an orchid that has been allowed to grow to great size and floriferousness instead of being divided; also refers to the species that typifies a genus.

Spike: Technically an unbranched flower stem with shortstalked or stalkless flowers. Long inflorescence having multiple florets.

Splash: Petal of an orchid flower that modifies its petals by duplicating the coloring found on the lip; a type of peloric condition.

Spore: A discrete sexual or asexual reproductive unit, usually enclosed by a rigid wall, capable of being disseminated.

Spot: A symptom of disease characterized by a limited necrotic area, as on leaves, flowers, and stems. Pots may be different in shape and size.

Spur: A hallow, sac-like or tubular extension of the base of the lip, usually bearing nectar.

Stamen: The pollen-producing reproductive organ.

Stegmata: Thickened wall adjacent to the underlying sclerenchyma cells, with progressively thinner lateral walls and thin outer walls.

Sterigma: A spike-like structure that bears basidiospores on the basidium.

Sterilant: A substance used for sterilization of seeds, tools etc.

Sterile fungus: A fungus that is not known to produce any kind of spores.

Sterilization: The elimination of pathogens and other living organisms from soil, containers, etc. by means of heat, chemicals and radiation.

Stigma: That part of the pistil of a flower which is receptive to pollen.

Stipe: The stalk-like support of a pistil, or other small organ such as pollinium.

Suberin: A complex of fatty substances present in the wall of cork tissue that waterproofs it and makes it resistant to decay.

Suberization: Deposition of suberin on the walls of plant cells and their subsequent conversion into cork tissue.

Suppressive soil: A soil in which certain disease(s) fail to develop due to presence of antagonistic microorganisms in the soil.

Suscept: Any plant that can be attacked by a given pathogen (a host plant).

Susceptibility: The inability of a host plant to resist the effect of a pathogen.

Swarmspore: Zoospore.

SWOT: Strength Weakness Opportunity Threat analysis.

Symbiosis: A mutual beneficial association between two or more disparate organisms.

Sympodial: One of the two forms of orchid growth (the other is monopodial), wherein each new growth arises from the rhizome of a previous growth, and each new growth is completely capable of bearing an inflorescence. (Sympodial means "with conjoined feet" and in biology is often used to refer to the outward morphology or mode of growth of organism.

Symptom: Visible sign of disease.

Symptomatology: The study of symptoms of disease and signs of pathogens for the purpose of diagnosis.

Synsepalum: An organ resulting from the fusion of the two lower sepals of a lady slipper orchid (*Paphiopedilum*). It is found below and behind the lip or pouch.

Systemic: Term used to describe pesticides or fungicides that are taken up by plant leaves and growths and then work from within the plant.

Systemic infection: Movement of pathogen / mycelium internally into the whole plant.

Systemic: In whole system of plant.

Teleomorph: The stage characterized by the production of asci / ascospores, basidia / basidiospores, teliospores, or other basidium-bearing organs.

Terrestrial: Growing in the ground or in the loose substrate atop the ground.

Tetraploid: Genetic aberration wherein the plant has twice as many chromosome sets as normal, often resulting in very vigorous, large plants and flowers.

Thallus: Any simple vegetative plant body that lacks roots, stems and leaves.

Throat: The inner portion of a tubular orchid lip.

Tissue culture: Artificial propagation of plants via laboratory mericloning, also known as meristernming.

Tissue: A group of cells having similar structure performs a special function.

Toxin: A compound produced by a microorganism and being toxic to a plant or animal.

Translocation: The movement of materials within the plant.

Transpiration: Loss of water from the plant, mainly from the leaves.

Unifoliate: Bearing one leaf per growth.

Variegation: It is the appearance of different coloured zones in the leaves and sometimes the stems of the plant.

Variety: A subdivision of a species that groups plants with a distinct form that is passed along to the progeny.

Vector: Living agent that transmits a pathogen from an infected to an uninfected plant.

Vein banding: A symptom of virus-infected leaves in which tissues along the veins are darker green than other leaf area.

Vein clearing: A symptom of virus-infected leaves in which veinal tissue is lighter green than that of healthy leaves.

Velamen: The thick corky layer of cells covering the aerial roots of epiphytic orchids, the function of which is to absorb moisture from the atmosphere and attachments to supports.

Virus: Infectious units composed of either RNA or DNA and enclosed by protein made protective coat.

Whorl: Three or more leaves, flowers or branches arranged in a circle at one point around the main stem.

Wilt: A disease symptom characterized by a loss of turgidity in a plant.

Wilting: Symptom characterized by loss of turgor pressure, which results in drooping of leaves, stems, and flowers.

Yellowing: A symptom characterized by the turning yellow of plant tissues that were once green.

Yellows: Any of a wide variety of plant diseases in which a major symptom is a uniform or non-uniform yellowing of leaves and/or other plant components. Yellows may be caused by fungi (e.g., celery yellows), viruses (e.g., sugar beet yellows virus), bacteria (e.g., coconut lethal yellowing), protozoa (e.g., hart rot), spiroplasmas or phytoplasmas.

Zoosporangium: A sac like structure having zoospores.

Zoospore: An asexual, motile spore that bears one or two flagella.

References

Aggarwal, S.Nirmala, C. 2012.Utilization of coir fibers as an eco-friendly substitute for costly gelling agents for *in vitro* orchid seed germination.*Sci Hortic.*; 133:89-92.

Albert, N. W. Arathoon, S. Collette, V. E. Schwinn, K. E. Jameson, P. E. Lewis, D. H. Zhang H.B. Davies, K. M.2010.Activation of anthocyanin synthesis in *Cymbidium* orchids: variability between known regulators.*Plant Cell, Tissue and Organ Culture*; 100 (3):355-360.

Apang O., Rao A. N.2008.Flowering period of some native orchid species in Arunachal Pradesh, India. *Indian Forester*;134 (8): 1116-1131.

Aravindhan, V. Sathiyadas, K. Rajendran, Thomas A. B. 2011.Some rare/endemic medicinal orchids of Velliangiri hills of Southern Western Ghats, Tamil Nadu, India.*Indian Forester*; 137(9): 1077-1081.

Arditti J, Ernst R.1993. In "Micropropagation of Orchids". J. Wiley and Sons Inc., New York. pp. 682.

Arditti, J.1992. Fundamentals of Orchid Biology (John Wiley and Sons, New York).

Arnon, D. I.1943.Mineral Nutrition of Plants, *Annual Review of Biochemistry*, 12: 493-528.

Attri, L. K. Kant, R.2011.Orchid pollination: an observation on pollination-pollinator interaction in *Cymbidium pendulum* (Sw.) Roxb.*Current Botany*; 2(7): 5-8.

Bag, T. K. 2006. Report of orchid wilt (*Sclerotium rolfsii*) on *Vanda* group of orchids.*Journal of Hill Research*;19(1): 44-45.

Bag, T. K.2003.Orchid wilt incited by *Sclerotium rolfsi* on some Indian orchids.*Indian Journal of Hill Farming*;16(1/2): 97-98.

Banziger, H. Pumikong, S. Srimuang, K. 2012.The missing link: bee pollination in wild lady slipper orchids *Paphiopedilum thaianum* and *P. niveum* (Orchidaceae) in Thailand.*Mitteilungen der Schweizerischen Entomologischen Gesellschaft*; 85(1/2): 1-26.

Banziger, H. S.H.Qin Luo Y.B. 2008.Pollination of wild lady slipper orchids *Cypripedium yunnanense* and *C. flavum* (Orchidaceae) in south-west China: why are there no hybrids. *Botanical Journal of the Linnean Society*; 156(1): 51-64.

Barbhuiya, H. A. Deori, C. Dutta, B. K. Das, A. K. Baishya, A. K. Schuiteman, A. 2012. *Ornithochilus cacharensis* (Orchidaceae), a new species from India (Assam).*Kew Bulletin;*. 67: 3, 511-516.

Barman, D. Bharathi, T. U. Medhi, R. P. 2012.Effect of media and nutrition on growth and flowering of *Cymbidium* hybrid 'H.C. Aurora'. *Indian Journal of Horticulture;* 69(3): 395-398.

Barman, D. Rajni, R. 2003.Studies on rooting in the keikis bamboo orchid (*Arundina gramminifolia* L.).*J Ornamental Hort.* (New Series); 6(3):260-263.

Barrett, C. F. Freudenstein, J. V. Taylor, D. L. Koljalg, U.2010.Rangewide analysis of fungal associations in the fully mycoheterotrophic *Corallorhiza striata* complex (Orchidaceae) reveals extreme specificity on ectomycorrhizal Tomentella (Thelephoraceae) across North America.*American Journal of Botany;* 97(4): 628-643.

Basumatary, N. Sarma, C. M. 2004.Epiphytic orchid flora of Chirang Reserve Forest.*Journal of Phytological Research;*17(1):33-37.

Bichsel, R. G. Starman, T. W. Wang, Y. T. 2008.Nitrogen, phosphorus, and potassium requirements for optimizing growth and flowering of the nobile *Dendrobium* as a potted orchid. *HortScience;*43(2):328-332.

Blanchard, M. G. Runkle, E. S. Lee, Y. I. 2010.First International Orchid Symposium, Taichung, Taiwan.*Acta Horti.;* 878:487.

Blanchard, M. G. Runkle, E. S. 2005.Temperature effects on flower induction of two *Phalaenopsis* orchid hybrids.Proceedings of the 18th World Orchid Conference, Dijon, France, 11-20 March; 120-123.

Blanchard, M. G. Runkle, E. S. 2006.Temperature during the day, but not during the night, controls flowering of *Phalaenopsis* orchids. *Journal of Experimental Botany;* 57(15): 4043-4049.

Blanchard, M. G. Runkle, E. S. 2008.Container opacity and media components influence rooting of potted *Phalaenopsis* and Doritaenopsis orchids.*Acta Horti.;* 788:115-120.

Borah, K.G.Sharma,R. L. Yonzone R.G. C.2012.Present status of orchid species diversity resources and distribution in Dibrugarh District of Assam of North East India.*International Journal of Modern Botany;* 2(2):19-34.

Borah, P. K. Modhusmita Kataky Bhagabati, K. N. Pathak, J. J. Bora, L. C. 2002.A new bacterial disease of orchid from India.*Journal of the Agricultural Science Society of North-East India;* 15(1):1-4.

Borah, R. L. Gogoi, K. Sharma, G. C. 2010.A taxonomic survey of the orchid diversity of Jokai reserve forest of Dibrugarh District of Assam, India.*Journal of Non-Timber Forest Products;*17(1):19-31.

Brochado, A. R., Matos, C., Moller, B. L., Hansen, J., Mortensen, U. H., Patil, K. R. 2010.Improved vanillin production in baker's yeast through in silico design.Food Science and Technology Abstracts Microbial Cell Factories. 9th , November; pp 15.

Brown JH, Lynch DV, Thompson J.E. 1987. Molecular species specificity of phospholipid breakdown in microsomal membranes of senescing carnation flowers. *Plant Physiol;*85: 679-683.

Brundrett, MC. 2002.Tansley review, Coevolution of roots and mycorrhizas of land plants, *New Phytologist;*154: 275 –304.

Bulpitt, C. J.2005. The uses and misuses of orchids in medicine QJM; 98(9):625-631.

Byung, D.L., Wan, G.K.Weon, D.C. and Jae, M.S. (2002). Occurrence of dry rot on *Cymbidium* Orchids Caused by *Fusarium* spp., *Plant Pathol. J.* ;18 (3):156-160.

C. J. Li Yan Bulpitt, P. F. Wang J.G. 2007.The use of orchids in Chinese medicine.*Journal of the Royal Society of Medicine*;100(12):558-563.

Cameron, K. M. Chase, M. W. Whitten, W. M. Kores, P. J. Jarrell, D. C. Albert, V. A. Yukawa, T. Hills, H. G. Goldman, D. H.A 1999.Phylogenetic analysis of the Orchidaceae: evidence from rbcL nucleotide sequences.*American Journal of Botany*; 86(2): 208-224.

Cardoso, J. C. Ono, E. O. Rodrigues, J. D. 2010.Gibberellic acid and water regime in the flowering induction of *BrassoCattleya* and *Cattleya* hybrid orchids.*Horticultura Brasileira*; 28(4): 395-398.

Cardoso, J. C. Ono, E. O. Rodrigues, J. D. 2012.Gibberellic acid in vegetative and reproductive development of *Phalaenopsis* orchid hybrid genus.*Horticultura Brasileira*; 30(1):71-74.

Carlquist, S. Schneider, E. L. 2006.Origins and nature of vessels in monocotyledons: 8. Orchidaceae.*American Journal of Botany*; 93(7): 963-971.

Carlsward, B.S., Stern, W.L. & Bytebier, B. 2006.Comparative vegetative anatomy and systematics of the angraecoids (Vandeae, Orchidaceae) with an emphasis on the leafless habit. *Botanical Journal of the Linnean Society*;151:165-218.

Carlsward, B.S., Whitten, W.M., Williams, N.H. & Bytebier, B. 2006.Molecular phylogenetics of Vandeae (Orchidaceae) and the evolution of leaflessness. *American Journal of Botany*;93:770-786.

Case, M. A. Mlodozeniec, H. T. Wallace, L. E. Weldy, T. W. 1998.Conservation genetics and taxonomic status of the rare Kentucky lady's slipper: *Cypripedium kentuckiense* (Orchidaceae). *American Journal of Botany*; 85(12):1779-1786.

Cating, R. A. Palmateer, A. J. McMillan, R. T. Dickstein, E. R. 2009. First report of a bacterial soft rot on *Tolumnia* orchids caused by a *Dickeya* sp. in the United States. *Plant Disease*; 93(12):1354.

Cating, R. A. Palmateer, A. J. Stiles, C. M. Rayside, P. A. 2010. Black rot of orchids caused by *Phytophthora cactorum* and *Phytophthora palmivora* in Florida. *Plant Health Progress*;. PHP-2010-0614-01-DG.

Cating, R. A. Palmateer, A. J. 2011. Bacterial soft rot of *Oncidium* orchids caused by a *Dickeya* sp. (*Pectobacterium chrysanthemi*) in Florida.*Plant Disease*; 95(1):74.

Catling, P. M. Kostiuk, B. 2011.Some observations on the pollination of round-leaf orchid, Galearis (Amerorchis) rotundifolia, near Jasper, Alberta. *Canadian Field-Naturalist*;125(1): 47-54.

Chadha, K.L. 2000. Post harvest management of flowers. *HandBook of Horticulture*. I.C.A.R, New Delhi; 969-981.

Chai, M.L., Xu, C.J., Senthil, K.K., Kim, J.Y. and Kim, D.H. 2002. Stable transformation of protocorm-like bodies in *Phalaenopsis* orchid mediated by *Agrobacterium tumefaciens*. *Sci Hortic.*; 96:213–224.

Chakrabarti K.1995. Indian orchid trade - a critical review.*Indian Forester*; 121(12):1095-1101.

Chang K.H. Tsai L.H. Wu R.Y. Dai T. Hsieh T.F.2007. Evaluation of bark as a growth medium for cultivation of *Phalaenopsis* orchids. *Journal of Taiwan Agricultural Research*;56(3):237-252.

Chang S.B. Chen W.H. Chen H.H.Fu Y.M. Lin Y.S.2000. RFLP and inheritance patterns of chloroplast DNA in intergeneric hybrids of *Phalaenopsis* and *Doritis.Botanical Bulletin of Academia Sinica*; 41(3): 219-223.

Chang Y.Y. Kao N.H. Li J.Y. Hsu W.H. Liang Y.L. Wu J.W. Yang C.H.2010. Characterization of the possible roles for B class MADS box genes in regulation of perianth formation in orchid. *Plant Physiology*;152(2):837-853.

Chaudhary A. S. , R. P. Achterberg C. van Heijerman, T. Lens, F. Dooren, T. J. M. van Gravendeel, B. 2011.Pollination and protection against herbivory of Nepalese *Coelogyninae* (Orchidaceae).*American Journal of Botany*; 98(7):1095-1103.

Chauhan N.S. 1999.Medicinal and Aromatic Plants of Himachal Pradesh. Indus Publishing Company, New Dehli.

Chauhan, N. S.1990.Medicinal orchids of Himachal Pradesh.*Journal of the Orchid Society of India*;4(1-2):99-105.

Chen L.J. Liu K.W. Xiao X.J. Tsai W.C. Hsiao YuYun Huang J. L. Zhong J.P.O.2012.The anther steps onto the stigma for self-fertilization in a slipper orchid.;7(5): e37478.

Chen L.L.Gao J.Y. 2011.Reproductive Ecology of *Bulbophyllum ambrosia* (Orchidaceae). *Chinese Journal of Plant Ecology*;35(11):1202-1208.

Chen L.R. Chen J.T. Chang W.C. 2002.Efficient production of protocorm-like bodies and plant regeneration from flower stalk explants of the sympodial orchid *Epidendrum radicans.In vitro Cellular & Developmental Biology - Plant*;38(5):441-445.

Chen, L., Hatano, T. and Niimi, Y.2002. High efficiency of *Agrobacterium*-mediated rhizome transformation in *Cymbidium*. *Lindleyana* 17:130–134.

Chen, W. H. Hsieh, R. M. Tsai, W. T. Wu, C. C. 1990.Application of biotechnology in the improvement of *Phalaenopsis* .*Taiwan Sugar*; 37(5): 20-24.

Chen, W. H. Tang, C. Y. Kao, Y. L. 2010.Polyploidy and variety improvement of *Phalaenopsis* orchids. *Acta Horti.*; 878:133-138.

Chen, W.H. Wang, Y.T.1996.*Phalaenopsis* orchid culture.*Taiwan Sugar*;43(5): 11-16.

Chen, W.H.Tang, C.Y. Kao, Y.L. 2009.Ploidy doubling by *In vitro* culture of excised protocorms or protocorm-like bodies in *Phalaenopsis* species.*Plant Cell, Tissue and Organ Culture*;98(2):229-238.

Cheng J. L.ShiYong H. R. Wei, X.L. L. Y. 2007.Food-deceptive pollination in *Cymbidium* lancifolium (Orchidaceae) in Guangxi, China. *Biodiversity Science*;15(6): 608-617.

Cheng, J. S. J. Shangguan F.Z. Dafni, A. Deng Z.H. Luo Y.B. 2009.The pollination of a self-incompatible, food-mimic orchid, *Coelogyne* fimbriata (Orchidaceae), by female Vespula wasps. (Special Issue: Orchid biology.) *Annals of Botany*; 104(3): 565-571.

Cheng, S.F. Chang, C. N. D. 2009.Growth responses and changes of active components as influenced by elevations and orchid mycorrhizae on *Anoectochilus formosanus* Hayata.*Botanical Studies*; 50(4): 459-466.

Chern L.L. Ann P.J. Wang I.T. 2011.*Cymbidium* black rot caused by an aberrant strain of *Phytophthora multivesiculata* in Taiwan. *Plant Pathology Bulletin*;20(1): 1-10.

Chiang, K.Y. Chen L.Y.2012.Effects of compound fertilizers on growth, flowering, and mineral nutrient concentrations of doritaenopsis orchids.*Journal of the Taiwan Society for Horticultural Science*;58(3):199-217.

Chiu, Y.T. Chang C.2011.Induction of *In vitro* flowering in the *Phalaenopsis* orchid.*Journal of the Taiwan Society for Horticultural Science*;57(1): 43-51.

Choi, S.H. Hong J.S. Ryu K.H.2010.Early detection of *Cymbidium* mosaic virus and odontoglossum ringspot virus in tissue-cultured *Phalaenopsis* plantlets.*Korean Journal of Horticultural Science & Technology*;28(1):103-107.

Chou, L.C. Chang, C. N. D. 2004.Asymbiotic and symbiotic seed germination of *Anoectochilus formosanus* and Haemaria discolor and their F1 hybrids. *Botanical Bulletin of Academia Sinica*; 45(2):143-147.

Chowdhury I. R., A. R. M. M. Islam, M. O. 2003.Matsui, S.Effects of plant growth regulators on callus proliferation, plantlet regeneration and growth of plantlets of *Doritaenopsis* orchid. *Biotechnology*; 2(3):214-221.

Chuang, H.T.Hsu S.T. Shen T.M. 2008.Breeding barriers in yellow *Phalaenopsis* orchids.*Journal of the Taiwan Society for Horticultural Science*; 54(1): 59-66.

Chung, MY Chung, MG. 2005. Pollination biology and breeding systems in the terrestrial orchid Bletilla striata.*Zoological Record Plant Systematics and Evolution*. 25April; 2(1-2): 1-9.

Coils, K. J. 2007.Securing the future of *Disa* orchids as a cut flower on the world market.*Acta Horti.*;755: 297-303.

Colombo, L. A. Faria, R. T. de Assis, A. M. de Fonseca, I. C. de B. 2005.*Cattleya* hybrid acclimatization in vegetal substrates under two irrigation systems. *Acta Scientiarum - Agronomy*; 27(1):145-150.

Colombo, L. A. Faria, R. T. de Carvalho, J. F. R. P. de Assis, A. M. de Fonseca, I. C. de B. 2004.The influence of the fungicide chlorothalonil on the vegetative and root development *In vitro* of two Brazilian orchid species. *Acta Scientiarum - Agronomy*;26(2): 253-258.

Combes, S. A. Dudley, R. 2009.Turbulence-driven instabilities limit insect flight performance. Proceedings of the National Academy of Sciences of the United States of America;106(22):9105-9108.

Conran, J. G. Bannister, J. M. Lee, D. E. 2009.Earliest orchid macrofossils: early Miocene *Dendrobium* and Earina (Orchidaceae: Epidendroideae) from New Zealand.*American Journal of Botany*; 96(2):466-474.

Cooke, T. (1901-1908). Flora of Presidency of Bombay, 3 Vols. Taylor and Francis, London.

Costa, P. A. Lima, A. L. da S. Zanella, F. Freitas, H. de (2010).Overcoming seed dormancy in *Adenanthera pavonina* L. *Pesquisa Agropecuaria Tropical*; 40(1):83-88.

Dai, T. E. Chang, K. H. Wang, Y. T. 2010.Growth and flowering of *Oncidium* sweet ears in an artificial textile fiber substrate. *Acta Horti.*;. 878: 361-366.

Dalayap, R. M. Torres, M. A. J. Demayo, C. G. 2011.Landmark and outline methods in describing petal, sepal and labellum shapes of the flower of Mokara orchid varieties. *International Journal of Agriculture and Biology*;13(5):652-658.

Darveau, C. A. Hochachka, P. W. Roubik, D. W. Suarez, R. K. 2005.Allometric scaling of flight energetics in orchid bees: evolution of flux capacities and flux rates.*Journal of Experimental Biology*; 208(18):3593-3602.

Das, M. C. S. Kumaria Tandon P.2007.Protocorm regeneration, multiple shoot induction and ex vitro establishment of *Cymbidium devonianum* Paxt.*Asian Journal of Plant Sciences*; 6(2):349-353.

Davies, K. M. Lewis, D. H. Boase, M. R. Zhang, H. Deroles, S. C. Schwinn, K. E. 2007.Biotechnological approaches for modifying flower colour in *Cymbidium* orchid.*Acta Horti.*;755:171-179.

De, K. K. Majumdar, S., Sharma R., Sharma, B.2006.Green pod culture and rapid micropropagation of *Dendrobium chrysanthum* Wall. - a horticultural and medicinal orchid. *Folia Horticulturae*;18(1):81-90.

Deb, C. R. Imchen, T. 2010.An efficient *In vitro* hardening technique of tissue culture raised plants.Biotechnology; 9(1):79-83.

Deb, C. R. Pongener, A. 2011.Asymbiotic seed germination and *In vitro* seedling development of *Cymbidium aloifolium* (L.) Sw.: a multipurpose orchid.*Journal of Plant Biochemistry and Biotechnology*;20(1):90-95.

Deb, C. Sungkumlong, R. 2009. Rapid multiplication and induction of early *In vitro* flowering in *Dendrobium primulinum* Lindl. *Journal of Plant Biochemistry and Biotechnology*; 18(2):241-244.

Debergh PC, Maene L. J. 1981. A scheme for commercial propagation of ornamental plants by tissue culture. *Sci Hortic.*;14: 335-345.

Delannoy, E. Fujii, S. Francs-Small, C. C. des Brundrett, M. Small, Rampant, I. 2011. Gene loss in the underground orchid *Rhizanthella gardneri* highlights evolutionary constraints on plastid genomes.*Molecular Biology and Evolution*; 28(7): 2077-2086.

Delle-Vedove, R. Juillet, N. Bessiere, J. M. Grison, C. Barthes, N. Pailler, T. Dormont, L. Schatz, B. 2011.Colour-scent associations in a tropical orchid: three colours but two odours.*Phytochemistry*;72(8):735-742.

Devadas, R. Medhi, R. P. Das, S. P. 2010.Interspecific hybrid developed in *Epidendrum* orchid from the cross *E. radicans* Pav. ex. Lindl. x *E. xanthinum* Lindl.*J Hort Sci.* ;5(2):144-147.

Devadas, R. Upadhyaya, R. C. Khatiwara, P. 2009.Characterization and evaluation of a rare orchid *Renanthera imschootiana* Rolfe from Manipur & Nagaland. *J Hort Sci.* ;4(2):181-183.

Devi, J. Jamir, C. Deka, P. C. 2005.*In vitro* pollination: its possible application in orchids for wide hybridization.Proceedings of the 18th World Orchid Conference, Dijon, France, 11-20 March; 173-176.

DeYong, G. Rowe, B. Runkle, E. 2011.Propagating orchids: multiplying and sharing helps conserve species and hybrids.*Orchids*; 80(8):486-489.

Dillon, M. E. Dudley, R.2004.Allometry of maximum vertical force production during hovering flight of neotropical orchid bees (Apidae: Euglossini).*Journal of Experimental Biology*; 207(3):417-425.

Dobriyal, M. J. 2002.Rare and endangered medicinal orchids of Asthavarga group in Uttaranchal - *Habenaria intermedia* (Virdhii) and *Habenaria edgeworthii* (Ridhii).*MFP News*; 12(4): 4-5.

Dobriyal, M. J. R. Bhojvaid, P. P. Gusain, M. S. 2003.Medicinal orchids - an important non-wood forest product.*International Journal of Forest Usufructs Management*; 4(1):12-21.

Doernenburg, H., Knorr, D. 1996.Production of phenolic flavor compounds with cultured cells and tissues of *Vanilla* species.Food Science and Technology Abstracts *Food Biotechnology*; 10(1): 75-92.

Dogane, Y. Ando, T. 1990.An estimation of carbon evolution during flowering and capsule development in a *Laeliocattleya* orchid. *Sci Hortic.*; 42(4):339-349.

Doorn, W. G. van 1997.Effects of pollination on floral attraction and longevity.*Journal of Experimental Botany*;48(314):1615-1622.

Douglas P. C. and Barry J. J. 2002. Optimizing a *Bacillus subtilis* isolate for biological control of sugar beet *Cercospora* leaf spot. *Biological Control*; 26 (2): 153-161.

Downie, D.G. 1957. *Corticium solani*-an orchid endophyte. *Nature*;179: 180.

Dressler, R. 1993.Phylogeny and classification of the orchid family.Cambridge University Press, Cambridge.

Dressler, R.L. 1981.The Orchids: Natural History and Classification. Harvard University Press, Cambridge, Mass.

Droll, P. W. 2005.Orchid watering decisions: moisture dynamics in orchid pots and media.Orchids;74(10):770-775.

Duan C.F. Li Z.L. Fang F.Yang G.H Chen S.S. Hong Q.Y. 2010.Isolation and identification research of mycorrhizal isolated from several orchids in Yunnan Province.*Southwest China Journal of Agricultural Sciences*; 23(3):756-759.

Duan, L.G. S, J. Chen, Z. L. Zeng, S. J. Jiang, Y. M. Joyce, D. C. 2006. KClO3 applications affect *Phalaenopsis* orchid flowering.*Sci Hortic.*; 110(4):362-365.

Dube, H.C.2007. An Introduction to fungi: 3rd edition, Vikash Publishing House Pvt. Ltd. Delhi.

Duff. J , Daly, A. 2002. Orchid diseases in the Northern Territory. *Agnote* ;13: 1-5.

Duncan DB 1955. Multiple range and multiple f-tests. *Biometrics*; 11:1-42

Durkee, S.2000.*Cypripedium acaule* and *Cypripedium reginae*: successful cultivation of two North American lady's-slipper orchids.*Orchids*;69(9):864-869.

Ejima, C. Kobayashi, Y. Honda, H. Shimizu, N. Kiyohara, S. Hamasaki, R. Sawa, S. A 2011.*Phalaenopsis* variety with floral organs showing C class homeotic transformation and its revertant may enable *Phalaenopsis* as a potential molecular genetic material. *Genes and Genetic Systems*;86(2):93-95.

Eken, C. Ercisli, S. Esitken, A. Demirci, E. Yuen, G. Y. 2005. First report of crown and stem rot of orchid (*Orchis palustris*) caused by *Sclerotinia* minor.*Plant Disease*; 89(8): 913.

Ernst R.1994. Effects of Thidiazuron on *In vitro* propagation of *Phalaenopsis* and *Doritaenopsis* (Orchidaceae). *Plant. Cell. Tissue Org*. 39: 273-275.

Eszeki, E. R. Tilly-Mandy, A. Forrai, M. 2009.The use of plant extract components in the *In vitro* propagation of some orchid species.Bulletin of University of Agricultural Sciences and Veterinary Medicine Cluj-Napoca. *Horticulture*;66(1):684.

Fadelah, A. A.2009.Field performance of tissue cultured micropropagated *Dendrobium* orchid hybrids. *Acta Horti.*;812: 499-506.

Falara, V. Amarasinghe, R. Poldy, J. Pichersky, E. Barrow, R. A. Peakall, R. 2013.The production of a key floral volatile is dependent on UV light in a sexually deceptive orchid.*Annals of Botany*; 111(1): 21-30.

Fan X.L. Barrett, S. C. H. Lin, Hua C. L.L., Zhou X. G. Jiang,Y 2012. Rain pollination provides reproductive assurance in a deceptive orchid.*Annals of Botany*;110(5):953-958.

Fay, M. F. Chase, M. W. 2009.Orchid biology: from Linnaeus via Darwin to the 21st century. (Special Issue: Orchid biology.) *Annals of Botany*;104(3):359-364.

Figueroa, D. M. Bass, H. W. 2010.A historical and modern perspective on plant cytogenetics. (Special Issue: Plant functional genomics.) *Briefings in Functional Genomics & Proteomics*; 9(2):95-102.

Finger, F. L. Barbosa, J. G. 2006.Postharvest physiology of cut flowers.*Advances in postharvest technologies for horticultural crops*;373-393.

Fischer, G. A. Sieder, A. Cribb, P. J. Kiehn, M. 2007.Description of two new species and one new section of *Bulbophyllum* (Orchidaceae) from Madagascar. *Adansonia*; 29(1):19-25.

Fischer, J. L. 2009.New world slipper orchids: 1. *Phragmipedium besseae* and *Phragmipedium fischeri* breeding.*Orchids*;78(12):700-709.

Fitch, C. M. 2003.*Brassias*: the spider orchid intrigues with bizarre flowers.*Orchids*;72(3): 186-189.

Freitag, D.Vanillin 2006.The essence of orchids.*Food Science and Technology Kleinbrennerei*;58(7): 7.

Freitas, F. C. L. Grossi, J. A. S. Barros, A. F. Mesquita, E. R. Ferreira, F. A. Barbosa, J. G. 2007.Chemical control of Pilea microphylla in orchid cultivation.*Planta Daninha*; 25(3):589-593.

Frenkel,H.D. French, J. C. Graft, N. M. Joel, D. M. Pak, F. E. Frenkel, C. 2004 Interrelation of curing and botany in *Vanilla* (*Vanilla* planifolia) bean.*Acta Horti.*;. 629, 93-102.

Frenkel,H.D. Frenkel, C. 2006.Postharvest handling and storage of cured *Vanilla* beans. *Stewart Postharvest Review*; 2: 4

Fu S.F. Tsai T.M. Chen Y.R. Liu C.P. Haiso L.J. Syue L.H. Yeh H.H. Huang H.J. 2012.Characterization of the early response of the orchid, *Phalaenopsis amabilis*, to *Erwinia chrysanthemi* infection using expression profiling.*Physiologia Plantarum*;145(3):406-425.

Gaba, V., Kathiravan, K., Amutha, S., Singer, S., Xiaodi, X. Ananthakrishnan, G.2006. The uses of ultrasound in plant tissue culture. p.417–426. In: G.S. Dutta and Y.Ibaraki (eds.), Focus on biotechnology, Vol. VI, *Plant tissue culture engineering.Springer, Dordrecht, The Netherlands.*

Gamble, J. S. 1915-1936. Flora of the Presidency of Madras, Adlard and Son, London.

Gavhale, R. V. Mahakhode, R. H. Bakale, V. L. 2010.Effect of herbicides on seed germination and seedling growth of weed *Malachra capitata* Linn. *Journal of Soils and Crops*;20(1):159-166.

Gawenda, I. Schroder-L, A. Debener, T. 2012.Markers for ornamental traits in *Phalaenopsis* orchids: population structure, linkage disequilibrium and association mapping.*Molecular Breeding*;30(1):305-316.

Gerlach, W. W. P. 1992.Flowering behaviour of ephemeral orchids of Western Samoa: I. Flowering periodicity.*Gartenbauwissenschaft*;. 57: 5, 219-222.

Gezgin, Y. Eltem, R. 2009.Diversity of endophytic fungi from various Aegean and Mediterranean orchids (saleps).*Turkish Journal of Botany*;33(6):439-445.

Ghanaksh, A. P. Kaushik 1999. Antibacterial effect of *Aerides multiflora* Roxb. A study *in vitro*, *Journ. Orchid Soc.India*; 13: 65-68.

Gigord, L. D. B. Macnair, M. R. Smithson, A.2001.Negative frequency-dependent selection maintains a dramatic flower color polymorphism in the rewardless orchid *Dactylorhiza sambucina* (L.) Soo.Proceedings of the National Academy of Sciences of the United States of America; 98(11):6253-6255.

Gigord, L. D. B. Macnair, M. R. Stritesky, M. Smithson, A. 2002.The potential for floral mimicry in rewardless orchids: an experimental study.*Proceedings of the Royal Society of London*. Series B, *Biological Sciences*;269(1498):1389-1395.

Giri, D. Tamta S. 2010.Combined effect of PGRs and soil facilitate early flowering of an endangered alpine orchid *Dactylorhiza hatagirea* at lower elevation.*Current Science*;99(1): 21-23.

Giri, L. Dhyani, P. Rawat S. Bhatt, I. D. Nandi, S. K. Rawal, R. S. Pande, V. 2012.*In vitro* production of phenolic compounds and antioxidant activity in callus suspension cultures of Habenaria edgeworthii: a rare himalayan medicinal orchid. *Industrial Crops and Products*;39: 1-6.

Gnasekaran, P. Poobathy, R. Mahmood, M. Samian, M. R. Subramaniam, S. 2012. Effects of complex organic additives on improving the growth of PLBs of *Vanda* Kasem's Delight. *Australian Journal of Crop Science;* 6(8): 1245-1248.

Goh, C. J. Wong, P. F.1990.Micropropagation of the monopodial orchid hybrid Aranda 'Deborah' using inflorescence explants.*Sci Hortic.;*. 44: 3-4, 315-321.

Goldman, D. H. Jansen, R. K. Berg, C. van den Leitch, I. J. Fay, M. F. Chase, M. W. 2004.Molecular and cytological examination of Calopogon (Orchidaceae, Epidendroideae): circumscription, phylogeny, polyploidy, and possible hybrid speciation. *American Journal of Botany;*91(5):707-723.

Gonzalez G., V. Portal Onco, M. A. Rubio S.V. 2006.Biology and systematics of the form genus *Rhizoctonia. Spanish Journal of Agricultural Research;*4(1):55-79.

Gouk S.S. He J. H. Choy S. 1999.Changes in photosynthetic capability and carbohydrate production in an epiphytic CAM orchid plantlet exposed to super-elevated CO_2.*Environmental and Experimental Botany;*41(3):219-230.

Govaerts, R., Campaccil, M.A., Baptista, D.H., Cribb, P.J., George, A., Kreuz, K. & Wood, J. 2010. *World Checklist of Orchidaceae*.Available at: http://www.kew.org/wcsp/ (accessed 5 March 2013).

Gow, W.P. Chen, J.T. Chang, W.C. 2008.Influence of growth regulators on direct embryo formation from leaf explants of *Phalaenopsis* orchids. *Acta Physiologiae Plantarum;*30(4):507-512.

Gowland, K. M. Wood, J. Clements, M. A. Nicotra, A. B. 2011.Significant phorophyte (substrate) bias is not explained by fitness benefits in three epiphytic orchid species.*American Journal of Botany;*98(2):197-206.

Grabiele, M. Honfi, A. I. Cerutti, J. C. Fernandez, V. Franco, D. Davina, J. R. 2010.Cytogenetic analysis in the terrestrial orchid *Sacoila argentina* (Griseb.) Garay from Paraguay. *Botanical Studies;*51(3):337-342.

Gravendeel, B. Smithson, A. Slik, F. J. W. Schuiteman, A. 2004.Epiphytism and pollinator specialization: drivers for orchid diversity? Philosophical Transactions of the *Royal Society of London*. Series B, *Biological Sciences;*359(1450):523-1535.

Gregg, K. B. 2011.Recovery from bud Disappearance explains prolonged dormancy in *Cleistes bifaria* (Orchidaceae). *American Journal of Botany;*98(2):326-330.

Greule, M. Tumino, L. D. Kronewald, T. Hener, U. Schleucher, J. Mosandl, A. Keppler, F.2010.Improved rapid authentication of vanillin using delta 13C and delta 2H values. *European Food Research and Technology;*. 231(6):933-941.

Griesbach, R. J. A.2005.Scientific approach to breeding blue orchids: exploring new frontiers in search of elusive flower colors.*Orchids;*74(5):376-379.

Griesbach, R.J. 1985. An orchid in every pot. *Florists' Rev.* ;176(4548):26–30.

Griffiths, K. E. Trueman, J. W. H. Brown, G. R. Peakall, R. 2011.Molecular genetic analysis and ecological evidence reveals multiple cryptic species among thynnine wasp pollinators of sexually deceptive orchids.*Molecular Phylogenetics and Evolution;* 59(1): 195-205.

Guo, Z.H. Zang, R.G. Qi, W.Q. Yu R.C.2003.The response and adaptation of two primitive orchids to a variable light environment in subtropical forests.Scientia Silvae Sinicae;. 39(3):23-29

Gupta S. Porwal, M. C. Roy, P. S. 2004.Orchid diversity of Great Nicobar Biosphere Reserve.*Current Science;* 86(10): 1372-1374.

Gustafsson, A. L. S. Verola, C. F. Antonelli, A. 2010.BMC Evolutionary Biology;. 10: 177, (14 June 2010). 54 ref.

Haddix, M. L. Glass, M. 2006.*In vitro* shoot and leaf proliferation of Encyclia tampensis (Lindl.) orchids.Proceedings of the 33rd Annual Meeting of the Plant Growth Regulation Society of America, Quebec City, Canada, 9-13 July; 201-204.

Hajra, P.K. 1983. New Species of Lister a from Nandadevi National Park,Chamoli, District, Uttar Pradesh. Bull.*Bot-Surv India*, 25: 181-182.

Halford, N. G. 2009.New insights on the effects of heat stress on crops.*Journal of Experimental Botany*; 60(15):4215-4216.

Harley J.L. Smith, S.E.1983. Mycorrhizal Symbiosis. Academic press Inc., London.

Harley, J.I. 1965. Mycorrhiza. *In: Ecology* of soil born plant pathogens. K.F Waker and W C Synder (eds.). University California Pres, Berkeley, Los Angles. 218-230.

Harrison, E.R. 1972.*Epiphytic orchids of Southern Africa; a field guide to the indigenous species*.Natal Branch, Wildlife Protection and Conservation Society of South Africa, Durban.

Hegde, S. N. 1997.Orchid wealth of India. Proceedings of the Indian National Science Academy. Part B, *Biological Sciences*;63(3):229-244.

Hegde, S. N. 2005.Orchid diversity in the Eastern Himalayas. *Journal of Hill Research*; 18(2):43-54.

Hermans, J. Hermans, C. 1995.Lilliputian adventures.*American Orchid Society Bulletin*; 64(10):1088-1097.

Hew CS, Yong JWH 2004. Recent advances in orchid tissue culture. In: *The Physiology of Tropical Orchids in Relation to the Industry* (2nd Edn) World Scientific Publishing, Singapore, 288-317.

Hijner, J.A., Arditti J.1973. Orchid mycorrhiza; vitamin requirements and production by the symbionts. *Amer.J.Bot.*, 60: 829-835.

Hillerman, F. & Holst, A. 1986.*An introduction to the Cultivated Angraecoid Orchids of Madagascar*. Timber Press, USA.

Hossain, M. M. R.Kant Pham T. V. Winarto, B. Zeng S.J. Silva, J. A. T. D .2013.The application of biotechnology to orchids.Critical Reviews in Plant Sciences; 32(2):69-139.

Hossain, M. M. 2012.Therapeutic orchids: traditional uses and recent advances - an overview. Fitoterapia; 82(2):102-140.

Hsu T.C. Chung, S.W. 2009.Nomenclature changes of some orchids in China and Taiwan.*Taiwan Journal of Forest Science*; 24(1):75-78.

Hsu, S. T. Chuang, H. T. Shen, T. M. 2010.Breeding barriers in red *Phalaenopsis* orchids.*Acta Horti.*; 878:145-152.

http://chemistry.about.com/od/chemistryhowtoguide/ht/makephpaper.htm .Retrieved on 2010-07-1

http://faq.gardenweb.com/faq/lists/orchids/2003011119000930.html. Retrieved on 2012-07-26

http://forum.terrorchid.org/viewtopic.php?t=900. Retrieved on 2012-08-28

http://fpdn.ifas.ufl.edu/. Retrieved on 2010-08-28

http://mycology.cornell.edu/listings.aspx?findex=Discussion. Retrieved on 2011-08-28

http://www.angelfire.com/on4/angelorchids/PropagatingOrchidsnew. Retrieved on 2011-08-9

http://www.beautyworlds.com

http://www.cotch.net/index.php/orchid_flowers

http://www.ehow.com/about_6617456_do-orchid-flowers-adapt-pollination.html #ixzz0z Ek5XFeA retrieved on 2012-06-17

http://www.indexfungorum.org/Names/NamesRecord.asp?RecordID. Retrieved on 2012-08-30

http://www.indg.in/agriculture/ruralemployment. Retrieved on 2010-08-28

http://www.ionopsis.com/sainsburyp.htm. Retrieved on 2010-098-28

http://www.orchidasia.com

http://www.orchidsusa.com/3Lightlevels.htm. Retrieved on 2010-09-27

http://www.pfc.cfs.nrcan.gc.ca/biodiversity/bcern/manual/index_e.html. Retrieved on 2010-08-3

Huang FY, Philosoph H. S, Meir S, Callahan DA, Sabato R, Zelcer A, Hepler PK.Ichikawa,K. and Aoki, T. 2000.New leaf spot disease of *Cymbidium* species caused by *Fusarium subglutinans* and *F. proliferatum. J.G.Pl. Path.*,66 (3): 213-218.

Huda, M. K. Wilcock, C. C. 2012.Rapid floral senescence following male function and breeding systems of some tropical orchids.*Plant Biology*;14(2):278-284.

Huetteman CA, Preece JE.1993. Thidiazuron: a potent cytokinin for woody plant tissue culture. *Plant Cell. Tissue Org. Cult.* 33: 105-119.

Hui,W. F. Chan, M.T. Liao D.C. Hsu, C.T. Lee, Y.W. Daniell, H. Duvall, M. R. Lin C.S. 2010.Complete chloroplast genome of *Oncidium* Gower Ramsey and evaluation of molecular markers for identification and breeding in Oncidiinae.*BMC Plant Biology*;10: 68.

Humeau, L. Micheneau, C. Jacquemyn, H. Gauvin-Bialecki, A. Fournel, J. Pailler, T. 2011.Sapromyiophily in the native orchid, *Bulbophyllum variegatum*, on Reunion (Mascarene Archipelago, Indian Ocean).*Journal of Tropical Ecology*;27(6):591-599.

Hwang, S.J. Jeong, B.R. 2007.Growth of *Phalaenopsis* plants in five different potting media.*Journal of the Japanese Society for Horticultural Science*; 76(4):319-326.

Ichihashi, S. Higuchi, T. Shibayama, H. Tesima, Y. Nishiwaki, Y. Ota, K. 2008.Aspects of CO2 uptake in the crassulacean acid metabolism orchid *Phalaenopsis .Acta Horti.*; 766: 245-256.

Ichikawa, S, Kawasaki, S., Tanaka, C and Tsuda, M.2003. Induced restance against Fusarium diseases of *Cymbidium* species by weakly virulent Strain HPF-1(*Fusarium* sp.).*Gen. Plant. Pathol.* 69: 400-405.

Jacquemyn, H. Brys, R. Jongejans, E. 2010.Seed limitation restricts population growth in shaded populations of a perennial woodland orchid.*Ecology*; 91(1):119-129.

Jacquemyn, H. Brys, R. 2010. Temporal and spatial variation in flower and fruit production in a food-deceptive orchid: a five-year study.*Plant Biology*;12(1):145-153.

Jaiphet, C. Rangsayatorn, N. 2010.Micropropagation of a rare orchid *Dendrobium gratiosissimum* using thin cell layers. *Acta Horti.*; 878:185-189.

Jakubska,B., A. Kadej, M. 2011.The pollination of Epipactis Zinn, 1757 (Orchidaceae) species in Central Europe - the significance of chemical attractants, floral morphology and concomitant insects.*Acta Societatis Botanicorum Poloniae*; 80(1):49-57.

Janarthanam, B. Gopalakrishnan, M. Sekar, T. 2008.A rapid *In vitro* propagation method for *Vanilla* planifolia.*Journal of Tropical Medicinal Plants*; 9(1):49-52.

Jha, L. K. Lalengmawia, R. 2008.Orchids: a review. *Indian Journal of Tropical Biodiversity;* 16(2):107-124.

Jianping ,Y. Webster, T. 2011.*Vanilla*: classic favourite.*Food Science and Technology Asia Pacific Food Industry*;46-47.

Johansson, D. 1974. *Ecology* of vascular epiphytes in West African rain forest.*Acta Phytogeographica Suecica*;59: 1-129.

Johansson, D. 1978. A method to register the distribution of epiphytes on the host tree.*American Orchid Society Bulletin*;47:901-904.

Johnson, S. D. Ellis, A. Dotterl, S. 2007.Specialization for pollination by beetles and wasps: the role of lollipop hairs and fragrance in *Satyrium microrrhynchum* (Orchidaceae). *American Journal of Botany*; 94(1):47-55.

Johnson, S. D. Hobbhahn, N. 2010.Generalized pollination, floral scent chemistry, and a possible case of hybridization in the African orchid *Disa* fragrans. (Special Issue: Chemical diversity and biological functions of plant volatiles.) *South African Journal of Botany*;76(4):739-748.

Johnson, S. D. Peter, C. I. Agren, J. 2004.The effects of nectar addition on pollen removal and geitonogamy in the non-rewarding orchid Anacamptis morio. *Proceedings of the Royal Society of London*. Series B, *Biological Sciences*; 271(1541):803-809.

Johnson, S. D. Peter, C. I. Ellis, A. G. Boberg, E. Botes, C. Niet, T. van D. 2011.Diverse pollination systems of the twin-spurred orchid genus Satyrium in African grasslands. *Plant Systematics and Evolution*; 292(1/2):95-103.

Johnson, S. D. Torninger, E. Agren, J. 2009.Relationships between population size and pollen fates in a moth-pollinated orchid.*Biology Letters*;.5(2):282-285.

Johnson, S. D. 2006.Pollination by long-proboscid flies in the endangered African orchid *Disa* scullyi. *South African Journal of Botany*; 72(1):24-27.

Johnson, S.D. 2000.Batesian mimicry in the non-rewarding orchid *Disa* pulchra, and its consequences for pollinator behaviour.*Zoological Record Biological Journal of the Linnean Society*; 71(1): 119-132.

Johnson, T. R. Kane, M. E. 2012.Effects of temperature and light on germination and early seedling development of the pine pink orchid (*Bletia purpurea*).*Plant Species Biology*; 27(2):174-179.

Junka, N. Wongs-A., C. Kanlayanarat, S. Thanomchit, K. 2007.Predominance of cyanidin found in flower from two species of Aerides orchids.*Acta Horti.*;755:549-556.

Kaewtrakulpong, K. Boonkorkaew, P. Tiampayotorn, J. Triboon, P. 2011.Development of the Thai orchid database system. Proceedings of the 49th Kasetsart University Annual Conference, Kasetsart University, Thailand, 1-4 February;361-367.

Kalimuthu, K. Senthilkumar, R. Murugalatha, N. 2006.Regeneration and mass multiplication of *Vanilla planifolia* Andr. - a tropical orchid.*Current Science*; 91(10): 1401-1403.

Kalimuthu, K. Senthilkumar, R. Vijayakumar, S. 2007.*In vitro* micropropagation of orchid, *Oncidium* sp. (Dancing Dolls). *African Journal of Biotechnology*;6(10):1171-1174.

Kan WS. 1986. Pharmaceutical botany. National Research Institute of Chinese Medicine, Taipei, Taiwan, p. 647.

Karuppaiah, P. Sendhilnathan, R. 2011.Studies on the effect of foliar application of organic nutrients on the growth, flowering and flower quality of *Dendrobium* orchid cv. Sakura pink.*Advances in Plant Sciences*;24(2):609-612.

Karuppusamy, S. Kumuthakalavalli, R. 2002.Medicinal orchids from the hilly tracts of Dindigul District in Tamil Nadu, India.Recent progress in medicinal plants. *Ethnomedicine and pharmacognosy*; 265-267.

Kataki, S. K. Abbas, S. L. 1992.Flowering calender of the North-East Indian orchids with a special reference to ornamental species.*Journal of the Orchid Society of India*; 6(1-2): 109-113.

Kauth, P. J. Johnson, T. R. Stewart, S. L. Kane, M. E. 2008.A classroom exercise in hand pollination and *In vitro* asymbiotic orchid seed germination.*Plant Cell, Tissue and Organ Culture*;93(2):223-230.

Keith, L. M. Sewake, K. T. Zee, F. T. 2005.Isolation and characterization of Burkholderia gladioli from orchids in Hawaii. *Plant Disease*; 89(12):1273-1278.

Khory,N. 1982.Materia Medica of India and their Therapeutics. Neeraj Publishing House.

Kirichenko, E. B. Chernyad'rev, N. I. Voronkova, T. V. Sokolova, R. S. Doman, N. G.1989. The activity of the photosynthetic apparatus in orchids during flowering. *Fiziologiya Rastenii*; 36(4):710-716.

Kizu, H., E.I. Kaneko E-I., Tomimori, T.1999. Studies on Nepalese crude drugs. XXVI. Chemical constituents of Panch aunle, the roots of *Dactylorhiza hatagirea* D. Don. *Chem. Pharm. Bull.*, 47(11): 1618-1625.

Klavina, D. Druva-Lusite, I. Gailite, A. 2009.Asymbiotic cultivation *In vitro* of the endangered orchid *Cypripedium* calceolus L. and some aspects of ex vitro growth.*Acta Horti.*; 812:539-544.

Klier, K. Leoschke, M. J. Wendel, J. F. 1991.Hybridization and introgression in white and yellow ladyslipper orchids (*Cypripedium candidum* and *C. pubescens*).*Journal of Heredity*; 82(4):305-318.

Knapp JE, Kausch AP, Chandlee J.M. 2000. Transformation of three genera of orchid using the *bar* gene as a selectable marker. *Plant Cell Reports*; 19:893-898.

Knudson L 1946.A new nutrient solution for germination of orchid seed.*American Orchid Society Bulletin*; 15:214-217.

Koch, G. W. Sillett, S. C. 2009.A response to: limitations within "The Limits to Tree Height". *American Journal of Botany*; 96(2):545-547.

Koike, S. T. Mayhew, D. E. 2001.Impatiens necrotic spot virus found in *Oncidium*: reporting a new ailment for orchids. *Orchids*;70(8):746-747.

Kolte, V. S. Raut, S. P. 2007. Efficacy of fungicides in management of collar rot of orchids incited by *Sclerotium rolfsii*. *Annals of Plant Protection Sciences*; 15(2):519-520.

Kuehnle AR, Sugii N .1992. Transformation of *Dendrobium* orchid using particle bombardment of protocorms. *Plant Cell Reports*, 11:484-488.

Kuehnle AR. 1997. Molecular biology of orchids. In: Arditti J, Pridgeon AM (Eds) *Orchid Biology: Reviews and Perspectives, VII*, Kluwer Academic Publishers,The Netherlands 75-115.

Kuehnle, A. R. Fujii, T. Mudalige, R. Alvarez, A. 2004.Gene and genome Melange in breeding of Anthurium and *Dendrobium* orchid. *Acta Horti.*;651(2):115-122.

Kumar SC, Shetty BV, Bennet SSR, Ananda Rao T, Sanjay Molur S.2001. Endemic Orchids of the Western Ghats. Conservation Assessment and Management Plan (CAMP) Workshop Report.

Kumar, A. Lal, J.1995.*Taeniophyllum crepidiforme* (King & Pantl.) King & Pantl.- a rare orchid.*Journal of Economic and Taxonomic Botany*; 19(3):545-546.

Kumar, K. M. Sheela, V. L. 2007.Status of breeding in orchids for improvement - a review.*J Ornamental Hort.* ; 10(4):199-208.

Kumari, I. P. George, T. S. 2010.Effect of polyploidy inducing agent colchicine on shoot regeneration in *Dendrobium* cv. Miss Snow White.*Current Biotica*; 4(2):218-228.

Kumari, I. P. George, T. S. 2011.*In vitro* clonal shoot morphogenesis of commercial *Dendrobium* orchid cultivars in polyamines supplemented medium.*Journal of Tropical Agriculture*;49(1/2):118-120.

Lakani, I. Suastika, G. Mattjik, N. Damayanti, T. A. 2010.Identification and molecular characterization of *Odontoglosum* Ringspot virus (ORSV) from Bogor, Indonesia.*Hayati Journal of Biosciences*; 17: 2 .

Lam E, Kato N, Lawton M .2001. Programmed cell death, mitochondria and the plant hypersensitive response. *Nature*;411: 848-853.

Lama, R.Y. , D. Bhujel, R. B. Khyanjeet G. Rai S.2012.Diversity, distribution and current ecological status of Zeuxine Lindl. (Orchidaceae) species of Darjeeling Himalaya of India. *Environment and Ecology*; 30(3A):646-648.

Lavine, B. K., Corona, D. T., Perera, U. D. N. T. 2012.Analysis of *Vanilla* extract by reversed phase liquid chromatography using water rich mobile phases.*Food Science and Technology Microchemical Journal*; 103:49-61.

Lawler, L. J. Rao, A. N. 2002.Medicinal uses of orchids in some Asian countries.*Journal of Tropical Medicinal Plants*;3(1):73-88.

Laws, N. 1995.Cut orchids in the world market.*FloraCulture International*;5(12):12-15.

Laws, N. 1996.The orchid cut-flower industry.*Orchids*;65(6):602-609.

Laws, N. 2004.The world's fascination with potted orchids.*Flora Culture International*; 14(12):26-27.

Lee B.D. Kim W.G. Cho W.D. Sung J.M.2002. Occurrence of dry rot on *Cymbidium* orchids caused by Fusarium spp. in Korea. *Plant Pathology Journal*; 18(3):156-160.

Lee H.C. Chiou, D. W. Chen W.H. Markhart, A. H. Chen Y.H. Lin T.Y. 2004.Dynamics of cell growth and endoreduplication during orchid flower development.*Plant Science*;166(3):659-667.

Lee S.C. Chang Y.C .2006.Multiplex RT-PCR detection of two orchid viruses with an internal control of plant nad5 mRNA. *Plant Pathology Bulletin*; 15(3):187-196.

Lee Y.H. 1995.Intergeneric hybridization in *Vanda*ceous orchid breeding.*Acta Horti.*;420:129-131.

Lee Y. Chang F.C. Chung M.C.2011.Chromosome pairing affinities in interspecific hybrids reflect phylogenetic distances among lady's slipper orchids (*Paphiopedilum*).*Annals of Botany*; 108(1):113-121.

Lee, Y. Yeung, E. C. 2012.Embryology of the lady's slipper orchid, *Paphiopedilum*delenatii: ovule development. *Botanical Studies*; 53(1):97-104.

Lee, L. Y. Wu F.H. Hsu C.T. Shen S.C. Yeh H.Y. Liao D.C.Fang M.J. Liu N.T. Yen, Y. C. Dokladal, L. Sykorova, E. Gelvin, S. B. Lin CS. 2012.Screening a cDNA library for protein-protein interactions directly in planta.*Plant Cell*; 24(5):1746-1759.

Lee, S. M. Lam-Chan, L. T. 1995.Study on the use of transparent polybags to culture orchid plantlets in vitro.Singapore. *Journal of Primary Industries*; 23: 53-54.

Lee, Y. H. Wong, S. M. Tan, W. K. Goh, C. J. 1996.Breeding *Vanda*ceous orchids for commercial cut-flowers in Singapore: an overview.*Euphytica*; 89(2):235-241.

Lev-Yadun, S. Neeman, G. 2012.Does bee or wasp mimicry by orchid flowers also deter herbivores?. *Arthropod - Plant Interactions*; 6(3):327-332.

Li B. S. JingY.G. ShunX.Z. G.2012.Screening and cloning of differentially expressed genes in *Dendrobium nobile* induced by orchid mycorrhizal fungus promoting the growth.*African Journal of Biotechnology*;11(51):11187-11192.

Li C.R. Zhang X.B. Hew C.S. 2003.Cloning of a sucrose-phosphate synthase gene highly expressed in flowers from the tropical epiphytic orchid *Oncidium* Goldiana.*Journal of Experimental Botany*; 54(390):2189-2191.

Li J.C. Matsui, S. Onwona A.S. 2001.Effects of cultural temperatures on antioxidant levels in leaves of *Cattleya* and *Cymbidium*.*Environment Control in Biology*; 39(1):1-7.

Li J.C. Zhao X.H. Matsui, S. 2001.Effect of temperatures on rate of nutrient concentrations in pseudobulbs and leaves and on flowering in a *Cattleya* and a *Cymbidium* hybrid.*Journal of Society of High Technology in Agriculture*;13(2):85-90.

Li P. Luo Y. 2009.Reproductive biology of an endemic orchid *Cypripedium* smithii in China and reproductive isolation between *C. smithii* and *C. tibeticum*. *Biodiversity Science*;17(4):406-413.

Li, P. Huang, B. Q. Pemberton, R. W. Luo, Y. B. Cheng, J. 2011.Floral display influences male and female reproductive success of the deceptive orchid *Phaius delavayi*.*Plant Systematics and Evolution*;296(1/2):21-27.

Liew C.F. Lim A.L. Loo M.Y. Swarup, S.A 1999.cDNA encoding the 16-kilodalton proteolipid subunit of the vacuolar H^+ ATP synthase from the orchid *Dendrobium crumenatum*. *Plant Physiology*;121(4):1384.

Lipow, S. R. Bernhardt, P. Vance, N. 2002.Comparative rates of pollination and fruit set in widely separated populations of a rare orchid (*Cypripedium fasciculatum*).*International Journal of Plant Sciences*;163(5):775-782.

Lokesha, R. Vasudeva, R. 1994.Do floral features determine the endangered status of Indian orchids? *Journal of the Orchid Society of India*; 8(1/2):53-54.

Lopez, R. G. Runkle, E. S. 2008.Effect of temperature and pseudobulb maturity on flowering of the orchid *Miltoniopsis augres* 'Trinity'. *Acta Horti.*; 766:273-278.

Lopez, R. G. Runkle, E. S. 2005.Environmental physiology of growth and flowering of orchids. *HortScience*;. 40(7):1969-1973.

Lopez, R. G. Runkle, E. S. 2004.The effect of temperature on leaf and flower development and flower longevity of *Zygopetalum* Redvale 'Fire Kiss' orchid.*HortScience*; 39(7):1630-1634.

Luyt, R. Johnson, S. D. 2001.Hawkmoth pollination of the African epiphytic orchid *Mystacidium venosum*, with special reference to flower and pollen longevity.*Plant Systematics and Evolution*; 228(1/2):49-62.

Mahmud S. Anand, V. K. Shah, N. H. 2008.Ethno-medico - botany of some important aquatic plants of Jammu province (J&K) India.*Journal of Phytological Research*; 21(2): 269-272.

Martins, D. J. Johnson, S. D. 2007.Hawkmoth pollination of aerangoid orchids in Kenya, with special reference to nectar sugar concentration gradients in the floral spurs. *American Journal of Botany*;94(4):650-659.

Matsumoto, T. K. 2006.Gibberellic acid and benzyladenine promote early flowering and vegetative growth of *Miltoniopsis* orchid hybrids. *HortScience*; 41(1):131-135.

Mayak, S. Dilley, D. 1976. Effect of sucrose on response of cut carnation flowers to kinetin, ethylene and abscisic acid. *J Am Soc Hort Sci.*; 101: 583-585.

Medhi, R. P. 2009.Syamali Chakrabarti Traditional knowledge of NE people on conservation of wild orchids. *Indian Journal of Traditional Knowledge;* 8(1):11-16.

Mercado ,S. S. A. 2012.Asymbiotic seed germination and *in vitro* seedling development of *Cattleya mendelii* Dombrain (Orchidaceae). *Acta Agronomica;* 61(1):69-78.

Miniraj, N.A.S. 2010.Medicinal orchids.*Indian Journal of Arecanut, Spices and Medicinal Plants;* 12(2): 15-23.

Misra, S. 1989.Orchid flora of Orissa. *Journal of the Orchid Society of India;* 3(1-2):61-72.

Mocan, C. Dragan-B.M. 2012.Mycorrhizal studies on some orchids from "Alexandru Borza" botanical garden.*Contributii Botanice;*47: 81-84.

Mohapatra S. Saravanan, S. 2006.Effect of different potting media on growth and flowering of *Dendrobium* orchid cv. Flamme.*Orissa Journal of Horticulture;* 34(2):65-68.

Moltmann, E. 1995.A bacterial disease causing leaf spots on orchids. *Nachrichtenblatt des Deutschen Pflanzenschutzdienstes;* 47(7):168-171.

Moniruzzaman, M. Zaman, M. A. Hossain, M. E. Bhuiyan, M. M. H. Rahman, M. Z. 2012. Genetic variability and character association in some native orchid species (*Dendrobium* spp.).*The Agriculturists;* 10(1):1-9.

Moore, S. 1877.Order XCIV. Orchidaceae.*Flora of Mauritius and the Seychelles*(ed. by J.G. Baker), pp. 317-363. L. Reeve & Co., London.

Moritz, E. Greifswald. Stuart G.A.1984.*Chinese Materia Medica.*Taipei. Southern Materials Centre. *A translation of an ancient Chinese herbal;* 31:751 759.

Mou Z.M. Yan N. L. Sh.Yun H. H. 2012.Nitrogen requirements for vegetative growth, flowering, seed production, and ramet growth of *Paphiopedilum armeniacum* (orchid).*HortScience;* 47(5):585-588.

Mroz, L. Kosiba, P. 2011.Variation in size-dependent fitness components in aterrestrial orchid, *Dactylorhiza majalis* (Rchb.) Hunt Et Summerh., in relation to environmental factors.*Acta Societatis Botanicorum Poloniae;* 80(2):129-138.

Mudalige, R. G. Kuehnle, A. R. 2004.Orchid biotechnology in production and improvement. *HortScience;* 39(1):11-17.

Murashige T, Skoog F. 1962. A revised medium for rapid growth and bioassays with tobacco tissue cultures. *Physiologia Plantarum;*15: 473-497.

Murren, C. J. Ellison, A. M. 1998.Seed dispersal characteristics of *Brassavola nodosa* (Orchidaceae).*American Journal of Botany;* 85(5):675-680.

Nadarajan, J. Wood, S. Marks, T. R. Seaton, P. T. Pritchard, H. W. 2011.Nutritional requirements for *In vitro* seed germination of 12 terrestrial, lithophytic and epiphytic orchids.*Journal of Tropical Forest Science;* 23(2):204-212.

Nagananda, G. S. Chandra, N. S. Kavyashree, R. 2012.Brick pieces soaked in liquid culture medium - a new matrix for seed germination and plantlet development for orchid *Flickingeria nodosa* (Dalz.) Seidenf.*Current Science;*102(8):1104-1107.

Nagaraju, V. Mani, S. K. 2005.Rapid *In vitro* propagation of orchid Zygopetalum intermedium. *Journal of Plant Biochemistry and Biotechnology;*14(1):27-32.

Nagtong, T. Thanonkeo, S. Klanrit, P. Thanonkeo, P. 2010.Cloning and differential expression of 1-aminocyclopropane-1-carboxylate synthase gene in different floral tissues of *Dendrobium* 'anna' flowers. *African Journal of Biotechnology;*9(15):2256-2266.

Naidu, M. M. Kumar, P. V. S. Shyamala, B. N. Guruguntla S. Prakash M. Thakur,M. S. 2012.Enzyme-assisted process for production of superior quality *Vanilla* extracts from green *Vanilla* pods using tea leaf enzymes.*Food and Bioprocess Technology;* 5(2):527-532.

Naik, S. K. Barman, D. Upadhyaya, R. C. 2006.Correlation studies on nutrient content of epiphytic orchids and associated substratum.*J Ornamental Hort* ; 9(2):133-135.

Naik, S. K. Barman, D. Upadhyaya, R. C. 2007.Correlation studies on nutrient contents of epiphytic orchids and their phorophytic substrata. *Journal of the Orchid Society of India*;. 21(1/2):61-64.

Naik, S. K. Barman, D. 2006.Response of foliar application of nitrogen on flowering in *Cymbidium* hybrid.*J Ornamental Hort*; 9(4):270-273.

Naik, S. K. Barman, D. 2007.Distribution of nitrogen, phosphorus, potassium, calcium, magnesium and sulphur in different orchid species of Sikkim. *J Ornamental Hort.* ;10(4): 255-259.

Naing A.H. Lim K.B.2011.Optimization of embryogenic callus induction and plant regeneration in orchid *Coelogyne cristata. Korean Journal of Horticultural Science & Technology*; 29(3):260-266.

Nair, S. A. Sujatha, K. 2010.Effect of varying levels of foliar nutrients on round the year production and quality of *Dendrobium* cv. Sonia 17.*J Ornamental Hort.*;13(2):87-94.

Nakazawa, Y. 1987.Characteristics and utilization of *Vanilla* flavours for food processing.*Food Science and Technology* Abstracts Koryo. 62 (12):49-61.

Nambiar, N. Siang T.C. Maziah M. 2012.Effect of 6-benzylaminopurine on flowering of a *Dendrobium* orchid.*Australian Journal of Crop Science*; 6(2):225-231.

Natarajan K. 2009.An *In vitro* study on micropropagation of *Cymbidium* orchids.*Current Biotica*; 3(2):244-250.

Navalinskiene, M. Raugalas, J. Samuitiene, M. 2005. Viral diseases of flower plants. Identification of viruses affecting orchids (*Cymbidium* Sw.).*Biologija*; 2:29-34.

Navalinskiene, M. Samuitiene, M. Koroteyeva, A. Polischuk, V. 2008.Detection and identification of viruses affecting orchids in *Lithuania* and Ukraine. *Botanica Lithuanica*; 14(3):191-198.

Nayak NR, Patnaik S, Rath SP.1997. Direct shoot regeneration from foliar explants of an epiphytic orchid *Acampe praemorsa* (Roxb.)Blatter and McCann. *Plant Cell Rep.* 16: 583-586.

Newman, E. Anderson, B. Johnson, S. D. 2012.Flower colour adaptation in a mimetic orchid.*Proceedings of the Royal Society of London*. Series B, *Biological Sciences*; 279: 1737: 2309-2313.

Newton, L. A. Runkle, E. S. 2010.Effects of paclobutrazol sprays on inflorescences of three potted moth orchid clones.*HortTechnology*; 20(5):892-895.

Newton, L. A. Runkle, E. S.2009.High-temperature inhibition of flowering of *Phalaenopsis* and *Doritaenopsis* orchids.*HortScience*; 44(5):1271-1276.

Nicholas D. J. D.1961.Minor Mineral Elements, *Annual Review of Plant Physiology*;12:63-90.

Nichols R.1973. Senescence and sugar status of the cut flower. *Acta Hortic.*; 41: 21-27.

Niet, T. van der H., D. M. Johnson, S. D. 2011.Carrion mimicry in a South African orchid: flowers attract a narrow subset of the fly assemblage on animal carcasses.*Annals of Botany*; 107(6):981-992.

Niknejad, A. Kadir, M. A. Kadzimin, S. B. 2011.*In vitro* plant regeneration from protocorms-like bodies (PLBs) and callus of *Phalaenopsis gigantea* (Epidendroideae: Orchidaceae). *African Journal of Biotechnology*;10(56):11808-11816.

Nongdam, P. Nirmala, C. Tewari, R. 2006.*In vitro* propropagation of *Cymbidium* lowianum Rchb.f using immature seeds: a floriculturally important orchid. Plant Cell Biotechnology and Molecular Biology;7(1/2):47-52.

Nongdam, P. C. 2011.*In vitro* rapid propagation of *Cymbidium aloifolium* (L.) Sw.: a medicinally important orchid via seed culture. *Journal of Biological Sciences*;11(3):254-260.

Nontachaiyapoom, S. Sasirat, S. Manoch, L. 2011.Symbiotic seed germination of *Grammatophyllum speciosum* Blume and *Dendrobium* draconis Rchb. f., native orchids of Thailand.*Sci Hortic.*; 130(1):303-308.

Norikane, A. Takamura, T. Morokuma, M. Tanaka, M. 2010. *In vitro* growth and single-leaf photosynthetic response of *Cymbidium* plantlets to super-elevated CO_2 under cold cathode fluorescent lamps. *Plant Cell Reports*; 29(3):273-283.

Ogura-Tsujita, Y. Gebauer, G. Hashimoto, T. Umata, H. Yukawa, T. 2009.Evidence for novel and specialized mycorrhizal parasitism: the orchid *Gastrodia confusa* gains carbon from saprotrophic Mycena.*Proceedings of the Royal Society of London*. Series B, *Biological Sciences*; 276(1657):761-767.

Ogura-Tsujita, Y. Yukawa, T. 2008.High mycorrhizal specificity in a widespread mycoheterotrophic plant, *Eulophia zollingeri* (Orchidaceae).*American Journal of Botany*; 95(1):93-97.

Orchard, W. 2000.Growing *Disa*s using hydroponics: methods to flower these orchids successfully.Orchids; 69(7):645-647.

Orlikowski, L. B. Szkuta, G. 2006.*Phytophthora* rot of some orchids - new disease in Poland.*Phytopathologia Polonica*; 40: 57-61.

Osorio-Gil, E. M. Forero-Montana, J. Otero, J. T. 2008.Variation in mycorrhizal infection of the epiphytic orchid *Ionopsis utriculariodes* (Orchidiaceae) on different substrata. *Caribbean Journal of Science*; 44(1):130-131.

Oyamada, T. Hiratsuka, A. Kurakake, S. 2011.Cultivation of an endangered orchid, *Cypripedium macranthos* var. speciosum using a mixture of rolled paper and vermiculite as culture medium support materials. *Horticultural Research (Japan)*; 10(3):315-320.

P.K. Kasera, Shukla J.K.2001.Bio-medicinal properties and cultivation of *Leptadaenia reticualta* (Jivanti)-an endangered plant of theThar Desert,India.*Curr. Sci.*;84(7):877-869.

Pal, R. Nagrare, V. S. Dayamma, M. 2010.Notes on growth and flowering behaviour of *Liparis cordifolia* J.D. Hook (Orchidaceae) a beautiful terrestrial orchid.Indian *Forester*;. 136(2):228-232.

Pandey, N. K. Joshi, G. C. Mudaiya, R. K. Tewari, V. P. Tewari, K. C. 2003.Management and conservation of medicinal orchids of Kumaon and Garhwal Himalaya.*Journal of Economic and Taxonomic Botany*;27(1):114-118.

Pant B., Thapa D.2012.*In vitro* mass propagation of an epiphytic orchid, *Dendrobium primulinum* Lindl. through shoot tip culture. *African Journal of Biotechnology*; 11(42):9970-9974.

Pant ,S.Tsewang R.2012.*Dactylorhiza hatagirea*: a high value medicinal orchid.*Journal of Medicinal Plants Research*; 6(19):3522-3524.

Pant, R. P. Mrinal Das Pun, K. B. Ramachandran, P. Medhi, R. P. 2010.Occurrence of *Cymbidium* mosaic and *Odontoglossum* ringspot viruses in orchid germplasm of Sikkim and Darjeeling hills, their identification and diagnosis. *Indian Phytopathology*;63(3):326-332.

Panwar, D. Kheta Ram H.Shekhawat, N. S. 2012.*In vitro* propagation of *Eulophia nuda* Lindl., an endangered orchid.*Sci Hortic.*;139: 46-52.

Parab, G. V. Krishnan, S. 2008.Assessment of genetic variation among populations of *Rhynchostylis retusa*, an epiphytic orchid from Goa, India using ISSR and RAPD markers.*Indian Journal of Biotechnology*;7(3):313-319.

Paradiso, R. Pascale, S. D.2012.Effect of light regime on growth and flowering of *Phalaenopsis* orchid. *Acta Horti.*; 952:329-334.

Pardio, V. T. Mariezcurrena, M. D. Waliszewski, K. N. Sanchez, V. Janczur, M. K. 2009. Effects of killing conditions of *Vanilla* (*Vanilla planifolia*, Andrews) pods during the curing process on aroma composition of pod ethanol extract.*International Journal of Food Science & Technology*; 44(12):2417-2423.

Parra-Tabla, V. Vargas, C. F. 2007.Flowering synchrony and floral display size affect pollination success in a deceit-pollinated tropical orchid.*Acta Oecologica*; 32(1):26-35.

Pathak ,P.Bhattacharya, A. Vij, S. P. Mahant, K. C. Dhillon, M. K. Piri, H. 2010.An update on the medicinal orchids of Himachal Pradesh with brief notes on their habit, distribution and flowering period.*Journal of Non-Timber Forest Products*;17(3):365-372.

Pathak, P. Vij, S. P. Verma, J. 2005.Ground orchids in North West Himalaya.Proceedings of the 18th World Orchid Conference, Dijon, France, 11-20 March; 383-388.

Patil, P. V. Nevkar, G. S. 2002.Standardization of exotic orchids for propagation through division. *South Indian Horticulture*;50(1/3):276-277.

Paulus, H. F.1997.Signals in pollinator attraction: imitation of females as a pollination principle in the Mediterranean orchid genus Ophrys. [German] *Verhandlungen der Zoologisch-Botanischen Gesellschaft in Osterreich*; 134: 133-176.

Pauw, A. 2006.Floral syndromes accurately predict pollination by a specialized oil-collecting bee (Rediviva peringueyi, Melittidae) in a guild of South African orchids (Coryciinae). *American Journal of Botany*;93(6):917-926.

Pellegrino, G. Bellusci, F. Musacchio, A. 2010.The effects of inflorescence size and flower position on female reproductive success in three deceptive orchids.*Botanical Studies*;51(3):351-356.

Perez-Hernandez, H. Damon, A. Valle-Mora, J. Sanchez-Guillen, D. 2011.Orchid pollination: specialization in chance?*Botanical Journal of the Linnean Society*;165(3): 251-266.

Pienaar, D. Combrink, N. J. J. 2007.The effects of N-source, shading and root zone cooling on two Disa hybrids. *South African Journal of Plant and Soil*; 24(3):166-171.

Pollet, B. Storme, E. Steppe, K. Lemeur, R. Dambre, P. Labeke, van M. C. 2011. Determining the optimal nighttime air temperature for *Phalaenopsis* during the vegetative stage. *Acta Horti.*; 893, 865-871.

Rai, I. D. P. K.Bharati, R. R. Adhikari, B. S. Rawat, G. S. 2010.*Corallorhiza trifida* Chatel.: a little known partial myco-heterotrophic orchid from the valley of flowers National Park, Western Himalaya, India.*Taiwania*; 55(4):391-395.

Ramakantha, V. 1995.A report on the threatened orchids of Manipur. *Journal of the Bombay Natural History Society*; 92(1):144-145.

Rao A.T., 1998. Conservation Of Wild Orchids Of Kodagu In The Western Ghats. The Karnataka Association for the Advancement of Science, Bangalore.

Rao U. S., V. R. Mohan, J. S. S. Reddy, A. S. 2007.*In vitro* propagation of *Dendrobium* microbulbon A. Rich - a rare ethnomedicinal herb.*Indian Journal of Biotechnology*;6(3): 381-384.

Ren ZongX. Wang Hong L.Y. 2012.Deceptive pollination of orchids.*Biodiversity Science*; 20(3):270-279.

Rey, S., Carbonel, F., Doorn, van M.1980.*Vanilla*; aromatic composition of different species.Food Science and Technology Abstracts *Annales des Falsifications et de l'Expertise Chimique*;73(788): 421-433.

Rhodehamel, W. A. 1994.Pollination of orchid flowers.*American Orchid Society Bulletin*; 63(5):534-541.

Rhodehamel, W. A. 2007.*Angraecum scottianum*: a compact orchid with fragrant flowers.*Orchids*; 76(8):582-583.

Rodrigues, D. T. Novais, R. F. Alvarez, V. V. H. Dias, J. M. M. Villani, E. M. de A. 2012.Chemical composition and component concentrations of culture medium for the *In vitro* growth of orchid. [Portuguese] Concentracoes e composicoes quimicas do meio nutritivo para o cultivo *In vitro* de orquidea. *Revista Ceres*; 59(1):1-8.

Rogerson, W. P. 1991.Hybridization of white *Paphiopedilums* - Part 1.*American Orchid Society Bulletin*;60(7):673-682.

Roy, A. R. Patel, R. S. Patel, V. V. Yadav, D. S. 2007.Medicinal orchids of Meghalaya. *Journal of the Orchid Society of India*;21(1/2):15-17.

Roychowdhury, N. Mandal, T. Munsi, P. S. 2004.Evaluation of different *Dendrobium* spp. under polyhouse in North-East Indian hills.*Acta Horti.*; 659(1):491-498.

Runkle, E. S. 2007.Innovative production systems for ornamental potted plants: a case study for *Phalaenopsis* orchids.*Acta Horti.*; 755: 55-59.

Runkle, E. S. 2010.Environmental and hormonal regulation of flowering in *Phalaenopsis* orchids: a mini review.*Acta Horti.*; 878:263-267.

Saikat G. Uma R.S. 2012.Rapid micropropagation of monopodial orchid hybrid (Aranda Wan Chark Kuan 'Blue' x *Vanda coerulea* Grifft. ex. Lindl.) through direct induction of protocorm-like bodies from leaf segments.*Plant Growth Regulation*;68(2):129-140.

Saiprasad, G. V. S. Raghuveer P. Anupam R. 2003.Effect of growth regulators on production of PLBs and multiple shoots in orchid - *Dendrobium* 'Sonia': assessment of role of methane and ethylene.*Phytomorphology*; 53(1):63-71.

Saiprasad, G. V. S. Raghuveer P. 2002.Effect of various nutrient media on the production of protocorm-like bodies (PLBs) and multiple shoots in orchids.*J Ornamental Hort.* (New Series); 5(2):72-73.

Saiprasad, G. V. S. Raghuveer, P. Khetarpal, S. Chandra, R. 2004.Effect of various polyamines on production of protocorm-like bodies in orchid - *Dendrobium* 'Sonia'.*Sci Hortic.*;100(1/4):161-168.

Saiprasad, G. V. S. Raghuveer, P. Khetrapal, S. Chandra, R. 2002.Effect of various growth regulators on the production of protocorm like bodies in three orchid genera.*Indian Journal of Plant Physiology*;7(1):35-39.

Saiprasad, G. V. S. Raghuveer, P. 2007.Influence of ethylene inhibitors and ethrel on production of protocorm like bodies in orchid - *Dendrobium* 'Sonia'. *J Hort Sci*; 2(1):13-18.

Sakai, W. S. Adams, C. Braun, G. 2000.Pseudobulb injected growth regulators as aids for year around production of Hawaiian *Dendrobium* orchid cutflowers.*Acta Horti.*;541:215-220.

Sakai, W. S. Ichihara, K. 2010.N6-benzyladenine induced flowering of potted nobile-type *Dendrobium* red emperor 'Prince' orchid plants.*Acta Horti.*; 878:317-320.

Sarmah, D. K. Borthakur, M. Borua, P. K. 2010.Artificial seed production from encapsulated PLBs regenerated from leaf base of *Vanda* coerulea Grifft. ex. Lindl. - an endangered orchid.*Current Science*; 98(5):686-690.

Satyanarayana, P. 2010.Orchid germplasm at National Orchidarium & Experimental Garden, Yercaud-notes on rare and endemics.*Journal of Economic and Taxonomic Botany*;34(1): 197-205.

Sazak, A. Ozdener, Y. 2006.Symbiotic and asymbiotic germination of endangered Spiranthes spiralis (L.) Chevall. and *Dactylorhiza osmanica* (Kl.) Soo var. osmanica (endemic).*Pakistan Journal of Biological Sciences*; 9(12):2222-2228.

Scheffknecht, S. Winkler, M. Hulber, K. Mata Rosas, M. Hietz, P. 2010.Seedling establishment of epiphytic orchids in forests and coffee plantations in Central Veracruz, Mexico.*Journal of Tropical Ecology*; 26(1):93-102.

Schelpe, E.A.C.L.E. 1976.*Growing orchids in South Africa*. Cape Orchid Society, Cape Town.

Schiestl, F. P. Cozzolino, S. 2008.Evolution of sexual mimicry in the orchid subtribe Orchidinae: the role of preadaptations in the attraction of male bees as pollinators. *BMC Evolutionary Biology*;8: 27.

Schiestl, F. P. Glaser, F. 2012.Specific ant-pollination in an alpine orchid and the role of floral scent in attracting pollinating ants.*Alpine Botany*;122(1):1-4.

Schluter, P. M. Xu S.Q. Gagliardini, V. Whittle, E. Shanklin, J. Grossniklaus, U. Schiest, F. P. 2011.Stearoyl-acyl carrier protein desaturases are associated with floral isolation in sexually deceptive orchids. *Proceedings of the National Academy of Sciences of the United States of America*;108(14):5696-5701.

Schultes, R. E. 1990.Medicinal orchids of the Indians of the Colombian Amazon.*American Orchid Society Bulletin*;59(2):159-161.

Sen, S. Rai, M. Acharya, R. Dasgupta, S. Saha, A. Acharya, K. 2009.Biological control of pathogens causing the *Cymbidium* Pseudobulb rot complex using fluorescent *Pseudomonas* strain BRL-1. *Journal of Plant Pathology*; 91(3):751-755.

Senthilkumar, S. Krishnamurthy, K. V. Britto, S. J. Arockiasamy, D. I. 2000.Visualization of orchid mycorrhizal fungal structures with fluorescence dye using epifluorescence microscopy.*Current Science*;79(11):1527-1528.

Seon , K. M., MyeongIl Lee Y.R.2010.*Cymbidium* hybrid 'Purple Princess' with dark purple flower. *Korean Journal of Horticultural Science & Technology*; 28(4):715-718.

Shangguan F.Z. Cheng J. X. YuanXin L. Y. 2008.Deceptive pollination of an autumn flowering orchid *Eria coronaria* (Orchidaceae). *Biodiversity Science*;. 16(5):477-483.

Sheela, V. L. Sarada, S. Anita, S. 2006. Micropropagation of orchids.*J Ornamental Hort.*; 9(1):1-11.

Shen, T. Chen, C. 2012.Misting cooling technique for protected culture of *Oncidium* orchids in subtropical regions. *African Journal of Biotechnology*; 11(80):14612-14620.

Sherif N. A., J. H. Franklin B., S. Muthukrishnan, T. Senthil Kumar and Rao M. V. 2012.Regeneration of plantlets from nodal and shoot tip explants of *Anoectochilus elatus* Lindley, an endangered terrestrial orchid. *African Journal of Biotechnology*;11(29): 7549-7553

Sherpa, A. R. Bag, T. K. Hallan, V. Zaidi, A. A. 2007.Incidence of *Cymbidium* mosaic virus (CymMV) in Sikkim. *Indian Phytopathology*; 60(1):133-136.

Shih,C.C. Y.W. Wu and W.C. Lin. 2002. Antihyperglycemic and antioxidant properties of *Anoectochilus formosanus* in diabetic rats.*Clinical and Experimental Pharmacology and Physiology*, 29(8): 684-688.

Shin Y.K. Baque, M. A. Salem E. Lee E.J. Paek K.Y. 2011.Effects of activated charcoal, plant growth regulators and ultrasonic pre-treatments on *In vitro* germination and protocorm formation of *Calanthe* hybrids.*Australian Journal of Crop Science*;5(5):582-588.

Shrestha, B.R., Chin, D.P., Tokuhara, K. and Mii, M. 2007. Efficient production of transgenic plants of *Vanda* through sonication-assisted *Agrobacterium*-mediated 465transformation of protocorm-like bodies. *Plant Biotechnol.* 24:429–434.

Shukla, B. K. Singh, P. Chauhan, A. S.1998.Orchid diversity in Sikkim and its conservation. *Journal of the Orchid Society of India*;12(1/2):53-76.

Silva, J. A. T. da Dong P. Chin P. T.Van M., M. 2011.Transgenic orchids.*Sci Hortic.*; 130(4):637-680.

Silvera, K. Santiago, L. S. Cushman, J. C. Winter, K. 2009.Crassulacean acid metabolism and epiphytism linked to adaptive radiations in the Orchidaceae.*Plant Physiology*;149(4): 1838-1847.

Singer, R. B. 2002.The pollination mechanism in *Trigonidium obtusum* Lindl (Orchidaceae: Maxillariinae): sexual mimicry and trap-flowers.*Annals of Botany*; 89(2):157-163.

Singh R.S.1994. Plant Disease. 7th edition, Oxford and IBH publishing Co. Pvt. New Delhi.

Singh R.S.1994. Plant Pathogens.The fungi. Third edition, Oxford and IBH publishing Co. Pvt. New Delhi.

Singh S. K. , J. Panj, F. G. Gupta, Sangwan, P. A.2011.Effect of chitosan on vegetative propagation and growth of orchid (*Dendrobium* cv. Sonia).*Annals of Biology*; 27(2):175-178.

Singh, D. R. Medhi, R. P.2006.Genetic resources of orchids in Andaman and Nicobar Islands. *Indian Forester*;132(6):700-706.

Singh, F.1990.Indian Orchids. *Indian Hort.*;35 (1): 14-16.

Singh, J. Singh, E. S. NK. Rajmuhon ,Singh, N. 2010.Biodiversity conservation and natural resources in north east India - with special reference to Manipur.*Zoological Record NeBIO.*; 1(1): 42-47.

Sinu, P. A. Giby K. Chandrashekara K. 2011.Epiphytic orchid diversity in farmer-managed Soppinabetta forests of Western Ghats: implications for conservation.*Current Science*; 101(10):1337-1346.

Sirisawat, S. Fukuda, N. Ezura, H. Handa, T. 2009.DMMADS4, a DEF-like gene from *Dendrobium* is required for floral organ identity and flower longevity of orchid.*Acta Horti.*;836:259-264.

Smith SE, Read DJ. 1997. Mycorrhizal Symbiosis, 2nd edn. San Diego, CA, USA: Academic Press.

Song, A. L. A. Abdullah, J. O. Mohd Puad Abdullah Norazizah S. Rahim, R. 2012.Functional expression of an orchid fragrance gene in *Lactococcus lactis*. *International Journal of Molecular Sciences*;13: 2, 1582-1597.

Soto-V., A.G. Torre-A., R. de la X.-Cazares, B. Ruiz-M..R. 2012. First report of *Cymbidium* mosaic virus and Odontoglossum ringspot virus in orchids in Mexico. *Plant Disease*;96(3):464.

Sreekumar, P. V. Veenakumari, K. Subramaniam, A. Prashanth M. 1997.On the orchid, *Bulbophyllum* crassipes Hook. f. in the Andaman Islands.*Current Science*;72(7):432.

Starin, D. 2012.Salepi extinction, Salepi survial: how a change in ingredients could help safeguard orchids.*Orchids*;81(8):490-493.

Steiner, K. E. Cruz, B. C. 2009.Hybridization between two oil-secreting orchids in South Africa.*Plant Systematics and Evolution*;277(3/4):233-243.

Steiner, K. E. Kaiser, R. Dotterl, S. 2011. Strong phylogenetic effects on floral scent variation of oil-secreting orchids in South Africa. *American Journal of Botany*; 98(10): 1663-1679.

Steiner, K. E. 2010.Twin oil sacs facilitate the evolution of a novel type of pollination unit (meranthium) in a South African orchid. *American Journal of Botany*; 97(2):311-323.

Stokl, J. Brodmann, J. Dafni, A. Ayasse, M. Hansson, B. S. 2011.Smells like aphids: orchid flowers mimic aphid alarm pheromones to attract hoverflies for pollination.*Proceedings of the Royal Society of London*. Series B, *Biological Sciences*; 278(1709):1216-1222.

Su, J. F. Lee, Y. C. Chen, C. W. Hsieh, T. F. Huang, J. H.2010.Sheath and root rot of *Phalaenopsis* caused by *Fusarium solani*. *Acta Horti.*; 878:389-394.

Subramaniam S. Rahman Z. A. 2010.Early GFP gene assessments influencing *Agrobacterium tumefaciens*-mediated transformation system in *Phalaenopsis violacea* orchid. *Emirates Journal of Food and Agriculture*;22(2):103-116.

Sun HaiQin Luo Yi B. G. S. 2003.A preliminary study on pollination biology of an endangered orchid, *Changnienia amoena*, in Shennongjia.*Acta Botanica Sinica*; 45: 9, 1019-1023.

Swartz O. 1799. Dianome Epidendri generis Linn. *Nova ActaRegiae Societatia Scientiarum Upsaliensia*, 6: 61–88.

Takahashi, K. Hatana, T. Ishikawa, H. Ogiwara, I.2012.Setting method of habitat protocorms inoculated with orchid mycorrhizal fungi isolated from habitat and covered with gel as a habitat for *Habenaria radiata* (Thunb.) K. Spreng. and seedling growth. *Horticultural Research (Japan)*;11(2):213-217.

Takahashi, K. Ishikawa, H. Ogino, T. Hatana, T. Ogiwara, I. 2007.Growth assay of daughter tubers from the tubers of Habenaria radiata (Thunb.) K. Spreng. seedlings gel covered and inoculated with orchid mycorrhizal fungi in habitat.*Horticultural Research*; 6(1):33-36.

Takahashi, Y. Takahashi, K. Sato, M. Watanabe, K. Kawano, T. 1997.Bacterial leaf rot of *Odontioda* orchids caused by *Enterobacter cloacae*. *Annals of the Phytopathological Society of Japan*;63(3):164-169.

Takahashi, Y. Takahashi, K. Watanabe, K. Kawano, T.2004.Bacterial black spot caused by Burkholderia andropogonis on Odontoglossum and intergeneric hybrid orchids. *Journal of General Plant Pathology*;70(5):284-287.

Taylor, D. L. Bruns, T. D. Hodges, S. A. 2004.Evidence for mycorrhizal races in a cheating orchid.*Proceedings of the Royal Society of London*. Series B, *Biological Sciences*; 271(1534):35-43.

Thakur, U. Ngarwar N. 2012. Artificial pollination and *In vitro* asymbiotic seed germination in garden orchid *Spathoglottis plicata* Blume (Orchidaceae).*Recent Research in Science and Technology*; 4(2):13-18.

Than, M. M. M. A.Pal Jha S..2011.Chromosome number and modal karyotype in a polysomatic endangered orchid, *Bulbophyllum* auricomum Lindl., the Royal Flower of Myanmar.*Plant Systematics and Evolution*;. 294(3/4):167-175.

Than, P.P., Haryudian, P. Sitthisack, P., Paul W.J. T and Kevin D. H. 2008. Chilli anthracnose disease caused by *Colletotrichum* species. *J Zhejiang Univ Sci B.* 9(10):764-778.

Thomas, M. 1937. Plant Hormones and their Possible Importance in Horticulture. *Scient. Hort.,* 5: 141.

Thompson, H. C. 1940. Effect of Temperature and Length of Day on Growth of Vegetables. *Minnesota Hort.* 68: 163.

Thornberry, H.H.Thompson, M.R., Izadpanah, K. and Canares, P.H. 1968. Orchid viruses: causality of *Cattleya* infection blossom necrosis. *J. Phyto. Path.,* 62: 305-310.

Trelka, T. Bres, W. Kozlowska, A. 2010.*Phalaenopsis* cultivation in different media. Part I. Growth and flowering.*Acta Scientiarum Polonorum - Hortorum Cultus*;9(3):85-94.

Trujillo, G. Hernandez, Y. 1999. Bacterial spot in orchid.*Fitopatologia Venezolana*; 12(1):5-8.

Tuyl, J. M. van Lim K.B. 2003.Interspecific hybri*Disa*tion and polyploi*Disa*tion as tools in ornamental plant breeding. *Acta Horti.*; 612:13-22.

Udomdee, W. Wen P.J. Chin S.W. Chen F.C. 2012.Shoot multiplication of *Paphiopedilum*orchid through *In vitro* cutting methods. *African Journal of Biotechnology*; 11(76):14077-14082.

Uniyal,M.R., M. Uniyal, M. 1977.Medicinal Plants and Minerals of Uttarakhand Himalaya. Baidyanath, Ayurved Shodh Sanstha,Patna 26 .

Upadhyaya, R. C. Nagaraju, V. Das, S. P. Barman, D. Pal R. 2006. Exploration, characterization and conservation of ornamental orchids.*J Ornamental Hort.*; 9(4):231-237.

Uthaichay, N. Ketsa, S. Doorn, van W. G. 2007.1-MCP pretreatment prevents bud and flower abscission in *Dendrobium* orchids.*Postharvest Biology and Technology*;43(3): 374-380.

Vacin EF, Went FW. 1949. Some pH changes in nutrient solutions. *Botanical Gazette,* 110: 605-613.

Vakhrameeva MG, Denisova, LV, Nikitina SV and Samsonov SK. 1991. Orkhidei nashei strany (orchids of our lands). *Moscow: Nauka.* 13: 27-39.

Valdivia, C. E. Cisternas, M. A. Verdugo, G. S. 2010.Reproductive biology aspects of two species of the genus *Gavilea* (Orchidaceae, Chloraeinae) in populations from Central Chile.*Gayana Botanica*; 67(1):44-51.

Vale, A. Navarro, L. Rojas, D. Alvarez, J. C. 2011.Breeding system and pollination by mimicry of the orchid *Tolumnia guibertiana* in Western Cuba.*Plant Species Biology*; 26(2):163-173.

Valpuesta V, Lange NE, Guerrero C, Reid M.S. 1995. Up-regulation of a cysteine protease accompanies the ethylene-insensitive senescence of daylily (*Hemerocallis*) flowers. *Plant Mol Biol* 28: 575-582.

Valsalakumari, P. K. Juliemol, T. Rajeevan, P. K. Geetha, C. K. 2007.Improvement in post harvest quality of orchid flowers (*Dendrobium* varieties).*Journal of the Orchid Society of India*; 21(1/2):51-55.

Valvi, R. J. Yadav, S. S. Varghese, M. 2006.New record of orchid species for the flora of West Khandesh Satpuda. *Plant Archives*; 6(2):753-755.

Van Doorn WG. 2001. Categories of petal senescence and abscission: a re-evaluation. *Ann Bot*, 87: 447-456.

van Doorn WG, Stead A.D., 1997. Abscission of flowers and floral parts. *J Exp Bot*, 309: 821-837.

Vaz, A. P. A. Kerbauy, G. B. 2000.Effects of mineral nutrients on *in vitro* growth and flower formation of Psygmorchis pusilla (Orchidaceae).*Acta Horti.*; 520:149-156.

Vejsadova, H. Sediva, J. Vlasinova, H. Havel, L. Mertlik, J. Kloudova, K. 2009.Organogenesis induction in horse chestnut (*Aesculus hippocastanum* L.). *Zpravy Lesnickeho Vyzkumu*;54(4):286-292.

Vejsadova, H. 2006.Factors affecting seed germination and seedling growth of terrestrial orchids cultured *in vitro*.*Acta Biologica Cracoviensia. Series Botanica*; 48: 109-113.

Vereecken, N. J. Cozzolino, S. Schiestl, F. P. 2010.Hybrid floral novelty drives pollinator shift in sexually deceptive orchids.*BMC Evolutionary Biology*; 10: 103.

Verma, N., MacDonald, L. and Punja, Z. K. 2006. Inoculum prevalence, host infection and biological control of *Colletotrichum acutatum* causal agent of blueberry anthracnose in British Columbia. *Plant Pathology*, 55: 442–450.

Vithoonjit, C. Paradornuwat, A. Klinkong, S. Piluek, C. 2005.Investigation on orchid dry rot: morphology of S. rolfsii Sacc., disease causing agent and attempt on antagonistic fungi. [Thai] Proceedings of 43rd Kasetsart University Annual Conference, Thailand, 1-4 February, 2005. Subject: Plants;. 664-670.

Wan G. K., Byung D. L., Woo S. K. and Weon D. C. 2002. Root Rot of Moth Orchid Caused by *Fusarium* spp. *Plant Pathol. J.*, 18 (4): 225-227.

Wang Y. T .2000.Impact of a high phosphorus fertilizer and timing of termination of fertilization on flowering of a hybrid moth orchid. *HortScience*;. 35: 1, 60-62

Wang, Y. T. Gregg, L. L. 1994.Medium and fertilizer affect the performance of *Phalaenopsis* orchids during two flowering cycles.*HortScience*; 29(4):269-271.

Warcup J.H. and Talbot P.H.B.1971. Perfect states of *Rhizoctonias* associated with orchids I-III. *New Phytologist*, 66: 631-641.

Webb, A. Webb, M. 1993.Australian orchids. *American Orchid Society Bulletin*;62(10): 1016-1017.

White J. 1990.Watering and Fertilizing by - American Orchid Society Bulletin.

Winter K, Wallace BJ, Stocker GC, Rocksandic Z. 1983. Crassulacean acid metabolism in Australian vascular epiphytes and some related species. *Oecologia (Berlin)* 57: 129–141.

Wu F.H. Chan M.T. Liao D.C. Hsu C.T. Lee Y.W. Daniell, H. Duvall, M. R. Lin C.S. 2010. Complete chloroplast genome of *Oncidium* Gower Ramsey and evaluation of molecular markers for identification and breeding in Oncidiinae. *BMC Plant Biology*;10: 68,

Wu H.H. Chen F.C. 2008.Effect of plant growth regulators on shoot multiplication from flower stalk nodal buds of *Phalaenopsis* and *Doritaenopsis*.*Journal of the Taiwan Society for Horticultural Science*;54(2):151-159.

Wu P.H. Chang, D. C. N. 2012.Cytokinin treatment and flower quality in *Phalaenopsis* orchids: comparing N-6-benzyladenine, kinetin and 2-isopentenyl adenine.*African Journal of Biotechnology*; 11(7):1592-1596.

Wu P.H. Huang D.D. Chang, D. C. N. 2011. Mycorrhizal symbiosis enhances *Phalaenopsis* orchid's growth and resistance to Erwinia chrysanthemi.*African Journal of Biotechnology*; 10(50):10095-10100.

Wu R.Z. Chakrabarty, D. Hahn E.J. Paek K.Y.2005.Growth of *Doritaenopsis* in peat-substituted growing medium.*Journal of the Korean Society for Horticultural Science*;. 46(1):76-81.

Wu R.Y. Lai Y.H.2010.Study on growing media for growth and development of Angraecum sesquipedale. *Journal of Taiwan Agricultural Research*;59(4):219-227.

Wu Z.Y. Fang Y. Chen X. Zhang H.L. Zhang M.Z. Li B.2010. Pathogen identification of bacterial soft rot in moth orchid in Zhejiang Province.*Journal of Zhejiang Forestry College*; 27(4):635-639.

Yadav, L. B. Tiwari, A. K. Maurya, R. P. Srivastav, R. M.2010. Occurrence of leaf blight and dry rot of *Cymbidium* orchids caused by Fusarium oxysporum under mid hill conditions of Uttarakhand.*Annals of Plant Protection Sciences* 18(2):541-542.

Yi S.X. Tan G.L. Li F. Wang Y.L. Ma Q.Xian K. L. 2008. Survey and pathogens identification of viral disease on partial orchids in Yunnan. *Journal of Yunnan Agricultural University*; 23(3):325-328.

Yong,D, G. Rowe, B. Runkle, E. 2011.Propagating orchids: multiplying and sharing helps conserve species and hybrids.*Orchids*;80(8):486-489.

Yoon Y.J. Mobin, M. Hahn E.J. Paek K.Y. 2009.Impact of *in vitro* CO2 enrichment and sugar deprivation on acclimatory responses of *Phalaenopsis* plantlets to *ex vitro* conditions.*Environmental and Experimental Botany*; 65(2/3):183-188.

Zakaria L. Hayati, N., Baharuddin M. Z. S, Maziah, Z. 2010. Identification and pathogenicity of Fusarium species associated with root rot and stem rot of *Dendrobium*. *Asian Journal of Plant Pathology*;4(2):103-110.

Zakaria L.Zain, N. H. M. D. Baharuddin S. Zakaria, M. 2009.Morphological and RAPD analysis of Fusarium species associated with root and stem rot of *Dendrobium* orchid in Northern Peninsula Malaysia. *Hayati Journal of Biosciences*; 16: 2

Zhao X.H. Li J.C. Matsui, S. Maezawa, S. 2005.Effect of near-ultraviolet radiation on the responsibility of antioxidative function of orchid leaves.*Journal of Science and High Technology in Agriculture*;17(4):192-198.

Zheng Y.X. Chen C.C. Chen Y.K. Jan F.J.2008. Identification and characterization of a potyvirus causing chlorotic spots on *Phalaenopsis* orchids.*European Journal of Plant Pathology*;121(1):87-95.

Zotz, G. 1999.Vegetative propagation in an epiphytic orchid. Occurrence and ecological relevance. *Ecotropica*;5(1):65-68.

Colour Figures

Fig.1 : Glimpse of Dendrobium flowers

Fig.2 : Packaging and marketing of Orchid flowers

Fig.3 : Packaging and marketing of Orchid flowers

Fig.4 : Packaging and marketing of Orchid flowers

Fig.5 : Packaging and marketing of Orchid flowers

Fig.6 : Packaging and marketing of Orchid flowers

Fig.7 : Packaging and marketing of Orchid flowers

Fig.8 : Oncidium

Fig.9 : Packaging and marketing of Orchid flowers

Fig.10 : Packaging and marketing of Orchid flowers

Fig.11 : Glimpse of Phalaenopsis orchid

Fig.12 : Field view of orchid

Fig.13 : Field view of orchid

Fig.14 : Field view of orchid

Fig.15 : Field view of orchid

Fig.16 : Glimpse of Dendrobium flowers

Fig.17 : Glimpse of Dendrobium flowers

Fig.18 : Glimpse of Dendrobium flowers

Fig.19 : Glimpse of Dendrobium flowers

Fig.20 : Glimpse of Dendrobium flowers

Fig.21 : Glimpse of Dendrobium flowers

Fig.22 : Glimpse of Dendrobium flowers

Fig.23 : Glimpse of Dendrobium flowers

Fig.24 : Glimpse of Dendrobium flowers

Fig.25 : Glimpse of Dendrobium flowers

Fig.26 : Glimpse of Phalaenopsis orchid

Fig.27 : Glimpse of Phalaenopsis orchid

Fig.28 : Glimpse of Phalaenopsis orchid

Fig.29 : Glimpse of Phalaenopsis orchid

Fig.30 : Glimpse of Phalaenopsis orchid

Fig.31 : Glimpse of Phalaenopsis orchid

Fig.32 : Glimpse of Phalaenopsis orchid

Fig.33 : Glimpse of Phalaenopsis orchid

Fig.34 : Glimpse of Phalaenopsis orchid

Fig.35 : Vanda orchids

Fig.36 : Vanda orchids

Fig.37 : Vanda orchids

Fig.38 : Vanda orchids

FSC
www.fsc.org
MIX
Papier | Fördert gute Waldnutzung
FSC® C083411

Zeitfracht Medien GmbH
Ferdinand-Jühlke-Straße 7
99095 Erfurt, Deutschland
produktsicherheit@kolibri360.de